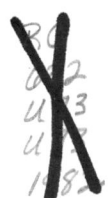  

WITHDRAWN

Date Due

UREA CYCLE DISEASES

ADVANCES IN EXPERIMENTAL MEDICINE AND BIOLOGY

Editorial Board:

NATHAN BACK, *State University of New York at Buffalo*

NICHOLAS R. DI LUZIO, *Tulane University School of Medicine*

EPHRAIM KATCHALSKI-KATZIR, *The Weizmann Institute of Science*

DAVID KRITCHEVSKY, *Wistar Institute*

ABEL LAJTHA, *Rockland Research Institute*

RODOLFO PAOLETTI, *University of Milan*

Recent Volumes in this Series

Volume 147
INTRAOVARIAN CONTROL MECHANISMS
Edited by Cornelia P. Channing and Sheldon J. Segal

Volume 148
STRUCTURE AND FUNCTION RELATIONSHIPS IN BIOCHEMICAL SYSTEMS
Edited by Francesco Bossa, Emilia Chiancone, Alessandro Finazzi-Agrò, and Roberto Strom

Volume 149
IN VIVO IMMUNOLOGY: Histophysiology of the Lymphoid System
Edited by Paul Nieuwenhuis, A. A. van den Broek, and M. G. Hanna, Jr.

Volume 150
IMMUNOBIOLOGY OF PROTEINS AND PEPTIDES—II
Edited by M. Z. Atassi

Volume 151
REGULATION OF PHOSPHATE AND MINERAL METABOLISM
Edited by Shaul G. Massry, Joseph M. Letteri, and Eberhard Ritz

Volume 152
NEW VISTAS IN GLYCOLIPID RESEARCH
Edited by Akira Makita, Shizuo Handa, Tamotsu Taketomi, and Yoshitaka Nagai

Volume 153
UREA CYCLE DISEASES
Edited by A. Lowenthal, A. Mori, and B. Marescau

Volume 154
GENETIC ANALYSIS OF THE X CHROMOSOME: Studies of Duchenne Muscular Dystrophy and Related Disorders
Edited by Henry F. Epstein and Stewart Wolf

Volume 155
MACROPHAGES AND NATURAL KILLER CELLS: Regulation and Function
Edited by Sigurd J. Normann and Ernst Sorkin

UREA CYCLE DISEASES

Edited by

A. Lowenthal
Born-Bunge Foundation
Universitaire Instelling Antwerpen
Wilrijk, Belgium

A. Mori
Institute for Neurobiology
Okayama University Medical School
Okayama, Japan

and

B. Marescau
Born-Bunge Foundation
Universitaire Instelling Antwerpen
Wilrijk, Belgium

PLENUM PRESS • NEW YORK AND LONDON

Library of Congress Cataloging in Publication Data

Main entry under title:

Urea cycle diseases.

(Advances in experimental medicine and biology; v. 153)
"Proceedings of the International Symposium on Urea Cycle Diseases, held September 27–29, 1981, in Okayama, Japan, which was a satellite symposium of the Twelfth World Congress of Neurology"—verso t.p.
Includes bibliographical references and indexes.
1. Urea—Metabolism—Disorders—Congresses. 2. Ammonia—Toxicology—Congresses. I. Lowenthal, A. (Armand) II. Mori, Akitane. III. Marescau, B. IV. International Symposium on Urea Cycle Diseases (1981: Okayama-shi, Japan) V. Series. [DNLM: 1. Urea—Metabolism—Congresses. W1 AD559 v. 153/QU 65 I61u 1981]
RC632.U73U73 1982   616.3'99   82-12302
ISBN 0-306-41037-0

Proceedings of the International Symposium on Urea Cycle Diseases,
held September 27–29, 1981, in Okayama, Japan, which was a
satellite symposium of the Twelfth World Congress of Neurology

©1982 Plenum Press, New York
A Division of Plenum Publishing Corporation
233 Spring Street, New York, N.Y. 10013

All rights reserved

No part of this book may be reproduced, stored in a retrieval system, or transmitted
in any form or by any means, electronic, mechanical, photocopying, microfilming,
recording, or otherwise, without written permission from the Publisher

Printed in the United States of America

ACKNOWLEDGEMENT

The Okayama meeting was sponsored in part by a grant of the Japanese Society for Promotion of Sciences.

The manuscripts were gathered by Dr. M. Hiramatsu and Dr. N. Ogawa of the Institute for Neurobiology, Okayama University Medical School, and Dr. D. Karcher and Dr. M. Noppe of the Laboratory of Neurochemistry of the Born-Bunge Foundation, Universitaire Instelling Antwerpen, Wilrijk, Belgium.

The editorial work was done by a team of the Laboratory of Neurochemistry at the Born-Bunge Foundation, Universitaire Instelling Antwerpen under the direction of Dr. D. Karcher.

To all these collaborators we wish to express our gratitude.

TABLE OF CONTENTS

Acknowledgement

I. INTRODUCTION

Introduction
    A. Lowenthal   1

New Facets in Urea Cycle Disorders
    A. Mori   5

II. DIAGNOSTIC, CLINICAL, PATHOLOGICAL AND BIOCHEMICAL ASPECTS OF THE DISEASES

1) Screening

Newborn screening for urea cycle disorders
    E.W. Naylor   9

A new method for screening of hyperammonemia
    K. Tada, H. Tateda, K. Metoki   19

Immobilization of multienzymes of urea cyle into fibrin membrane : an approach to an artificial liver
    H. Hagiwara, T. Nagasaki, Y. Saito, Y. Inada   29

2) Urea Cycle Enzyme Deficiencies

N-acetylglutamate synthetase (NAGS) deficiency: diagnosis, clinical observations and treatment
    C. Bachmann, J.P. Colombo, K. Jaggi   39

Ornithine transcarbamylase (OTC) in white blood cells and jejunal mucosa
    N. Nagata, I. Akaboshi, J. Yamamoto, F. Endo, I. Matsuda, T. Katsuki   47

Immunochemical assay in 16 boys with ornithine transcarbamylase deficiency
  B. François, P. Briand, L. Cathelineau   53

Enzymatic analysis of citrullinemia (12 cases) in Japan
  T. Saheki, A. Ueda, M. Hosoya, M. Sase,
  K. Nakano, T. Katsunuma   63

Qualitative abnormality of liver argininosuccinate synthetase in a patient with citrullinemia
  Y. Matsuda, A. Tsuji, N. Katunuma   77

Argininosuccinic aciduria in adult : a clinical, electrophysiological and biochemical study
  T. Grisar   83

First case of argininosuccinic aciduria in Japan : clinical observations and treatment
  T. Sakiyama, T. Suzuki, M. Owada, T. Kitagawa   95

Complementation in argininosuccinate synthetase and argininosuccinate lyase deficiencies in human fibroblasts
  L. Cathelineau, D. Pham Dinh, P. Briand,
  P. Kamoun   101

Clinical and biochemical findings in argininemia
  H.G. Terheggen, A. Lowenthal, J.P. Colombo   111

Argininemia : report of a new case and mechanisms of orotic aciduria and hyperammonemia
  M. Yoshino, K. Kubota, I. Yoshida, T. Murakami,
  F. Yamashita   121

Therapy of neonatal onset urea cycle enzymopathies (UCE)
  M. Batshaw, S. Brusilow   127

3) Secondary and Transient Hyperammonemia

Introduction to the Session on Secondary and Transient Hyperammonemia
  H.G. Terheggen   131

Hyperammonemia secondary to hereditary organic acidurias : a study of 29 cases
  J.M. Saudubray, F.X. Coudé, H. Ogier,
  L. Cathelineau, P. Briand, C. Charpentier   135

CONTENTS

Neonatal isovaleric acidemia associated with hyperammonemia
    M. Yoshino, I. Yoshida, F. Yamashita, M. Mori,
    C. Uchiyama, M. Tatibana   141

Hyperammonemia in the neonate with hypoxia
    Y. Sakaguchi, K. Yuge, M. Yoshino, F. Yamashita,
    T. Hashimoto   147

A mechanism for valproate-induced hyperammonemia
    F.X. Coudé, D. Rabier, L. Cathelineau, G. Grimber,
    P. Parvy, P. Kamoun   153

Reduced activity of OTC in the liver of patient with Reye's syndrome
    S. Arashima, Y. Takekoshi, M. Anakura, H. Nanbu,
    I. Matsuda   163

4) Animal Models

Spontaneous animal models of ornithine transcarbamylase deficiency studies on serum and urinary nitrogenous metabolites
    I.A. Qureshi, J. Letarte, R. Ouellet   173

Sparse-fur mutation : a model for some human ornithine transcarbamylase deficiencies
    P. Briand, L. Cathelineau   185

III. BASIC BIOCHEMISTRY

1) Regulation

Regulation of the N-acetylglutamate content of rat hepatocytes by the glutamate concentration
    H. Zollner   197

Enzyme regulation of N-acetylglutamate synthesis in mouse and rat liver
    M. Tatibana, S. Kawamoto, T. Sonoda, M. Mori   207

Acute glucagon treatment in rats fed various protein diets. Effect on N-acetylglutamate concentration
    L. Cathelineau, D. Rabier, F.X. Coudé   217

The relation between the developmental timing of birth and developmental increases in urea cycle enzymes
    W.H. Lamers, P.G. Mooren, W. Oosterhuis,
    H. Lunstroo, A. De Graaf, R. Charles   229

Studies on the enzymes of urea cycle intermediates
in normal and infarcted myocardial tissue of rat
    B. Sadasivudu, M. Swamy, G.N. Roa   241

Effects of arginine-free meals on ureagenesis in cats
    P.M. Stewart, M. Batshaw, D. Valle, M. Walser   243

Dynamism of rat liver ornithine metabolisms in relation to
high-protein stimulation of the urea cycle
    T. Matsuzawa, N. Sugimoto, I. Ishiguro   245

Regulation of urea synthesis : changes in the concentration of ornithine in the liver corresponding to changes in urea synthesis
    T. Saheki, M. Hosoya, S. Fujinami, T. Katsunuma   255

2) Enzymes

Synthesis and intracellular transport of mitochondrial carbamylphosphate synthetase I and ornithine transcarbamylase
    M. Mori, S. Miura, T. Morita, M. Tatibana   267

Isolation of argininosuccinase from bovine brain : catalytic, physical and chemical properties compared to liver and kidney enzymes
    K. Murakami-Murofushi, S. Ratner   277

On the mechanism of the alterations of rat kidney transamidinase activities by diet and hormones
    J.F. Van Pilsum, D.M. McGuire, H. Towle   291

Sciatectomic stimulation of muscle arginase and its implications
    V. Mohanachari, P. Neeraja, K. Indira, K.S. Swami   299

Enzymes of arginine metabolism in the lizard, calotes versicolor
    T.G. Baby, S.R.R. Reddy   303

3) Relation of Urea Cycle to Organic Acid Metabolism

Orotic acid in urine and hyperammonemia
    C. Bachmann, J.P. Colombo   313

Comparison between amino acids and orotic acid analysis in the detection of urea cycle disorders in the Quebec urinary screening program
    B. Lemieux, Ch. Auray-Blais, R. Giguère   321

Transient hyperammonemias in infants with and without
organic acidemia
    W.L. Nyhan, V. Rubio, A. Jordá, S. Grisolia,
    F. Gutierez, C. Canosa   331

Mechanism of hyperammonemia in an experimental model
of propionic acidemia
    P.M. Stewart, M. Walser   339

The study of organic acid metabolism in a patient with
ornithine transcarbamylase (OTC) deficiency
    H. Kodama, O. Nose, S. Okada, H. Yabuuchi   341

4) Relation of Urea Cycle to Proline Metabolism

Hyperornithinemia, gyrate atrophy and ornithine
keto-acid transaminase
    S. Hayasaka, T. Shiono, K. Mizuno, T. Saito,
    K. Tada, T. Matsuzawa, I. Ishiguro   353

Gyrate atrophy of the choroid and retina (GA) : toxic
effects of ornithine and long-term therapy with an
arginine-restricted diet
    D. Valle, M. Walser, S. Brusilow, M.I. Kaiser-
    Kupfer   359

Disease of ornithine-proline pathway : an Δ'-pyrro-
line-5-carboxylate reductase deficiency in the retina
of retinal degeneration mice
    T. Matsuzawa, K. Iwasaki, N. Hiraiwa, E. Inagaki,
    I. Ishiguro   361

Toxic effects of ornithine and its related compounds
on the retina
    Y. Ishikawa, T. Kuwabara, M.I. Kaiser-Kupfer   371

5) Relation of Urea Cycle to Guanidino Compounds

Quantitative determination of guanidino compounds :
the excellent preparation of biological samples
    A. Ando, T. Kikuchi, H. Mikami, M. Fujii,
    K. Yoshihara, Y. Orita, H. Abe   381

Recommended deproteinizing methods for plasma guanidino compounds analysis by liquid chromatography
    T. Hoshino   391

Evolutionary relationship between arginine and
creatine in muscle
    F.J.R. Hird, S.P. Davuluri, R.M. McLean   401

Metabolism of arginine in invertebrates : relation to urea cycle and to other guanidine derivates
    Y. Robin   407

α Guanidinoglutaric acid and epilepsy
    A. Mori, Y. Watanabe, S. Shindo, M. Akagi, M. Hiramatsu   419

Guanidino compounds in hyperargininemia
    B. Marescau, A. Lowenthal, H.G. Terheggen, E. Esmans, F. Alderweireldt   427

Guanidinosuccinic acid and the alternate urea cycle
    B.D. Cohen, H. Patel   435

Guanidinosuccinic acid excretion in argininosuccinic aciduria
    H. Böhles, B.D. Cohen, D. Michalk   443

Metabolic pathway of guanidino compounds in chronic renal failure
    H. Mikami, Y. Orita, A. Ando, M. Fujii, T. Kikuchi, K. Yoshihara, A. Okada, H. Abe   449

Guanidino compounds and hemodialysis
    Y. Ochiai, S. Abe, T. Yamada, K. Tada, F. Kosaka   459

Effect of guanidino compounds on hen egg development
    S. Seki, N. Yuyama, M. Hiramatsu   465

IV. FREE COMMUNICATIONS

α-ketoglutarate induced transamination during ischemic exercise
    C. Cerri, F. Fici, G. Scarlato   473

Effect of pyridoxine-2-oxoglutarate administration in patients with advanced cirrhosis : control of ammonia pyruvate and lactate high plasma concentrations
    F. Salerno, M.C. Lorenzini, M. Conti, R. Abbiati, F. Fici   479

Treatment of pyruvic and lactic acidaemia in ophthalmoplegia plus
    G. Scarlato, M. Moggio, F. Fici, C. Cerri   487

Intracerebral pH regulation and ammonia detoxification
    N.M. Van Gelder   501

CONTENTS xiii

CONCLUDING REMARKS
   A. Lowenthal 509

PARTICIPANTS 513

INDEX 521

I. INTRODUCTION

Introduction

New Facets in Urea Cycle Disorders

INTRODUCTION

A. Lowenthal

Laboratory of Neurochemistry, Born-Bunge Foundation,
Universitaire Instelling Antwerpen, Wilrijk, Belgium

  This occasion is by no means the first meeting devoted to
urea cycle diseases. It has been preceeded by meetings held in
the Netherlands and in Spain. Accordingly the justification for
a further meeting is not immediately evident. The reason for it
is that the problems related to urea cycle diseases are developing
fast, as instanced inter alia by :

1) the relation observed between hyperornithinemia and gyrate
  atrophy and the therapeutic acquisitions which result from it ;

2) the treatment of hyperammonemias.

  If the diagnosis of urea cycle disease is easily established
by following standard principles and techniques, i.e. by means of
amino acid analysis, with or without prior loading tests, by
ammonemia measurement and by enzyme determination, also if these
operations produce clear and precise conclusions in the matter of
genetics and preventive medicine, yet many physiopathological
questions remain unanswered and a number of therapeutic problems
of these remain unsolved.

  Thus I come to mention here some of the problems which, in
my opinion, could well be discussed during this meeting :

1) given that the different urea cycle diseases occur in the same
  cyclus is there a physiopathological relationship between
  these diseases ? What is their relation to renal diseases ?
  Could changes of the urea concentration of hyperammonemia have
  an effect on the functioning of the kidneys or any other organ ?

2) admitting that urea cycle diseases are caused by an intoxication due to the accumulation of one or other amino acid, what then is the effect of hyperammonemia in the urea cycle diseases ? What about deficiencies in urea synthesis ? Does the production of less urea have a physiopathological influence ? Again, if a block is to be found in the urea cycle, where is the urea present in such patients coming from ?

3) how can we explain the complex biochemical anomalies seen in some patients such as hyperargininemia where we note

   a) an increase of arginine,
   b) an exaggerated elimination of cystine, lysine, arginine and ornithine in the urines,
   c) an increase of the ammonia content.

4) then the fact that an excess of arginine in hyperargininemia is eliminated in part as monoguanidino compounds in the urines, is a problem in itself. In connection with these compounds, it may be asked whether they do not have a part in the physiopathology of the disease. We know that some of these compounds have an epileptogeneous activity. Should this not apply also in hyperargininemia ?

   Furthermore we have the intriguing indication that one of the guanidino compounds does not seem to originate from arginine : the guanidinosuccinic acid. Up to the present no explanation has been found for this anomaly. Let it be said on this point that a decreased elimination of guanidinosuccinic acid is also seen in other diseases of the urea cycle, such as citrullinemia ?

   The appearance of these guanidino compounds in the urine of patients affected with hyperargininemia causes the question to be asked as to what becomes of the amino acids accumulated (ornithine, citrulline, argininosuccinic acid) in other diseases of the urea cycle.

5) as far as we are concerned we will also wish to know the treatment to be applied for urea cycle diseases. Is a diet poor in proteins satisfactory ? A diet poor in amino acids such as arginine or ornithine, and is this diet applicable or have we to consider other therapeutics such as those applied to hyperammonemias ?

   These are some of the reasons why we thought it could be of interest to organize a symposium devoted to urea cycle diseases and since our colleague Mori is studying the physiopathological activity of the guanidino compounds in epilepsy in particular, we believed that this symposium could take place here, and since

INTRODUCTION

it is connected with neurological problems, we also thought it could become a satellite symposium of the World Congress of Neurology. In this connection we have asked and obtained the sponsorship of the World Federation of Neurology, which we thank very sincerely.

NEW FACETS IN UREA CYCLE DISORDERS

A. Mori

Department of Neurochemistry, Institute for Neurobiology
Okayama University Medical School, Okayama, 700 Japan

It is a great pleasure to have so many participants at the symposium on Urea Cycle Diseases here in Okayama. Clinical, pathological and biochemical studies on metabolic disorders in the urea cycle, i.e., argininosuccinic aciduria, citrullinemia, ornithinemia, argininemia and hyperammonia have been pursued vigorously in the recent decade, and we are beginning now to at least understand some of the mechanisms of these disorders. In this symposium, additional new ideas on the origin of these diseases will be reported by many participants, and we are expecting to gather more exact information to clarify the pathogenesis. In addition to these disorders, a new clinical entity, N-acetyl-glutamate synthetase deficiency, will be introduced in detail by the discoverers themselves. Sparse-fur mutant mice have become a model for human ornithine transcarbamylase deficiency, and will contribute extensively to investigations of human diseases in the future.

In this symposium, many papers regarding basic biochemistry will also be presented. N-acetyl-glutamate metabolism as a regulatory mechanism involved in the urea cycle will be one of the topics. The special properties of other enzyme systems and regulating factors in the urea cycle, as well as the relationship of the urea cycle to organic acids, proline and guanidino compounds, will be discussed in detail.

Guanidino compounds have been suggested as a possible "uremic toxin" in chronic renal failure. Personally, I will discuss an epileptic effect of α-guanidinoglutaric acid, which we first isolated from the cobalt-induced epileptogenic focus in the rat.

Therapeutic problems will also be discussed, especially with respect to trials of treatment based on biochemical pathogenesis.

In conclusion, I would like to introduce a traditional story from the Okayama district. You may see a statue of Momotaro, the peach boy, at the Okayama station. He was born from a peach, which came from the upper stream of a river. He grew up to become a superman, and one day went on an expedition against demons who lived on Onigashima, the Demon Island. Peach boy brought 3 faithful companions. They were a dog, a monkey and a pheasant. The dog is brave and can bite a demon directly. The monkey is clever and can attack with intelligence. The pheasant can fly and can therefore attack from an unexpected direction.

The same principles apply to our research on the Urea Cycle Diseases. We are using clinical, pathological or biochemical techniques, while the ultimate goal remains the same. Such a diversity of techniques should in the end lead us to conquer these diseases of the urea cycle. Now, the demon is right here! Release the dog, release the monkey and release the pheasant! Ladies and gentlemen, now it is up to you!

II. DIAGNOSTIC, CLINICAL, PATHOLOGICAL AND
    BIOCHEMICAL ASPECTS OF THE DISEASES

    1) Screening

NEWBORN SCREENING FOR UREA CYCLE DISORDERS

E.W. Naylor

Department of Pediatrics, State University of New York
at Buffalo, 352 Acheson Hall, 3435 Main Street
Buffalo, New York 14214, U.S.A.

Neonatal screening for inherited metabolic disorders is one of the most exciting advances in modern preventive pediatrics. By definition, it is a search in the newborn population for individuals possessing certain genotypes that are either already associated with disease or predisposed to disease. Important preconditions for the establishment of such programs are the availability of effective therapy or improved medical management and the need for early, pre-symptomatic diagnosis. The first successful newborn screening program had its beginnings in the early 1960's with the development of the Guthrie bacterial inhibition assay for the detection of phenylketonuria (PKU)[1]. This simple screening test was the first to utilize dried filter paper blood specimens collected by heel prick from newborn infants during the first week of life.

The field of neonatal screening, however, has not remained limited to the search for patients with PKU. Instead, it has evolved into an exciting area of preventive medicine, which has seen the establishment of widespread screening for congenital hypothyroidism, the introduction of additional bacterial assays for other inborn errors of metabolism, automation of screening laboratories, and regionalization of screening programs[2]. It has also seen the introduction of other specimens for mass screening, including cord blood, 4-6 week follow-up blood specimens, and urine specimens collected on filter paper.

In recent years, a number of screening methods have been developed which are potentially capable of detecting urea cycle disorders using dried blood specimens on filter paper. These include: paper or thin-layer chromatography, bacterial inhibition assays, multiple auxotroph assays, an enzyme-auxotroph assay, an enzyme-multiple auxotroph assay, and a fluorescent spot test.

The first screening methods to be developed involved the use of paper chromatography of dried blood specimens. This approach as well as the use of thin-layer chromatography has been used in a number of blood screening programs for the detection of PKU and other inborn errors. Although this approach to PKU screening has generally been replaced by the bacterial inhibition assay, it continues to be used in a few laboratories and is capable of detecting infants with citrullinemia and argininemia.

A number of bacterial assays for the detection of urea cycle disorders have been developed in our laboratory. The earliest of these involved the isolation of Bacillus subtilis auxotrophs which require ornithine, citrulline, or arginine for growth. The placement of a blood disc containing an elevated amount of any one of these substrates on the surface of an agar plate, containing appropriate spores and minimal growth media, results in a large growth zone around the disc. This multiple auxotroph assay was used by the St. Joseph Hospital Laboratory in Los Angeles, California in a brief field trial during 1979 and 1980. A total of 63,847 newborns were screened with no confirmed cases detected and only four initial elevations.

In 1972, Murphey et al[3] used one of these auxotrophs as the basis for a unique enzyme-auxotroph assay capable of screening for argininosuccinic acid (ASA) lyase deficiency. In this screening test, ASA is added to the agar plate containing minimal media and an arginine requiring B. subtilis auxotroph. The ASA lyase present in a normal blood spot converts the ASA into arginine which then is used by the auxotroph to produce a measurable growth zone. In the absence of the enzyme, no arginine is produced and there is no detectable growth. This test was used in a large scale pilot screening program in the early 1970's, but it was never widely accepted as a routine screening test. The results of that field trial are summarized in Table 1. Of approximately 1 million infants tested, only four cases were confirmed. In addition, one presumptive positive case from one of the Pacific Islands was detected, but was lost to follow-up confirmation. Because of a number of technical problems and the relatively low yield, this test has been discontinued by most laboratories. The only exception is the laboratory in Vienna, Austria which continues to screen.

A considerable amount of time and energy was also spent by us in an attempt to develop a bacterial inhibition assay for glutamine. A successful assay was developed using O-carbamyl-DL-serine as the metabolic inhibitor. Unfortunately, when this assay was tested in an actual screening program, it was discovered that the variability of glutamine in normal newborns overlapped with the levels in hyperammonemic patients. This has severely limited the use of this assay

Table 1. Blood Enzyme-Auxotroph Screening
for ASA Lyase Deficiency
(through 1980)

Laboratory[a]	No. Tested	Cases
Austria	659,775	3
California	139,684	0
New Zealand	96,464	0[b]
Oregon	80,192	0
Western New York	57,052	1
Pacific Islands	15,473	0[c]
Totals	1,048,640	4

[a] All laboratories have stopped screening, except the one in Austria
[b] One case detected in an institution, not a newborn
[c] One presumptive positive case, not confirmed

for routine screening.

A simple fluorescent spot test has also been developed for the identification of individuals with arginase deficiency[4]. The assay is based on the conversion of arginine to ornithine and urea by the arginase present in a 3.2 mm disc of dried blood on filter paper. The enzyme activity is visually estimated by the oxidation of NAD·H to NAD$^+$ in a coupled kinetic reaction (Figure 1). In the absence of the enzyme, there is no oxidation of the NAD·H and consequently no loss of fluorescence. This screening test has been used to identify successfully both heterozygous arginase-deficient crabeater macaques as well as a number of patients with argininemia. This test has only been used for screening by two laboratories. Our laboratory in Buffalo has screened approximately 5,000 mentally retarded residents of institutions in New York and California with no cases detected. Dr. Toshiaki Oura in Osaka, Japan has also used this on 35,000 new born and high-risk infants with no cases detected to date.

Ideally, the best approach to screening for urea cycle disorders is the direct measurement of blood ammonia. Unfortunately, this is technically impossible in a dried blood specimen. Dr. Keiya Tada, however, will present an exciting screening test for rapid ammonia determinations using a drop of liquid blood collected in the newborn nursery. This approach has the advantage of also picking up patients with carbamyl phosphate synthetase and ornithine transcarbamylase deficiency, but because of the need to collect an additional specimen, it should be considered a high-risk screening test rather than a mass screening test.

$$\text{L-Arginine} \xrightarrow[\text{Mn}^{2+},\ 37^\circ\text{C}]{\text{arginase}} \text{L-Ornithine} + \text{Urea}$$

$$\text{Urea} \xrightarrow[\text{Room Temp.}]{\text{urease}} 2\text{NH}_4^+$$

$$\alpha\text{-Ketoglutarate} + \text{NH}_4^+ + \text{NAD·H} \xrightarrow[\text{Room Temp.}]{\text{glutamate dehydrogenase}} \text{L-Glutamate} + \text{NAD}^+ + \text{H}_2\text{O}$$

Figure 1. Enzymatic reactions involved in the qualitative arginase screening assay

If we choose to test a urine specimen collected on filter paper as the means of screening for disorders in the urea cycle, there are a number of approaches available. The most widely used method is that of paper or thin-layer chromatography. Descending one-dimensional chromatography is used by the Massachusetts program[5] on specimens collected a 3 to 4 weeks of age, whereas ascending one-dimensional chromatography on specimens collected at 6 weeks of age was carried out in New South Wales[6]. Other urine screening programs in British Columbia[7] and Wales[8] also use one dimensional chromatography on specimens collected between 2 to 4 weeks of age. Thin-layer chromatography is carried out on specimens collected at 2 weeks of age in Quebec[9]. These programs have developed a considerable body of experience with these methods and are capable of diagnosing those cases of citrullinemia, argininosuccinic aciduria, and argininemia that survive the immediate neonatal period. The results of the major urine screening programs are summarized in Table 2.

Over 2 1/2 million infants have been screened using paper or thin-layer chromatography. Twenty patients with argininosuccinic aciduria, two with citrullinemia, and one case each with argininemia and ornithinemia have been diagnosed. Two observations are worthy of note regarding these programs. The first is the rather low yield from the large Australian program compared to the Massachusetts and Canadian programs. The second is the realization that nearly all of the cases of ASA and citrullinemia detected are the more mild or later onset forms. The patients with the severe neonatal onset forms are generally missed because of the age at which the specimens are collected (2-6 weeks of age).

Two bacterial tests have been developed by our laboratory for use in a urine screening program. The first is a multiple auxotroph (MB 1047) which responds to ornithine, citrulline, and arginine. The second is a bacterial inhibition assay for uracil which utilizes 5-fluorouracil as the specific inhibitor. This test is based on the observation that a block in any of the steps of the urea cycle will result in hyperammonemia and the stimulation of other pathways for the detoxification of ammonia. One of these alternate pathways involves the biosynthesis of pyrimidines. With the exception of carbamylphosphate synthetase, a block in any step of the urea cycle will result in marked excretion in the urine of orotic acid, uridine and uracil. This uracil assay was used in a pilot urine screening program in Western New York as part of a battery of bacterial inhibition assays for various amino acids, purines and pyrimidines[11]. One infant with citrullinemia was identified out of 18,400 screened.

Recently, Dr. Henry Talbot, a postdoctoral fellow working in our laboratory, developed the most comprehensive and most easily implemented mass screening test to date.[10] Starting with Murphey

Table 2. Urine Chromatography Screening for Urea Cycle Disorders[a]

Laboratory	Number Tested	Cases
New South Wales[b]	1,000,000 (through 9/78)	2 (ASA)
Massachusetts	799,234 (through 6/81)	10 (ASA)[c]
		1 (Orn)
Quebec	550,500 (through 7/81)	7 (ASA)
		2 (Cit)
		1 (Arg)
British Columbia	155,763 (through 8/81)	1 (ASA)
Wales[b]	135,295 (through 12/73)	0
Totals	2,640,792	20 (ASA)
		2 (Cit)
		1 (Arg)
		1 (Orn)

[a] Abbreviations used are: ASA, argininosuccinic aciduria; Cit, citrullinemia; Arg, argininemia; Orn, ornithinemia
[b] Program has stopped
[c] Plus 1 atypical case and 1 case diagnosed in a sibling

et al's enzyme-auxotroph assay for ASA lyase deficiency, he developed an improved enzyme-multiple auxotroph assay which is capable of detecting three urea cycle disorders (citrullinemia, ASA lyase deficiency, and argininemia) as well as the several types of ornithinemia which can result in gyrate atrophy of the retina or in mental retardation. The most important modification in this assay has been the replacement of the arginine-requiring auxotroph with a multiple auxotroph which can utilize either citrulline, arginine, or ornithine for growth. There have also been a number of modifications which were designed to eliminate problems in the correct interpretation of test results. This assay is illustrated in Figure 2.

Around normal specimens, a dense growth zone (stained red) of about 10-12 mm in diameter is seen. This indicates normal enzyme activity. If abnormal amounts of free arginine, citrulline, or ornithine are present in the specimen, another less dense growth zone of 20 mm or greater in diameter will appear. Specimens with added arginine, citrulline, and ornithine are included on each plate as standard controls. ASA lyase deficiency is recognized in a different manner. ASA cannot be assayed directly using this auxotroph, since it cannot be utilized by this organism. However, the activity of the enzyme can be assessed. ASA lyase, which is active in normal erythrocytes but inactive in cases of ASA lyase deficiency, catalyzes the cleavage of ASA to arginine and fumaric acid. Since ASA is included in the assay medium, the auxotroph will produce a small dense

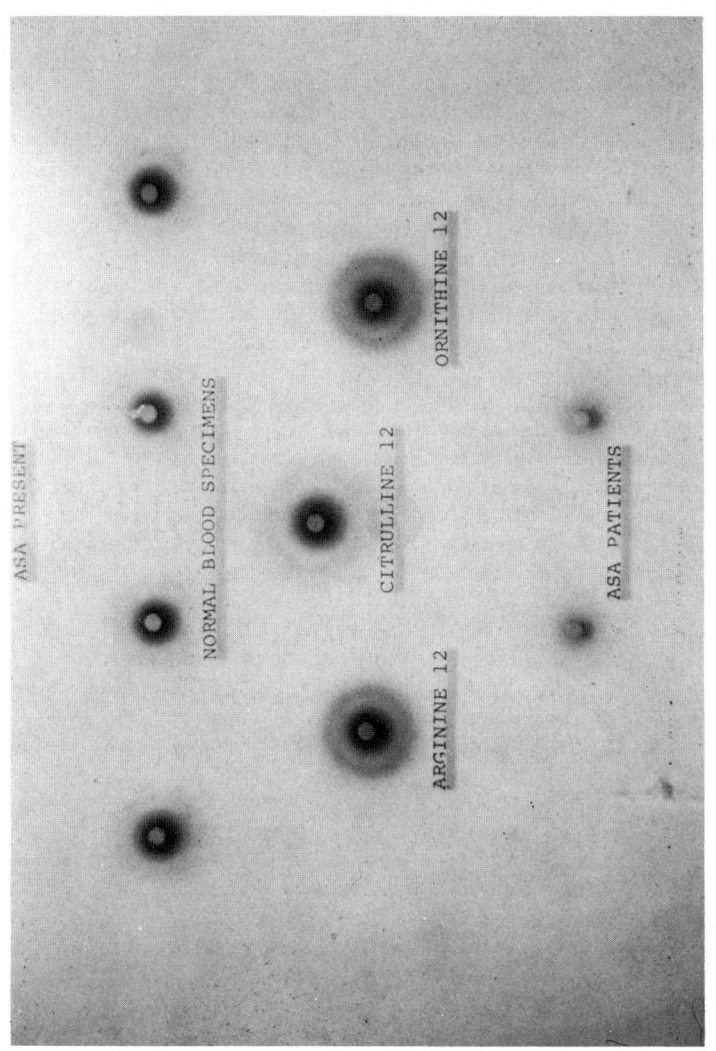

Figure 2. Sample assay plate with ASA present in the medium. The upper row illustrates normal blood specimens with the dense 10-12 mm diameter inner growth zones which result from arginine produced by normal activity of ASA lyase in the discs. The middle row illustrates the increased density of the outer growth zones when abnormal amounts (12 mg/dl) of arginine, citrulline or ornithine are added. The bottom row illustrates the absence of the inner growth zone in specimens from patients with ASA lyase deficiency.

growth zone around blood specimens with normal enzyme activity in response to the released arginine. In this case, the presence of a growth zone indicates a normal condition, in contrast to most other bacterial assays for metabolic disease. This dense inner growth zone is absent in ASA lyase deficient patients.

The reliability of this assay was determined by collecting and testing a number of dried blood specimens from patients with known urea cycle disorders. The results of these studies are summarized in Table 3. The assay meets many of the applicable basic criteria for the evaluation of new screening tests set forth by the World Health Organization[12] and the USA National Academy of Sciences Committee for the Study of Inborn Errors of Metabolism[13]. Its reliability, repeatability and accuracy have been reasonably demonstrated. Preliminary data suggest that the test is both sensitive and specific. It has the advantage of being able to utilize filter paper blood specimens already being routinely collected for PKU and congenital hypothyroidism screening. It is designed for use with 3.2 mm blood discs automatically punched onto the agar surface of test trays by the quadratic punch-index machine, which is in use in regional newborn screening laboratories carrying out multiple testing procedures[14]. The stability of the ASA lyase at room temperature is as great as that of galactose-1-phosphate uridyl transferase, the basis of a well established screening test for galactosemia[15].

Pilot screening programs to evaluate the effectiveness of this new assay are just getting underway. To date, only about 30,000 newborn have been screened (Table 4). The only positive result was an infant with a significantly elevated level of homocitrulline. This patient remains clinically asymptomatic, but is being evaluated further.

In summary, the experience with newborn screening for urea cycle disorders on a mass population basis is still somewhat limited. The technology for mass screening, however, does exist. Therefore, we are at the point where large scale screening programs are feasible for full evaluation of existing or newly developed screening tests.

Talbot's enzyme-multiple auxotroph assay appears to be the most comprehensive single test available for mass screening and is ready for large-scale field evaluation. It has the advantages of being able to utilize blood specimens already being routinely collected and it can readily be added to a battery of tests being carried out in an automated multi-test regional laboratory. This single test should identify patients with citrullinemia, ASA lyase deficiency, argininemia, and ornithinemia.

The use of paper or thin-layer chromatography or a combination of the multiple auxotroph assay and the 5-fluorouracil bacterial

Table 3. Results of Specimens from Patients with Known Urea Cycle Disorders Using the Enzyme-Multiple Auxotroph Assay

Disorder	Age of Specimen Patient Status	Auxotroph Response (mg/dl)	Enzyme Response
Citrullinemia	3 w.-untreated	10-12	+
Argininosuccinic Aciduria[a]	3 d.-untreated	<2	-
Argininosuccinic Aciduria[a]	?	<2	-
Argininosuccinic Aciduria[a]	3 d.-untreated	<2	-
Argininemia[b]	8 y.-low protein	12	+
Argininemia[b]	15 y.-low protein	10-12	+
Argininemia[c]	4 y.-untreated	12	+

[a] Specimens provided by Dr. Vivian Shih and Dr. Harvey Levy. Not severe neonatal onset forms.
[b] Specimens provided by Dr. Stephen Cederbaum
[c] Specimen provided by Dr. Makoto Yoshino

Table 4. Newborn Screening Results Using the Enzyme-Multiple Auxotroph Assay

Laboratory	Number Tested	Cases
Oregon	14,200	0
Western New York	9,038	0[a]
Albany, New York	5,712	0
Totals	28,950	0

[a] One patient with elevated homocitrulline.

inhibition assay appear to be the most promising and productive approaches to urine screening for urea cycle disorders.

The primary justification for mass newborn screening for urea cycle disorders at the present time is to gather frequency data on these disorders, to elucidate their natural history, to detect individuals who will be severely affected clinically at a time when therapy may have a beneficial effect, and to identify mildly affected or asymptomatic variants. Early detection should permit better medical management of the patient and accurate early diagnosis should permit more effective genetic counseling for family members at risk.

REFERENCES

1. R. Guthrie and A. Susi, A simple phenylalanine method for detecting phenylketonuria in large populations of newborn infants, Pediatrics 32: 338 (1963).
2. H. Bickel, R. Guthrie, and G. Hammersen, "Neonatal Screening for Inborn Errors of Metabolism", Springer-Verlag, Heidelberg (1980).
3. W.H. Murphey, L. Patchen, and R. Guthrie, Screening tests for argininosuccinic aciduria, orotic aciduria, and other inherited enzyme deficiencies using dried blood specimens, Biochem. Genet. 6: 51 (1972).
4. E.W. Naylor, A.P. Orfanos, and R. Guthrie, A simple screening test for arginase deficiency (hyperargininemia), J. Lab. Clin. Med. 89: 987 (1977).
5. H. Levy, P.M. Madigan, and V.E. Shih, Massachusetts metabolic disorders screening program. 1. Technics and results of urine screening, Pediatrics 49: 825 (1972).
6. B. Turner and D.A. Brown, Amino acid excretion in infants and early childhood. A survey of 200,000 infants, Med. J. Aust. 1: 62 (1972).
7. D.A. Applegarth, D.F. Hardwick, S. Israels, and P.M. Ross, Results of a screening program for aminoacidopathies in British Columbia, British Columbia Med. J. 12: 129 (1970).
8. D.M. Bradley, Screening for inherited metabolic disease in Wales using urine-impregnated filter paper, Arch. Dis. Child. 50: 264 (1975).
9. D. Shapcott, B. Lemieux, and A. Sahapoglu, A semi-automated device for multiple sample applications to thin-layer chromatography plates, J. Chromatogr. 70: 174 (1972).
10. T.D. Paul, E.W. Naylor, and R. Guthrie, Urine screening for metabolic disease in newborn infants, J. Pediatrics 96 : 653 (1980).
11. H.W. Talbot, A.B. Sumlin, E.W. Naylor, and R. Guthrie, A neonatal screening test for argininosuccinic acid lyase deficiency and other urea cycle disorders, Pediatrics, in press (1981).
12. World Health Organization, "Screening for inborn errors of metabolism", Report of a WHO scientific group, WHO Tech. Rep. Ser. No. 401, Geneva (1968).
13. National Academy of Sciences, "Genetic Screening - Programs, Principles, and Research", Report of the Committee for the Study of Inborn Errors of Metabolism, Washington (1975).
14. R. Guthrie, Mass screening for genetic disease, Hosp. Practice 7: 93 (1972).
15. E.Beutler and M.D. Baluda, A simple spot screening test for galactosemia, J. Lab. Clin. Med. 68 : 137-141 (1966).

A NEW METHOD FOR SCREENING OF HYPERAMMONEMIA

K. Tada, H. Tateda and K. Metoki

Department of Pediatrics, Tohoku University
Medical School, Sendai 980, Japan

A number of enzymatic defects are known such as congenital hyperammonemia, citrullinemia, argininosuccinic aciduria or hyperargininemia, which involve the urea cycle. Patients with these disorders exhibit mental retardation, presumably resulting from intoxication with ammonia. Mental retardation can be prevented by early treatment with a low protein diet and/or use of sodium benzoate to decrease the blood ammonia level.

Furthermore, neonatal transient hyperammonemia associated with perinatal asphyxia or respiratory distress is now being considered as a possible cause of neonatal death and damage of the central nervous system in the surviving cases. Exchange transfusion or peritoneal dialysis is know to be an effective treatment for neonatal hyperammonemia (Goldberg et al. 1979).

Therefore, the early diagnosis of hyperammonemia in newborn period is very important in pediatric practice. However, there has been no simple method available for screening of hyperammonemia. We developed a new and simple screening kit to detect hyperammonemia (Tada et al. 1979).

The principle of the kit is based on microdiffusion of ammonia (Fig. 1). Ammonia, liberated from blood when it is mixed with alkali, diffuses through a polyethylene film which is permeable to gas but not to liquid. It then reacts with an indicator (bromcresol green) and a color change develops which is related to the ammonia concentration.

The instructions for the kit are as follows :

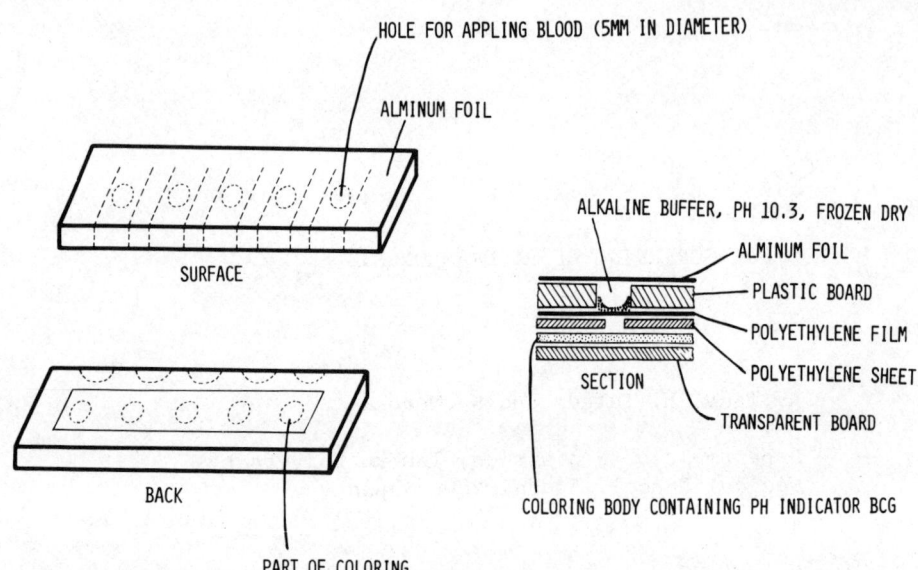

Fig. 1. Kit for screening of hyperammonemia.

1. Tear off the aluminum foil with a finger.
2. Place a drop of freshly collected blood (0.02 ml).
3. Seal the hole with the foil immediately after applying the blood.
4. Leave the kit for 15 minutes at room temperature.
5. Compare the color intensity at the back of the kit with the standard.

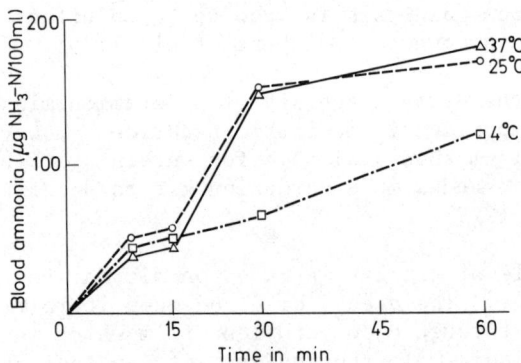

Fig. 2. Time course of the reaction at various temperatures (Whole blood sample : original concentration of ammonia : 51 N-μg/100 ml).

Table 1. Recovery experiment using the present method

(NH₄)₂SO₄ added (μg NH₃-N/100 ml)	Value expected	Value measured (μg NH₃-N/100 ml)	Recovery (%)
0	55		
50	105	100	95.2
100	155	150	96.8
200	255	250	98.0
300	355	350	98.6

The optimal conditions for the reaction were investigated using different temperatures and reaction times. Satisfactory results were obtained with a 15 min reaction time which gave no significant variation with temperature, as shown in Fig. 2. Therefore, the experiments described below were made using 15 min incubation at room-temperature.

Recovery : known amounts of $(NH_4)_2SO_4$ were added to a blood specimen and recovery was investigated. Results were considered adequate, averaging 95.2 to 98.6 % over a wide range of concentrations, as shown in Table 1.

Reproducibility : reproducibility was assessed in two blood specimens with known amounts of ammonia. Results were satisfactory, giving a mean concentration of 227.14 ± 18.58 for 238 μg NH_3-N/100 ml and 62.86 ± 8.25 for 70 μg NH_3-N/100 ml, respectively (cf. Table 2).

A comparison was made between the present method and the enzymatic method for quantitative determination of blood ammonia (Mondzac et al., 1965), using many blood specimens. The results indicated that the values obtained by the two methods were in good agreement, as shown in Fig. 5. This was particularly observed in the higher range of blood ammonia, and agreement was not quite so good in the lower range of less than 100 μg NH_3-N/100 ml. As blood ammonia levels are normally less than 150 μg NH_3-N/100 ml (Shih, 1978), this method is sufficiently accurate for detecting hyperammonemia.

Random blood samples from normal newborn infants and from patients known to have congenital hyperammonemia due to ornithine transcarbamylase deficiency were tested (without informing the tester) by the present method. No false positive results were found among 300 samples from the newborn infants whereas the

Table 2. Reproducibility experiments using the present method.

No.	Sample[a] (μg NH_3-N/100 ml)	
	A	B
1	220	70
2	240	60
3	240	50
4	250	70
5	200	80
6	220	50
7	220	70
8	220	60
9	250	60
10	200	60
11	240	60
12	200	70
13	250	60
14	230	60
$\bar{x}$	227.14	62.86
$\pm$ S.D.	18.58	8.25
C.V.	8.18%	13.1%

[a] Sample A = 238 μg NH_3-N/100 ml
Sample B = 70 μg NH_3-N/100 ml
by enzymatic method

samples from the two patients with hyperammonemia were clearly identified (cf. Table 3).

With this method, we checked blood ammonia in newborns admitted to NICU, in collaboration with Dr. Fujiwara, Akita and Dr. Baba, Tokyo.

Table 4 shows the results of blood ammonia monitoring. In 29 out of 85 cases monitored, hyperammonemia, ranging from 150 to 996 µg/100 ml, was found, which consisted of 3 cases associated with perinatal asphyxia, 11 cases with IRDS, 6 cases with dyspnea except IRDS, 2 cases with convulsions, 4 cases with miscellaneous disorders, 2 cases with low birth weight infants with no

Table 3. Blood ammonia screening by the present method

	Blood ammonia (μg NH$_3$-N/100 ml)	
	present method	enzymatic method
Normal newborns		
$n = 298$	50 ~ 100	< 100
$n = 2$	100 ~ 150	127, 140
Congenital hyperammonemia		
Case 1	350	387
	250	270
Case 2	400	358
	200	190

Table 4. Results of blood ammonia monitoring in newborns admitted to NICU.

	TOTAL	NORMAL BIRTH WEIGHT	LOW BIRTH WEIGHT*
PERINATAL ASPHYXIA	16 (3)	12 (2)	4 (1)
IRDS	18 (11)		18 (11)
DYSPNEA (EXCEPT IRDS)	14 (6)	8 (2)	6 (4)
CONVULSIONS	6 (2)	4 (2)	2 (0)
MISCELLANEOUS DISORDERS	21 (4)	13 (2)	8 (2)
NO COMPLICATION	9 (2)		9 (2)
ORNITHINE TRANSCARBAMYLASE DEFICIENCY	1 (1)	1 (1)	
	85 (29)	38 (9)	47 (20)

LOW BIRTH WEIGHT : BELOW 2500 G
PARENTHESES : NUMBERS OF HYPERAMMONEMIA BEYOND 150 μG%

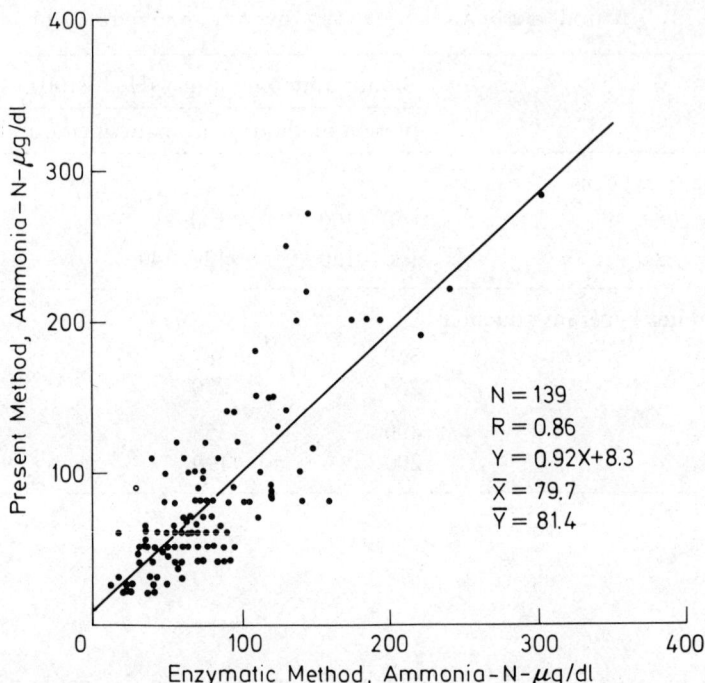

Fig. 3. Comparison of results for blood ammonia concentration by the present method and by enzymatic method.

complications and a case with ornithine transcarbamylase deficiency. Higher frequency of hyperammonemia was found in IRDS (61 %) and dyspnea except IRDS (43 %). In normal birth weight group hyperammonemia was found in 9 out of 38 cases (24 %), whereas it was found in 20 out of 47 cases (43 %) in low birth weight group, especially 14 out of 24 cases (58 %) in low birth weight infants below 1500 g. Nine out of 11 cases (82 %), which showed above 300 µg % of blood ammonia, died in neonatal period, as compared with 21 % of mortality rate of the newborns subjected to this study (Table 5).

Table 5. Mortality rate.

Total	18/85	(21.2 %)
Cases with hyperammonemia	11/29	(37.9 %)
> 300 µg %	9/11	(81.8 %)
Cases without hyperammonemia	9/56	(16.1 %)

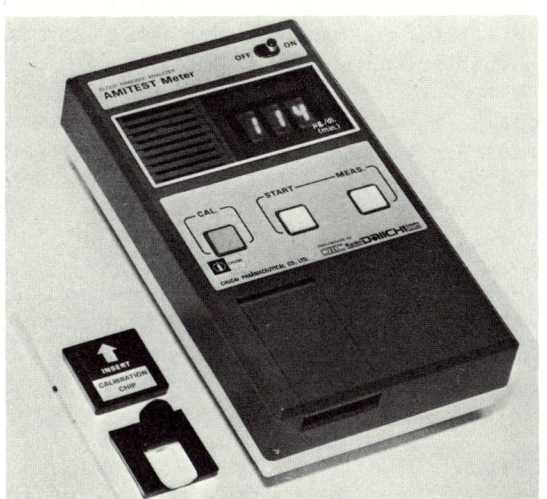

Fig. 4. A reflectance-meter for blood ammonia determination.

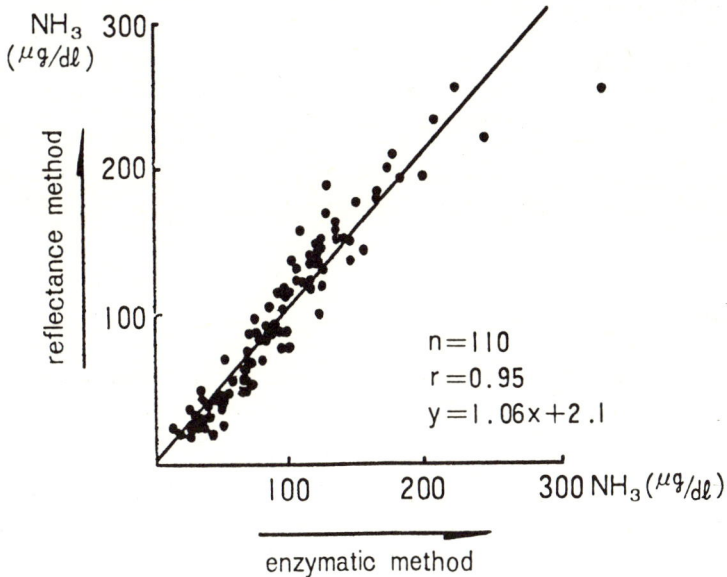

Fig. 5. Correlation between reflectance method and enzymatic method.

These results indicate a high incidence of hyperammonemia in newborns with perinatal asphyxia or respiratory distress, especially in low birth weight infants and a high mortality in newborns with hyperammonemia.

Only recently we made a reflectance-meter as shown in Fig. 4. In this case we use a kit with a single hole. The kit is put in the meter immediately after applying the blood and pushing the start button. In fifteen minutes, the concentration of blood ammonia is automatically indicated.

Fig. 5. shows the correlation between enzymatic method and reflectance method. A good correlation (R = 0.95) was obtained for both values.

CONCLUSION

So far, there has been no simple method for the detection of hyperammonemia because of the volatility of ammonia. The present method is simple and requires only a drop of blood. This is thought to be useful for bedside monitoring of hyperammonemia especially for newborns.

ACKNOWLEDGEMENT

We are indebted to Dr. T. Fujiwara, Department of Pediatrics Akita University and Dr. K. Baba, Department of Pediatrics Nihon University for kind cooperation.

This work was supported by grants from the Ministry of Education, the Ministry of Public Welfare, Japan and Chugai Seiyaku Co., Ltd.

REFERENCES

Goldberg, R. N., Cabal, L. A., Sinatra, F. R., Plajstek, C. E., and Hodgman, J. E., 1979, Hyperammonemia associated with perinatal asphyxia, Pediatrics, 64:336.
Johnson, J. D., Albritton, W. L., and Sunshine, P., 1972, Hyperammonemia accompanying parenteral nutrition in newborn infants, J. Pediatr., 81:154.
Mondzak, A., Ehrlich, G. E., and Seegmiller, J. E., 1965, An enzymatic determination of ammonia in biological fluid, J. Lab. Clin. Med., 66:526.
Shih, V. E., 1978, Urea cycle disorders and other congenital hyperammonemic syndromes, in : "The metabolic basis of inherited diseases," J. B. Stanbury, J. B. Wyngaarden, D. S. Fredrick-

son, eds., 362, New York : McGraw-Hill.

Snyderman, S. E., Sansaricq, C., Phansalkar, S. V., Schacht, R. G., and Norton, P. M., 1975, The therapy of hyperammonemia in a male neonate, Pediatrics, 56:65.

Tada, K., Okuda, K., Watanabe, K., Limura, Y., and Yamada, S., 1979, A new method for screening for hyperammonemia, Europ. J. Pediatr., 130:105.

IMMOBILIZATION OF MULTIENZYMES OF UREA CYCLE INTO FIBRIN MEMBRANE:

AN APPROACH TO AN ARTIFICIAL LIVER

H. Hagiwara, T. Nagasaki, Y. Saito and Y. Inada

Laboratory of Biological Chemistry, Tokyo Institute of
Technology, Ookayama, Meguroku, Tokyo 152, Japan

INTRODUCTION
   The final obligatory step for the metabolism of nitrogeneous
compounds in human body is the conversion of ammonia to urea through
carbamyl phosphate and urea cycle with the aid of ATP. Accumulation
of ammonia in human blood is observed for patients with liver
failure and hepatic coma and gives rise to a serious damage to
brain. The metabolic pathway for the urea synthesis from ammonia
is shown below, which is mainly conducted in liver.
Embedding these enzymes shown in the figure below into a matrix
may make an artificial liver for the patients with ammoniemia.
The authors utilized as the matrix fibrin polymer formed from fibrinogen in the final step of blood coagulation cascade. Fibrinogen is converted to fibrin monomer by the enzymic action of thrombin,

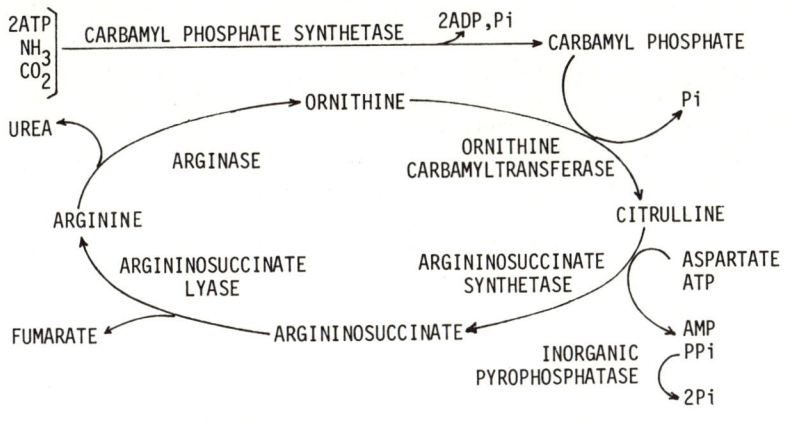

Conversion of ammonia to urea

and fibrin monomers are spontaneously associated with each other to
form fibrin polymer. Blood coagulation factor XIII catalyzes cross-
linking reactions between α chains or γ chains of the fibrin monomer
molecules to stabilize fibrin polymer. The above reaction proceeds
under mild conditions(at neutral pH, room temperature and in aque-
ous solution), and therefore, embedding of enzymes into fibrin
polymer may be performed without imparing their functions, even if
enzymes have subunits and are rather unstable. Another advantage
of the matrix is that this is a biological material from human
blood and scarcely causes the excitation of immune system.

MATERIALS AND METHODS
    Fibrinogen(93% clottable) and thrombin(12.5 units/mg) from
human blood were supplied by Green Cross Co. The fibrinogen preparation contains a small amount of blood coagulation factor XIII.
The enzymes embedded in the fibrin polymer are as follows;
carbamyl phosphate synthetase I(EC 6.3.4.16), ornithine carbamyl-
transferase(EC 2.1.3.3), argininosuccinate synthetase(EC 6.3.4.5),
argininosuccinate lyase(EC 4.3.2.1), arginase(EC 3.5.3.1) and
inorganic pyrophosphatase(EC 3.6.1.1). Inorganic pyrophosphatase
was used to remove pyrophosphate formed by argininosuccinate
synthetase, which strongly inhibits the latter enzyme[1].
Carbamyl phosphate synthetase I and ornithine carbamyltransferase
were partially purified from rat liver by the method of Raijman
and Jones[2]. Argininosuccinate synthetase and argininosuccinate
lyase were also partially purified from beef liver by the methods
of Rochovansky and Ratner[1] and Havir et al.[3], respectively.
Arginase(55 IU/mg protein) from bovine liver was purchased from
Sigma Chemical Co. Inorganic pyrophosphatase(200 IU/mg protein)
from yeast was obtained from Boehringer Co.

    Immobilization of enzymes with fibrin membrane was carried out
as follows; to the mixture of fibrinogen solution and the enzyme
solution was added thrombin and incubated for 2 h to complete the
reaction. During the incubation, the enzymes were trapped in
fibrin polymer which was further stabilized by blood coagulation
factor XIII. Fibrin polymer thus obtained was pressed by a weight
of 300 g/cm^2 and was washed completely with water to remove free
enzymes. Although the real amount of each enzyme embedded in
fibrin membrane was unclear, the apparent enzymic activity in the
membrane was approximate 10% of that initially applied. The
amounts of products such as citrulline, AMP and urea were determined
by the methods of Nakamura and Jones[4], Rochovansky and Ratner[1] and
Schimke[5], respectively.

RESULTS AND DISCUSSION
Urea synthesis from carbamyl phosphate
    All four enzymes in urea cycle together with inorganic pyro-
phosphatase were immobilized in a single fibrin membrane. Following amounts of each enzyme were used; 60 units of ornithine

carbamyltransferase, 1.7 units of argininosuccinate synthetase, 1.0 unit of argininosuccinate lyase, 150 units of arginase and 50 units of inorganic pyrophosphatase. It was verified that those enzymes were completely embedded in the fibrin membrane since there was no enzymic reaction after the membrane was removed from the reaction mixture. The most difficult point we encountered in this study was that the optimal pH values for all of these enzymes were not the same. We have set the pH of the reaction mixture at 7.5, because it is the optimal pH for two enzymes in the cycle, argininosuccinate synthetase and argininosuccinate lyase[1,3]. As far as those enzymes of which optimal pH deviates greatly from 7.5 were concerned, we were compelled to use rather larger amounts of enzymes. The optimal pH for arginase, for example, is reported to be 9.7(ref. 5). The immobilized multienzyme complex was put into a reaction medium containing 1.25 mM L-ornithine, 12.5 mM carbamyl phosphate(as the starting substrate), 6.25 mM L-aspartate, 6.25 mM ATP, 6.25 mM $MgCl_2$ and 30 mM KCl in 12.5 mM phosphate buffer(pH 7.5). The reaction was carried out at 37°C with gentle stirring. The formation of citrulline, AMP and urea were shown in Fig. 1. The formation of other reaction products and intermediates were not able to be determined due to the lack of appropriate procedures. The formation of citrulline preceded that of AMP(and urea). The former, however, stopped after 60 min of

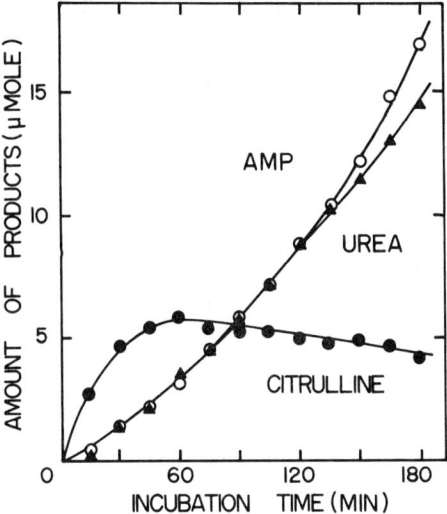

Fig. 1. Urea synthesis from carbamyl phosphate by immobilized enzyme system. The fibrin membrane containing ornithine carbamyltransferase, argininosuccinate synthetase, argininosuccinate lyase, arginase and inorganic pyrophosphatase was introduced to a reaction medium containing 1.25 mM ornithine, 12.5 mM carbamyl phosphate, 6.25 mM aspartate, 6.25 mM ATP, 6.25 mM $MgCl_2$ and 30 mM KCl in 12.5 mM phosphate buffer(pH 7.5).

incubation and the latter continued steadily even after 3 h . The constant level of citrulline, therefore, was obtained due to the balanced formation from ornithine(and carbamyl phosphate) and degradation to urea via argininosuccinate and arginine. The sum of AMP(or urea) and citrulline formed after 3 h was greater than the amount of ornithine introduced initially, which could never happen unless the cyclic system was operative to supply the starting substrate, ornithine, back to the cycle. These results all point to the conclusion that this immobilized cyclic enzyme system is functioning well.

The next series of experiments were carried out to see whether or not the cyclic enzyme complex in the fibrin membrane thus prepared has an improved efficiency over enzymes in solution. The solutions of these five enzymes were mixed in such a way that the activities of each enzyme would be equal between those in membrane and in solution by the following procedure. Apparent activity of the first enzyme in the cycle, ornithine carbamyltransferase, in the membrane was measured in the absence of aspartate and ATP. Knowing this value and the relative amounts of enzymes embedded in the membrane, we estimated the activities of other enzymes to be used in the solution. Fig. 2 represents the change in concentration of citrulline and urea formed from the starting substrate

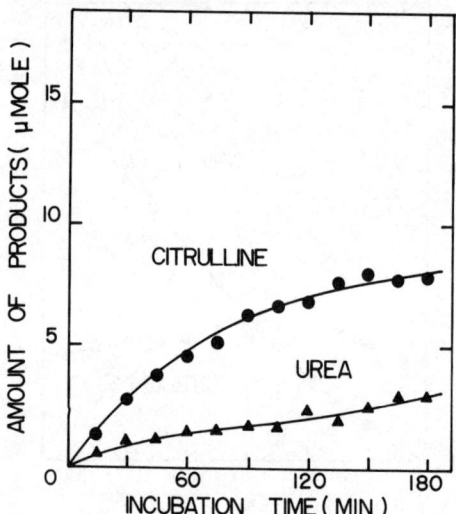

Fig. 2. Urea synthesis from carbamyl phosphate by soluble enzyme system. The soluble enzyme system composed of ornithine carbamyltransferase, argininosuccinate synthetase, argininosuccinate lyase, arginase and inorganic pyrophosphatase in a reaction medium containing 1.25 mM ornithine, 12.5 mM carbamyl phosphate, 6.25 mM aspartate, 6.25 mM ATP, 6.25 mM $MgCl_2$ and 30 mM KCl in 12.5 mM phosphate buffer(pH 7.5).

ornithine by the mixture of the five enzymes in solution. The measurement of AMP production was not feasible here due to a technical reason. The amount of urea produced by the reaction for 3 h was lower than that of citrulline or of starting substrate, indicating the considerably lower efficiency in solution.

The product of the first reaction, citrulline, which is also an intermediate in the cyclic system, was found a little more in the soluble system than in the immobilized system. On the other hand, a product of the last reaction, and also an end product of the cycle, urea, was produced by the immobilized system as much as five times more than by the soluble system. This high efficiency of immobilized system can be expected by the proximity of the enzymes which might reflect the compartmentalization of enzymes in vivo such as the electron transfer system in mitochondrion as well as urea cycle.

Urea synthesis from ammonia

Carbamyl phosphate synthetase I which catalyzes the formation of carbamyl phosphate from ammonia and carbon dioxide with the aid of ATP was embedded with four enzymes of urea cycle and inorganic pyrophosphatase. Ten ml of the enzyme solution containing 12.5 units of carbamyl phosphate synthetase I and ornithine carbamyl-

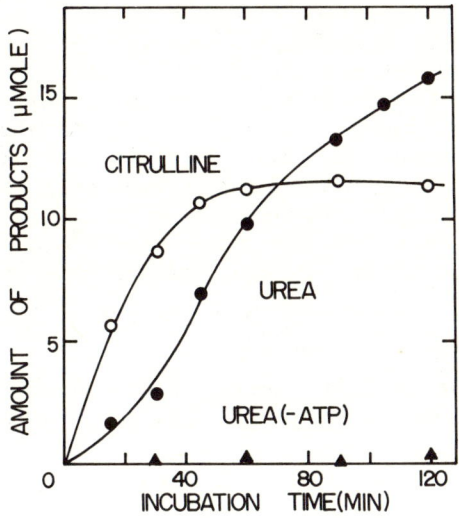

Fig. 3. Urea synthesis from ammonia by immobilized enzyme system. The fibrin membrane containing carbamyl phosphate synthetase I, ornithine carbamyltransferase, argininosuccinate synthetase, argininosuccinate lyase, arginase and inorganic pyrophosphatase was introduced to a reaction medium containing 50 mM NH_4HCO_3, 25 mM ATP, 10 mM N-acetyl-L-glutamate, 2 mM ornithine, 10 mM aspartate, 30 mM KCl, 20 mM $MgCl_2$ and 1 mM $MnCl_2$ in 20 mM PIPES buffer(pH 7.5).

transferase, 15.0 units of argininosuccinate synthetase, 58.4 units of argininosuccinate lyase, 440 units of arginase and 200 units of inorganic pyrophosphatase was added to 10 ml of 5% fibrinogen in 0.4 M glucose and 0.1 M Na-citrate buffer(pH 6.9) and thrombin(40 units) was added to the mixture. The fibrin membrane (6 x 5.5 cm) was prepared by the method described in Materials and Methods. The fibrin membrane(one-fourth of the total) was introduced to a medium containing 50 mM NH_4HCO_3, 25 mM ATP, 10 mM N-acetyl-L-glutamate, 2 mM ornithine, 10 mM aspartate, 30 mM KCl, 20 mM $MgCl_2$ and 1 mM $MnCl_2$ in 20 mM PIPES buffer(pH 7.5) and the amount of urea and of citrulline formed were measured at a given time. The reaction was carried out at 37°C with gentle stirring. The result is shown in Fig. 3. The amount of urea increased linearly with incubation-time, and that of citrulline, an intermediate of urea cycle, increased for the first 60 min of incubation and tended to approach a constant level. Since this reaction is endothermic, production of citrulline or urea was not observed at all in the absence of ATP. These results indicate that ammonia is continuously and efficiently converted to urea by these immobilized enzymes.

As a more practical approach to an artificial liver, a similar experiment was carried out in human plasma including aforementioned substrates and activators instead of buffer solutions. As is seen in Table 1, the fibrin membrane had an ability to synthesize urea from ammonia through urea cycle in plasma. Neither urea nor citrulline was formed without the fibrin membrane containing enzymes. In order to prevent fibrinolysis in plasma, approtinine (Trasylol 100 U/ml of plasma) was added to human plasma anticoagulated with 0.4% Na-citrate buffer(pH 7.4) in this experiment.

We have successfully immobilized living cells such as Chlorella cells and sea urchin eggs[6], as well as enzymes such as asparaginase[7], chloroplast ATPase[8], lysozyme and ribonuclease[9], and multienzymes including uricase, catalase, allantoinase and allantoicase[10].

Table 1. Urea synthesis from ammonia by immobilized enzymes in human plasma

Incubation Time (hour)	Amount produced (μmoles)	
	Citrulline	Urea
0	0	0
1	7.9	10.0
2	7.7	15.5
3	6.9	19.7

The amount of each enzyme added to the fibrinogen solution was as follows; 14.4 units of carbamyl phosphate synthetase I and ornithine carbamyltransferase, 33 units of argininosuccinate synthetase, 78 units of argininosuccinate lyase, 600 units of arginase and 150 units of inorganic pyrophosphatase.

In the present study, we have successfully synthesized urea from ammonia with fibrin membrane containing six enzymes. As far as we know, the removal of toxic ammonia in plasma of patients with ammoniemia has been done clinically by the injection of sodium glutamate or by the treatment with active charcoal. Under the circumstances, the present study was designed so as to give a clue to the development of an artificial liver for therapy.

SUMMARY

In order to give a clue to the development of an artificial liver, fibrin polymer was endowed with a biological function of urea synthesis from ammonia. Four enzymes participating in the urea cycle together with carbamyl phosphate synthetase I and inorganic pyrophosphatase were embedded in a single fibrin membrane formed from fibrinogen by the enzymic action of thrombin and blood coagulation factor XIII. The immobilized enzyme system thus prepared had the ability not only in buffer solutions but also in human plasma.

REFERENCES

1. O. Rochovansky and S. Ratner, J. Biol. Chem., 242:3839 (1967)
2. L. Raijman and M. E. Jones, Arch. Biochem. Biophys., 175:270 (1976)
3. E. A. Havir, H. Tamir, S. Ratner and R. C. Warner, J. Biol. Chem., 240:3079 (1965)
4. M. Nakamura and M. E. Jones, Methods in Enzymology, Vol. XVII-A, Academic Press, New York, (1970) p.286.
5. R. T. Schimke, Methods in Enzymology, Vol. XVII-A, Academic Press, New York, (1970) p.313
6. Y. Inada, S. Hirose, A. Matsushima, H. Mihama and Y. Hiramoto, Experientia, 33:1257 (1977)
7. Y. Inada, S. Hirose, M. Okada, and H. Mihama, Enzyme, 20:188 (1975)
8. Y. Inada, S. Yamazaki, S. Miyake, S. Hirose, M. Okada and H. Mihama, Biochem. Biophys. Res. Commun., 67:1275 (1975)
9. H. Okamoto, S. Kanai, P. Tipayang and Y. Inada, Enzyme, 24:273 (1979)
10. H. Okamoto, P. Tipayang and Y. Inada, Biochim. Biophys. Acta, 611:35 (1980)

DIAGNOSTIC, CLINICAL, PATHOLOGICAL AND
BIOCHEMICAL ASPECTS OF THE DISEASES

2) <u>Urea Cycle Enzyme Deficiencies</u>

N-ACETYLGLUTAMATE SYNTHETASE (NAGS) DEFICIENCY :

DIAGNOSIS, CLINICAL OBSERVATIONS AND TREATMENT

C. Bachmann, J.P. Colombo and K. Jaggi

Dept. of Clinical Chemistry, Inselspital
University of Berne, and
Children's Hospital Lucerne, Switzerland

NAGS plays an important role in the short term regulation of urea synthesis. This enzyme has extensively been studied by Shigesada and Tatibana in rodents [1,2,3]. We have purified human NAGS in order to find out what substrate concentrations should be used in an assay for human liver. The kinetic results including product inhibition studies are compatible with a rapid equilibrium random bibi mechanism. In the absence of arginine the K_m are for Acetyl CoA 4.4 mmol/l, for glutamate 8.2 mmol/l. The K_i of N-acetylglutamate is at 0.5 mmol/l. Arginine approximately triples the activity (half maximum activation at about 30 µmol/l), and appears to lower the K_m of acetyl CoA, but not of glutamate[4].

Based on these data we developed an enzyme assay for human liver biopsies. The sample should immediately be frozen and rapidly homogenized and sonicated on ice. The supernatant will be stable for at least 4 days at -20°C. The incubation (37°C) is done in Tris-HCl 125 mmol/l (printing error in Ref. 5), 1.5 mmol/l EDTA, pH 8.7, glutamate 20 mmol/l, acetyl CoA 6 mmol/l, with or without arginine 1 mmol/l. The assay is linear up to 20 min at 37°C and up to 10 mg tissue/assay. Details are given elsewhere[5].

We encountered a patient with a deficiency of NAGS. A short report has been published[5]. The patient was closely followed because of his family history. Relevant biochemical data are shown in Fig. 1 and table 1.

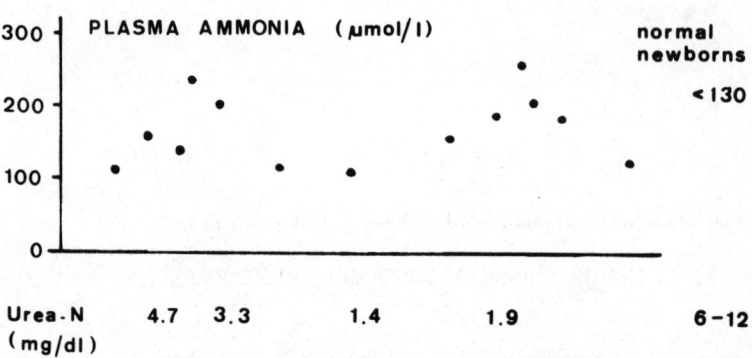
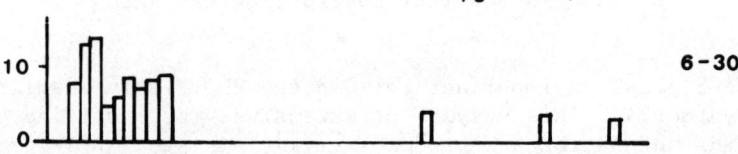
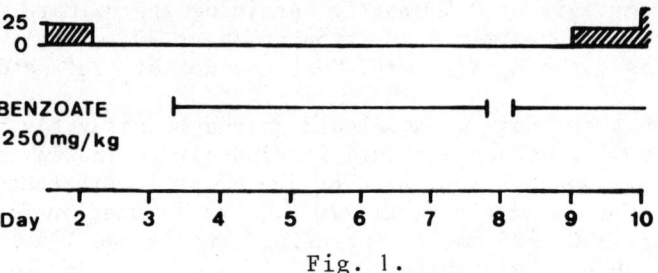

Fig. 1.

Hyperammonemia increased despite a stop of protein supply (100 -120 kCal/kg were given as glucose). Plasma urea nitrogen was low, as well as urinary orotic acid excretion. In addition hypocalcemia (as in a sibling who died in the neonatal period) and coagulation problems were present. We found no organic aciduria (except for lactic aciduria). The aminoacid pattern exhibited nonspecific changes secondary to hyperammonemia.

Table 1. Aminoacids in plasma and urine during the neonatal period

	Plasma			
	Normal newborns mmol/l	Patient Day 3	7	10
GLN	0.24 - 0.82	1.02	0.49	0.22
GLU	0.03 - 0.10	0.23	0.09	0.06
ALA	0.13 - 0.45	0.36	0.57	0.23
ORN	0.04 - 0.21	0.14	0.11	0.09
LYS	0.07 - 0.27	0.28	0.16	0.18
ARG	0.02 - 0.12	0.12	0.04	0.14
	Urine (mmol/g Creat.)			
		Day 2	3	9
GLN	< 0.66	0.52	0.42	0.57
GLU	< 0.14	0.35	0.33	1.19
ALA	< 0.40	0.63	0.72	6.41
ORN	< 0.02	0.30	1.31	0.69
LYS	< 0.12	0.22	0.68	6.14
ARG	< 0.08	0.06	0.50	0.27
CIT	< 0.04	0.08	0.30	1.10

The increased citrulline excretion before and after benzoate should be noted. Benzoate treatment[6] was effective in reducing plasma ammonia as illustrated by its decrease upon starting the treatment and after an accidental withdrawal on day 8. Liver biopsy was done on the fifth day, immediately put on dry ice and homogenized within three hours. The major part was used for CPS determination, since we expected its deficiency. CPS was normal, but as described, we found no activity above blank level of NAGS done in triplicate, while a biopsy from an infant had an activity of 23 nmol min^{-1}g^{-1} tissue (150 nmol min^{-1}g protein), (blank 84 dpm; sample 239 dpm). Reference values (table 2) had been obtained for adult livers ; the biopsies were taken during colonic or sigmoid cancer operations ; the patients, having received only tea or bouillon for the two days preceeding the intervention, were thus on a low protein intake comparable to that of the hyperammonemic patient.

Table 2. Reference values of NAGS (adult liver biopsies, n = 12)

Median	Range	
14.5	(5.6 - 33) nmol min^{-1}g^{-1}	
35.7	(22.5 - 52) nmol min^{-1}g^{-1}	with arginine
92	(34 - 203) nmol min^{-1}g^{-1}	protein
233	(144 - 320) nmol min^{-1}g^{-1}	protein with arginine

We had not enough material to measure the arginine stimulated NAGS in the patient. Surprisingly OTC was reduced. The conditions used were pH 7.7 triethanolamine buffer, ornithine 2.35 mmol/l and carbamylphosphate 5 mmol/l final concentration. As reported[5] using autopsy material we found an inhibition of OTC when pre-incubating the 1:300 homogenate with benzoic acid for 10 minutes at 37°C before doing the final dilution 50/650 with triethanolamine buffer for the OTC assay. Whether this inhibition is due to accumulating phosphate (a known inhibitor of OTC) produced for the activation of benzoic acid or nonspecific action e.g. activation of phosphatases is not clear. There was not enough material of the patient for evaluating the type of inhibition in a kinetic study of OTC. Thus we cannot prove directly that OTC was inhibited in the patient. A further tempting hypothesis would be that NAG plays a role in activating the OTC precursor found extramitochondrially, since Rivas et al.[7] have found that NAG stimulates markedly proteolysis at neutral pH. We realize however that the increased citrulline excretion found also while the patient was on benzoate treatment renders an _in vivo_ inhibition of OTC unlikely. Citrulline was present in plasma, but could not be quantitated because of bad resolution from alanine.

Because Brown et al.[8] had advocated the use of carbamylglutamate (CG) and arginine in hepatic coma and Kim et al.[9] actually used it in rats, we tried it on the patient. ✶ This was furthermore necessary because while on benzoate the patient had oedemas,

---

✶ We thank S. Grisolia for sending us the first CG sample for initiating treatment.

abdominal distension, and respiratory insufficiency. Similar symptoms reappeared later when benzoate was used to control acute hyperammonemia which occured after increasing protein supply during a trial of total withdrawal of CG.

CG was administered perorally. We gave a first dose of 300 mg. The aminoacid changes (especially the decrease of arginine) are of interest (table 3) :

Table 3. Changes of selected aminoacids in plasma (mmol/l) after a single dose of CG

	GLN	GLU	ORN	ARG	AMMONIA (μmol/l)
0	0.22	0.06	0.09	0.14	157
1 h	0.30	0.05	0.11	0.05	154
2 h	0.24	0.09	0.21	0.03	146
4 h	0.02	0.13	0.11	0.02	151

We increased the dose of CG gradually up to 3600 mg/day (750 mg/kg). Symptoms of intoxication occured which can be characterised as a sympathicomimetic reaction : tachycardia, profuse sweating, increased bronchial secretion, increased temperature and persistent crying.

At that time orotic acid was increased (5890 μmoles/g creatinine), plasma arginine was normal (0.08 mmol/l) and ornithine (0.29 mmol/l) certainly not reduced. Glutamine was markedly increased to 1.0 mmol/l ; there was no hyperammonemia (49 μmol/l). In consequence we kept the arginine supply constant, but stopped CG administration for 24 hours. Ammonia increased to 173 μmol/l. 6 Hours after reinstituting CG (1800 mg/day) ammonia values. were at 92 μmol/l. The orotic acid excretion went down to 80 μmol/g creatine within 36 hours, and normalized completely when we increased the arginine supply to 2 mmol/kg/day.

The main problem in the treatment of this patient has been to find the right dosage of CG and arginine. The demand further depends on the amount of ammonia to be detoxified. Reducing GC led to an increase of ammonia, while a decrease of arginine supply increased orotic acid excretion and in consequence produced hyperammonemia.

The dependence on arginine is not clear. Peroral application of arginine leads to the changes in plasma shown in table 4.

Table 4. Changes of plasma arginine, ornithine and leucine after a peroral dose of 0.8 mmol/kg of arginine-HCl

mmol/l	0	30'	60'	90'	120'
ARG	0.05	0.21	0.25	0.27	0.70
ORN	0.04	0.11	0.15	0.18	0.21
LEU	0.10	0.09	0.08	0.07	0.08

Arginine is thus certainly resorbed and increases in plasma as well as ornithine. Its excretion does not account for a net deficit, albeit not all the urines have been processed yet. Urea formation is stimulated both by CG and arginine. Since urea is formed ornithine should be available for carbamylphosphate fixation. Creatinine excretion was normal. We did not have a chance, so far, to investigate the excretion of guanidino compounds or of polyamines. We recognize that plasma concentrations of aminoacids are perhaps not reflecting the intramitochondrial concentrations. We are tempted to speculate that CG and perhaps N-acetylglutamate affect the transport of ornithine or citrulline at the mitochondrial membrane.

We further ignore the metabolic fate of CG. As measured by GC/MS only 12-15 % of the perorally administered CG is excreted unchanged. Studies with stable isotopes are needed to get a better insight.

The patient is actually slighty retarded. He is ataxic, but lost the muscle rigidity which was a problem during his first few months of life. Clinically episodes of hyperammonemia are manifested first by hyperpnea. If ammonia increases further, hypotonicity and loss of consciousness occur. CG intoxication leads to the symptoms mentioned above, and is best recognized by the profuse sweating.

ACKNOWLEDGEMENT

We thank the nurses of the Lucerne Children's hospital for their dedication, Mrs. Pfister, Gradwohl, Kokorovic and Mr. Krähenbühl, Bühlmann and Sansano for their technical assistance and Mrs. Krähenbühl for the careful preparation of the manuscript. This work was supported by Grant N° 3.551.-0 79 of the Swiss National Research Foundation.

REFERENCES

1. K. Shigesada and M. Tabibana, J. Biol. Chem. 246:5588 (1971).
2. K. Shigesada and M. Tatibana, BBRC 44:1117 (1971).
3. C. Uchiyama, M. Mori and M. Tatibana, J. Biochem. 89:1777 (1981).
4. C. Bachmann, St. Krähenbühl and J. P. Colombo (in preparation).
5. J. P. Colombo, St. Krähenbühl, C. Bachmann and P. Aeberhard, (submitted for publication).
6. M. C. Batshaw and S. W. Brusilow, J. Pediat. 97:893 (1980).
7. J. Rivas, A. Reglero and S. Grisolia, in "The Urea Cycle", S. Grisolia, R. Baguena, F. Mayor, eds. Wiley, New York (1976).
8. R. Brown, R. Manning, M. Delp and S. Grisolia, Lancet I:591 (1958).
9. S. Kim, W. U. Pain and P. P. Cohen, Proc. Nat. Acad. Sci. (USA) 69:3530 (1972.)

ORNITHINE TRANSCARBAMYLASE (OTC) IN WHITE BLOOD CELLS AND JEJUNAL MUCOSA

N. Nagata, I. Akaboshi, J. Yamamoto, F. Endo,
I. Matsuda, T. Katsuki[*]

Department of Pediatrics, Department of Microbiology[*]
Kumamoto University Medical School, 860, Japan

INTRODUCTION

The presence of OTC in white blood cells was demonstrated by Wolfe and Gatfield(1) and Snodgrass et al.(2), but was denied by Rabier et al (3). OTC deficiency was diagnosed using white blood cells by Wolfe and Gatfield (1) and Krieger et al (4). On the other hand, Snodgrass et al (2) reported that OTC deficiency in the liver can not be inferred from the measurements of the enzyme's activity in peripheral white blood cells, because the latter parameter was normal in their patients. The presence of OTC in jejunal mucosa and reliability of this material for detecting OTC deficiency were demonstrated by Levin et al (5) and Cathelineau et al (6). However, the nature of OTC in intestinal mucosa was not clearly defined. We would like to describe some kinetic studies on OTC in white blood cells and jejunal mucosa.

MATERIAL AND METHOD

L-ornithine hydrochloride and carbamylphosphate (dilithium salt) used were purchased from Sigma Chemical Co, St. Louis, ^{14}C-ornithine was purchased from New England Nuclear, Boston. All other chemicals were of reagent quality.
White blood cells : white blood cells were prepared by the method of Wolfe and Gatfield(1). Separation of mononuclear cells from the granulocytes was performed by differential filtration using a modified method of Boyüm (7).
Establishment of lymphoid cell line : lymphoid cell lines were established from normal subject and OTC-deficient infant after incubation with Epstein-Barr (EB) virus (8). Liver and jejunum : post mortem specimens of the liver and jejunum were obtained from

a patient dead by accident. Materials were stored at -60°C for three months before being studied.

Measurement of OTC activity : the radiochemical method of Sinatra et al (9) was used for OTC assay in white blood cells. White cell pellet was suspended in 0.1% cetylpridium chloride, frozen and thawed three times, and centrifuged at 12,000 X g for 25 min. One hundred μl of the supernatant, ^{14}C-carbamylphosphate (final concentration 0.87 mM; specific activity 1.3×10^3 cpm/μmoles) and ornithine (final concentration. 5 mM) were mixed and 0.05 M triethanolamine acetic acid buffer pH 8.5 was added. The mixture was incubated at 37°C for 60 min. The reaction was then terminated by adding 100 μl of 3N formic acid. After heating 5 min in boiling water; the excess $^{14}CO_2$ was removed by adding crushed dry ice. The supernatant, after centrifugation for 10 min at 2,000 X g, was transferred to a vial and counted by liquid scintillation spectrometer.

OTC in liver was measured by both radiochemical method and conventional colorimetric method (10) (11), and OTC in jejunal mucosa by colorimetric method only.

Autoradiography : logarithmically growing cells, each 5×10^5 in number were washed with arginine-free RPMI 1640 and incubated for 48 h at 37°C in RPMI 1460, which contained ^{14}C-ornithine or ^{14}C-citrulline instead of arginine. After the incubation, slides were dipped in Kodak NTB 3 emulsion, exposed for 5 days, developed in D-19 for 3 min, and stained with Giemsa.

Growth curve of lymphoid cells : lymphoid cells were cultured in two different media : RPMI 1640 with 1 mM arginine (ornithine-, arginine+); and RPMI 1640 with 1 or 5 mM ornithine (ornithine+, arginine-).

RESULT

OTC activity in white blood cells and cultured lymphoid cells : a positive linear relationship was found between incubation time (40 and 60 min) and conversion of ^{14}C-carbamylphosphate. The relationship between the conversion rate of radioactive substrate and the amount of cell lysate, measured after 60 min incubation, was linear. The results obtained are summarized in Table 1. Apparent Km of OTC for ornithine and carbamyl phosphate are shown in Table 3.

Growth curves of cultured lymphoid cells : lymphoid cells from both the controls and the OTC-deficient infant showed normal logarithmic growths in the arginine (+), ornithine (-) medium, whereas the cells failed to grow in the arginine (-), ornithine (+) medium.

Autoradiography : when in the presence of ^{14}C-ornithine, few grains were found on lymphoid cells from both the normal subjects and the OTC-deficient patient.

Table 1   OTC in white blood cells and cultured lymphoid cells

	mean $\pm$ S.D.
White blood cells (N=10)	1.32 $\pm$ 0.95 (nmoles/mg protein/h)
Granulocyte (N=2)	1.0
Mononuclear cells (N=2)	0.4
Cultured lymphoid cells * (N=4)	0

* including a sample derived from an OTC - deficient patient.

However, when ^{14}C-citrulline was substituted for ^{14}C-ornithine more than 80% of cultured lymphoid cells contained more than 20 grains per cell.
OTC activity in jejunal mucosa : six samples (70 mg of wet weight) of jejunal mucosa were taken out from the different area of stored jejunum at random. As shown in Table 2, OTC activities obtained were homologous in these samples. Apparent Km of OTC in jejunum are shown in Table 3.
Heat stability of OTC in jejunal mucosa and liver : heat stability was tested by the incubation in water both for one min at given temperature. As shown in Fig 1, a very similar profile was obtained between hepatic and jejunal samples.
pH profile of OTC in jejunal mucosa and liver : the obtained results indicated that pH profile of OTC was similar between these two organs and the maximum activity was found at pH 8.0 in both samples.

Table 2   OTC in jejunal mucosa and liver.

Jejunal mucosa (sample numbers)		
	1.	9.94 (μmoles/mg protein/h)
	2.	10.35
	3.	10.75
	4.	9.00
	5.	9.20
	6.	10.72
mean $\pm$ S.D.		9.99 $\pm$ 0.75
liver (N=5)		43.0 $\pm$ 8.0 (μmoles/mg protein/h)

Table 3 Apparent Km of OTC in white blood cells, jejunal mucosa and liver.

	Km for ornithine	Km for carbamylphosphate
White blood	6.7, 6.2 mM	0.6 mM
Liver	0.6 mM	0.12 mM
Jejunum	0.6 mM	0.2 mM

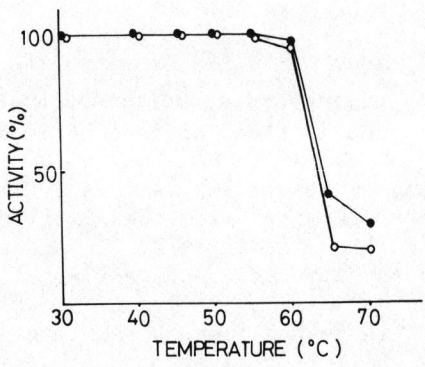

Fig. 1. Heat stability of OTC in liver (o) and jejunal mucosa (●).

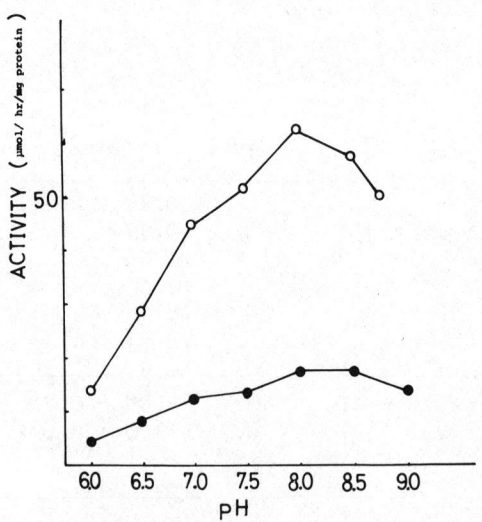

Fig. 2. pH profile of OTC in liver (o) and jejunal mucosa (●).

COMMENT

In the present study, the presence of OTC in white blood cells is clearly demonstrated, as was found by Wolfe and Gatfield (1) and Snodgrass et al. (2), whereas the enzyme activity in established lymphoid cells is absent. This is further confirmed by cell growth curve and autoradiography.

Our study demonstrated that OTC activity in the granulocytes is higher than that in mononuclear cells. Wide variation of OTC activities in white blood cells found in the present study and those by Wolfe and Gatfield (1) and Snodgrass et al (2), may be related to individual difference of two cell populations. Apparent Km of OTC for carbamyl phosphate and for ornithine are approximately 5 and 10 times, respectively, higher in white cells than those in liver. On the other hand, apparent Km of OTC in jejunal mucosa are close to those of liver. Other parameters including heat stability and pH profile are again close to those of liver. These results suggest that OTC in white cells is different from that in liver, as was suspected by Snodgrass et al (2) and the enzyme in jejunal mucosa might be of the same genetic origin as that in liver In addition, the fact that OTC activities are homologous in jejunal samples taken out from different area, indicates that jejunal mucosa obtained by peroral biopsy may be useful for detecting OTC deficiency as alternate to liver tissue. On the other hand, OTC in white blood cells could not be used as a reliable marker of genetic OTC deficiency.

REFERENCES

1. D.M. Wolfe, and Gatfield, P.D., Leukocyte urea cycle enzymes in hyperammonemia. Pediatr. Res. 9 : 531. (1975).
2. P.J. Snodgrass, P.S. Wappner, and I.K. Brandt., Letter to the Editor : White cell ornithine transcarbamylase activity cannot detect the liver enzyme deficiency. Pediatr. Res. 12 : 873 (1978).
3. P. Rabier, L. Cathelineau, and P. Kamoum., Letter to the Editor : Lack of mitochondrial enzymes of urea cycle in human white blood cells. Pediatr. Res. 13 : 207 (1979).
4. F. Krieger, C. Bachmann, W. Gronemeyer, and J. Cejka., Propionic acidemia and hyperlysinemia in a case with ornithine transcarbamylase deficiency. J. Clin. Endcrinol. Metab. 43 : 796 (1976).
5. B. Levin , V.G. Oherholzer, and L. Sinclair., Biochemical investigations of hyperammonemia. Lancet 2 : 120 (1969).
6. L. Cathelineau, J. Saudubray, and C. Polonsvski., Heterogenous mutations of the structural gene of human ornithine carbamyltransferase as observed in five personal cases. Enzyme 18 : 103 (1974).
7. A. Boyüm., Separation of leukocytes from blood and bone marrow. Scan. J. Clin. Lab. Invest. Suppl. 21 : 77 (1968).

8. I. Matsuda, J. Yamamoto, N. Nagata, N. Ninomiya, I. Akaboshi, H. Ohtsuka, and I. Katsuki., Lysosomal enzyme activity in cultured lymphoid cell lines. Clin. Chim. Acta. 80 : 483 (1977).
9. F. Sinatra, T. Yoshida, M. Applebaum, W. Mason, J. Hoogenraad, and P. Sunshine, K., Abnormalities of carbamyl phosphate synthetase and ornithine transcarbamylase activity in liver of patients with Reye's syndrome. Pediatr. Res. 9 : 829 (1975).
10. G.W. Jr. Brown, and P.P. Cohen., Comparative biochemistry of urea cycle synthesis. I. Methods for the quantitative assay of urea cycle enzymes in liver. J. Biol. Chem. 239 : 1770 (1959).
11. I. Matsuda, S. Arashima, H. Nambu, Y. Takekoshi, and M. Anakura., Hyperammonemia due to a mutant enzyme of ornithine transcarbamylase. Pediatrics. 48 : 595 (1971).

IMMUNOCHEMICAL ASSAY IN 16 BOYS WITH ORNITHINE TRANSCARBAMYLASE DEFICIENCY

B. Francois, P. Briand,[*] L. Cathelineau[*]

Neuropediatric Department, Louvain University, Brussels
Belgium and Biochemical Genetics Laboratory[*], Hôpital
Enfants Malades, Paris, France

SUMMARY

Enzymatic activities and kinetics of liver ornithine transcarbamylase (OTC) were studied in sixteen human male infants with an OTC mutation. In the same liver fragments, cross reacting material (CRM) was measured with anti-OTC monospecific antibody. These studies allow us to describe five groups of mutations.

INTRODUCTION

Deficiency of ornithine transcarbamylase (OTC) (EC.2.1.3.3.) a mitochondrial enzyme of the urea cycle, appears to be one of the most frequent causes of inherited ammonia intoxication. Recent studies provided strong evidences for a X-linked mode of inheritance of this disease[1,2]. Different kinds of mutations leading to different degrees of enzymatic dysfunction in affected hemizygous males or heterozygous females have been reported with increasing frequency during this last decade[3].

In the classical form of the disease, OTC activity varied widely in heterozygous females and was generally undetectable (less than 1% of control) in male infants who died during the neonatal period. But, in addition, several male atypical cases have been reported with a mild clinical course and partial OTC deficiency apparently due to a variant enzyme[4,5,6,7,8,9,10,11,12,13,14]. However, the molecular basis of the underlying enzyme defect leading to the heterogeneity of the disease is poorly understood. In order to define a better correlation between the clinical course and the biochemical characteristics, we carried out detailed studies on the kinetic and immunochemical properties of the liver enzyme from normal individuals and sixteen affected male infants.

MATERIAL AND METHOD

The livers from sixteen male patients who suffered from hyperammonemia due to an OTC deficiency have been studied. Among these patients, 8 (6 non-related, and 2 related) had a neonatal onset of symptoms, and died soon after birth whatever the treatment which was performed. The diagnosis of OTC deficiency was confirmed on post-mortem samples kept at -80°C. Eight patients from five families had a later onset of symptoms: periodic attacks of lethargy and vomiting were the most common features observed in the clinical course of the disease. Unfortunately, in 5 cases the diagnosis of OTC mutations was established on post-mortem specimens only after the last and fatal hyperammoniaemic episode. For the remaining three, the diagnosis was performed on tissue samples obtained by liver biopsy. Two of them are still alive.

Preparation of liver extracts

The patients' liver tissue was removed shortly after death, immediately frozen and stored at -80°C. The livers of children who died from another cause than urea cycle deficiency were used as control. Tissue fragments were weighed and homogeneized in 14 volumes of cold 0.1% cetyltrimethylammonium bromide.

Enzyme assays

OTC assay was performed according to the method of Snodgrass[15]. In the kinetic studies, one of the two substrates (ornithine or carbamylphosphate) was at the final concentration of 5 mM while the other ranged from 0.04 mM to 5 mM. The effects of pHs variation were studied in 0.2 M triethanolamine HCl buffer (pH 6.5-10). Homogenates were diluted in 0.5% of bovine serum albumin according to their activities previously determined with saturating concentration of the two substrates. The final volume of 1 ml containing 100 µl of the diluted homogenate was incubated for 10 min. at 37°C and the reaction was stopped by addition of 500 µl of 10% trichloracetic acid. The citrulline formed was assayed in the supernatant after centrifugation by an automated colorimetric method[16,17]. Protein contents were determined according to the method of Lowry[18]. Velocity of each substrate concentration was plotted on a graph to calculate Km and Vmax values on the basis of double reciprocal plots.

Preparation of antibodies

A monospecific antiserum against purified bovine hepatic OTC (kindly supplied by Dr. Margaret Marshall) has been raised in rabbits by injecting 0.8 mg of the purified enzyme mixed with Freund's complete adjuvant subcutaneously three times, once a week and boosted twice every 6 weeks. The antiserum was used without any further purification.

Immunodiffusion

Ouchterlony double immunodiffusion[19] and Mancini radial immunodiffusion[20] were performed in agarose 1% containing veronal buffer (25 mM pH 8.6) and sodium azide (3 mM). For the Mancini assays, antibovine OTC antiserum was diluted (1/30) before adding to agarose. After diffusion for two days at room temperature the layer was washed with 0.1 M NaCl, dried, stained with Coomassie blue and destained[19].

RESULTS

OTC activities and kinetics

All urea cycle enzymes activities but OTC are found normal in patients' liver. Normal control values for OTC are (mean ± SD) : 35.5 ± 8.0 M/mg protein/h at pH 8.0 and 30.0 ± 7.0 at pH 9.0. Normal Km for ornithine is : 0.22 ± 0.10 mM at pH 8.0 and for carbamylphosphate : 0.12 ± 0.05 at pH 8.0 (n = 10).

Residual activities of each patient are reported in tables 1 and 2 as percentage of the mean obtained in control livers. The lowest activity detected by the assay is 0.1% of the control values. In eight patients who died in the neonatal period no residual enzymatic activity could be detected. On the other hand, all children with later onset of symptoms have partial activity at pH 8.0 (the optimum pH for the enzyme) ranging from 5% to 30% of the control values. Among these patients we were able to distinguish three kinds of abnormalities in the enzymatic activity: four patients from two non-related families had 5% of the control activity at all pHs (fig. 1) and normal Km for the two substrates, ornithine and carbamylphosphate. Another group, with three patients from two non-related families exhibited all three the same kinetic abnormalities: an increased Km for ornithine, which was ten times the Km of the normal enzyme (table 2) while the Km for carbamylphosphate was normal, and an abnormal curve of activity versus pH (fig. 1), with 10% of normal activity at pH 8.0 and 60% of the control activity at pH 9.0. A third group with one patient had an increased Km for ornithine (5 times the normal value) and an abnormal curve of activity versus pH with 30% of the normal activity at pH 8.0 and 150% at pH 9.5 (fig. 1).

Determination in CRM in human patients

Previous studies had shown that monospecific antibody against

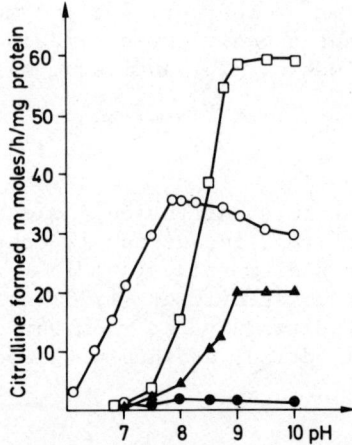

Fig. 1. Variation of the activities of human OTC as function of % of control in three different groups of OTC mutations. o-o normal human OTC (mean of ten experiments). •-• first group (see table 2 patients 9 to 12). ▲-▲ second group (patients 13 to 15). □-□ third group (patient 16).

the purified bovine hepatic OTC can completely precipitate human OTC. Furthermore the two enzymes, human and bovine are equally cross reactive with the antiserum using the Ouchterlony double immunodiffusion[21]. Using this antiserum, the amount of CRM was determined by Mancini radial immunodiffusion. Among the 8 patients who died in the neonatal period no CRM was detectable in all but the two related patients (table 1 - fig. 2). In these two last children's livers the amount of CRM approximated that of control liver.

Among the patients with later onset of symptoms, three different groups of data were obtained, data which are closely related to the kinetic characteristics of the mutant enzyme.

In the first group four patients showed an amount of CRM which approximated that of their enzymatic activity (table 2 - fig. 2) 5% of normal values. In the second group, the CRM amount was 50 to 60% of the normal values, exactly the same percentage as the higher value of activity observed (at pH 9.5 see table 2 and fig. 2). And the last patient has a CRM concentration higher than in normal liver (table 2).

Since monospecific antiserum against purified mouse OTC was also available (as discussed later in this book), radial immunodiffusion was performed using these antibodies and the date presented as comparison in table 1 and 2.

Table 1. OTC activities and OTC cross reactive material (CRM)

Patients	Residual activity pH 8.0 & pH 9.5	C.R.M. (% of control) With bovine antiserum	With mouse antiserum
1-G.O.	0°	-°°	-
2-D.T.	0	-	-
3-L.A.	0	-	-
4-C.G.	0	-	-
5-P.S.	0	-	-
6-A.A.	0	-	-
7-J.JM.	0	48	80
8-J.V.	0	ND°°°	80

° is indicated for <0.01% of normal activity
°° undetectable CRM
°°° Not determined

Table 2. OTC activities, kinetics and cross reactive material (CRM) in patients with partial OTC deficiency.

PATIENTS	RESIDUAL ACTIVITY % of control pH 8.0	pH 9.5	Km for Ornithine	C. R. M. (% of control) With bovine antiserum	With mouse antiserum	ANIMAL MODEL
9-B.J.[14]	5	5	normal	6	ND°	
10-B.Jo.	5	5	normal	6	ND	spf-ash
11-B.C.	5	5	normal	5	ND	
12-M.B.	5	5	normal	30	10	
13-V.V.[8]	10	60	X 10	63	50	
14-H.A.	10	60	X 10	ND	52	
15-B.N.	10	60	X 10	53	52	
16-D.L.[12]	30	150	X 5	> 100	ND	spf

° ND : Not determined

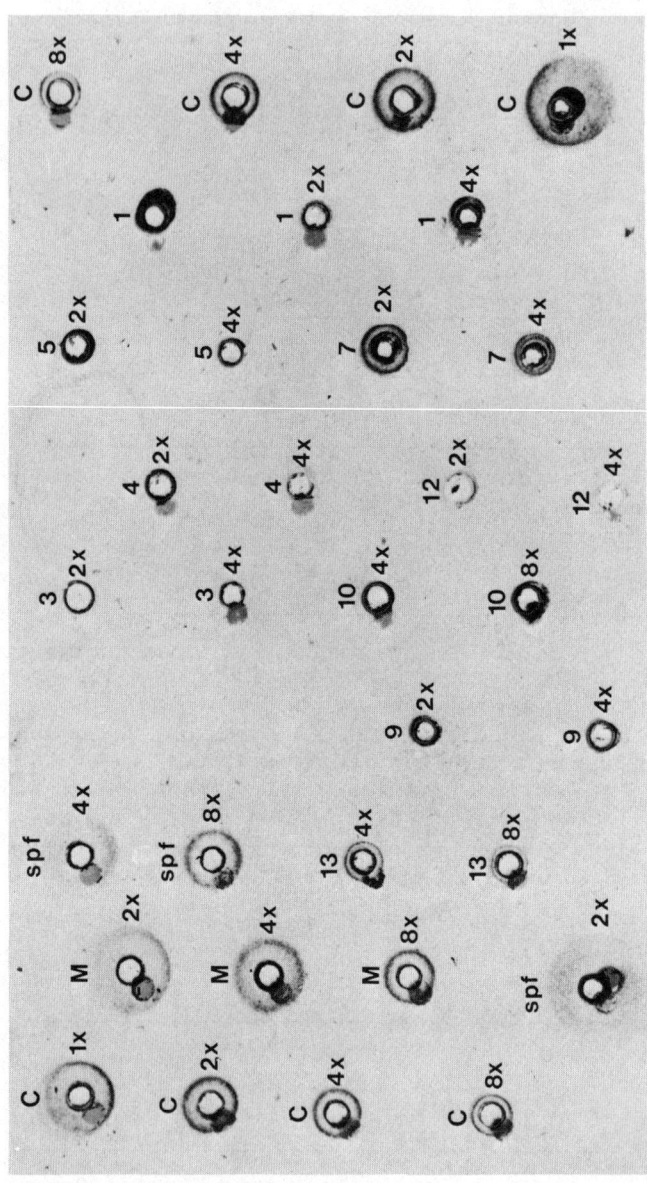

Fig. 2. Radial immunodiffusion of serial dilutions of liver extracts from human controls (C), mouse controls (M), spf mutant mice as well as some patients with OTC deficiency (numbers as in table 1 and 2).

DISCUSSION

In a previous work[8] a tentative attempt was made to clarify the heterogeneity of OTC deficiencies according to their kinetic characteristics. The possibility to use immunoassay in relationship with kinetic studies of the mutant enzyme on a larger number of male infants with OTC deficiency provides a better opportunity to define the biochemical phenotype of the OTC mutations.

Among the 16 patients presented in this study, our data suggest that at least five different classes of enzymatic defect could be distinguished.

All the patients with neonatal onset of symptoms presented no or very poor residual activity. In all but two no cross reactive material could be detectable. The observed agreement between the lethal clinical outcome, the absence of residual activity and no reactive immuno material may reflect in these patients a structural gene mutation which alters the rate of synthesis of the OTC protein as a punctual non-sense mutation or delation. The possibility of a regulatory mutation controlling the OTC synthesis or degradation could not be ruled out in such cases. The pedigrees being always in favour of a x-linked transmission, we should have to assume that the presumed gene of regulation for the OTC is located on the X chromosome, this remaining to be demonstrated.

The two related patients with no residual activity but normal CRM who died a few days after birth must represent a second class. This could possibly be due to a gene mutation controlling the active site of the enzyme (mis-sense) while it does not change the total synthesis of the polypeptide.

In the group of patients with later onset, the findings of boys with residual CRM which approximates that of their enzymatic activities suggests a well defined group. Although the same picture was documented in three individuals of a same family, both clinical data and enzyme studies suggested that this mutation is related to one previously described[6,9] in two other non-related patients in whom no immunological study was performed. Therefore this mutation probably represents an additional class. Moreover a parallelism should be drawn between this enzyme abnormality found in human and the one found in an animal model, the spf-ash mouse, as discussed elsewhere in this book. This mutation could be understood in the light of the most recent work concerning the translocation of mitochondrial enzymes across both mitochondrial membranes to the matrix. It has been shown that OTC as some other enzymes[22,23,24] is synthetized in the cytosol as a larger precursor which has higher molecular weight than the mature enzyme. This peptide is translocated into the mitochondria in association with a proteolytic processing to form the mature OTC. In this class, it must be

speculated that the translocated system is impaired leading to a
low amount of normal protein in the mitochondria.

Among patients with partial OTC deficiency, two additional
classes were identified. Three patients from two non-related families exhibited a similar decrease of enzyme activity and immunological material as compared to controls, in addition to an abnormal Km for ornithine. We may speculate that this picture represents
a mutation of the structural gene affecting the active site of the
OTC enzyme.

The last class is defined only by one patient 16-D in whom
abnormal Km for ornithine and an increased activity at pH 9.5
as compared to that obtained in control livers at the optimum
pH 8.0 were detected. Another non related infant with the same
kinetic abnormalities has been observed recently and he is still
under investigation. The amount of CRM found in the patient's
liver was here increased as compared to control. The fact that
these data are similar to that found in another mice mutant "spf"
allow us to differentiate this mutant enzyme from the latter. This
mutation may suggest that the immunological protein accumulation is
an adaptative advantage compared to the mutant enzyme.

Our study brings additional support in the concept of the
heterogeneity of the OTC deficiency. However the number of cases
in some classes is limited, and these preliminary data needs further
confirmation.

AKNOWLEDGMENTS

This work has been supported by a grant from INSERM n°803004.
We are grateful to the following for sending us samples from their
patients : Pr. J.M. Saudubray (Paris, France), Pr. Farriaux (Lille,
France), Pr. Vildaihet (Nancy, France), Dr. C. Bachmann (Berne,
Switzerland).

REFERENCES

1. E.M. Short, H.O. Conn, P.J. Snodgrass, A.G.M. Campbell and
   L.E. Rosenberg, Evidence for x-linked dominant inheritance
   of ornithine transcarbamylase deficiency, N. Engl.J.Med.
   288:7 (1976).
2. F.C. Ricciuti, T.D. Gelchrler and L.E. Rosenberg, X chromosome
   inactivation in human liver; confirmation of X linkage of
   ornithine transcarbamylase, Am.J.Hum.Genet. 28:332 (1976).
3. V.E. Shih, Urea cycle disorders and other congenital hyperammonemic syndromes. In: "The metabolic basis of inherited
   disease."J.B. Stanbury, J.B. Wyngaarden and D.S.Fredrickson,
   eds., Mc Graw-Hill, New-York (1978).

4. B. Levin, R.H. Dobbs, E.A. Burgess and T. Palmer, Hyperammonemia. A variant type of deficiency of liver ornithine transcarbamylase, Arch.Dis.Childr. 44:162 (1969).
5. P. Mc Leod, S. Mackenzie and C.R. Scriver, Partial ornithine carbamyl transferase deficiency : an inborn error of the urea cycle presenting as orotic aciduria in a male infant. Canad. Med. Ass. J. 197:405 (1972).
6. J. P. Farriaux, Le déficit en ornithine carbamyl transferase ou hyperammoniémie de type II. In: "Le cycle de l'urée et ses anomalies", G. Fontaine et J.P. Farriaux, eds., Doin, Paris (1978).
7. M. Thaler, N. Hoogenraad and M. Boswell, Reye's syndrome due to a novel protein-tolerant variant of ornithine, Lancet II:438 (1974).
8. L. Cathelineau, J.M. Saudubray and C. Polonovski, Heterogenous mutations of the structural gene of human ornithine carbamyl transferase as observed in five personal cases, Enzyme 41:103 (1974).
9. J. M. Saudubray, L. Cathelineau, J. M. Laugier, C. Charpentier, J.A. Lejeune and P. Mozziconacci, Hereditary ornithine transcarbamylase deficiency, Acta Pediat.Scand. 64:464 (1975).
10. C. Van der Heiden, J. Desplanques and H.D. Bakker, Some kinetic properties of liver ornithine carbamyl transferase (OTC) in a patient with OTC deficiency, Clin.Chim.Acta. 80:519 (1977).
11. L. Krieger, J.P. Snodgrass, J. Roskanys, A typical course of ornithine transcarbamylase deficiency due to a new mutant (comparison with Reye's disease), J.Clin.Endocrinol.Metab. 48:388 (1979).
12. L. Cathelineau, P. Briand, F. Petit, J.P. Nuyts, J.P. Farriaux and P. Kamoun, Kinetic analysis of a new human ornithine carbamoyl transferase variant, Biochem.Biophys.Acta 614:40 (1980).
13. M. Yudkoff, W. Yang, P.J. Snodgrass and S. Segal, Ornithine transcarbamylase deficiency in a boy with normal development, J. Pediatr. 96:441 (1980).
14. P. Landrieu, B. François, G. Lyon and F. Van Hoof, Liver peroxisomal damage during acute hepatic failure in partial ornithine transcarbamylase deficiency, Submitted to Ped. Res. (1981).
15. P. J. Snodgrass, The effects of pH on the kinetics of human liver ornithine carbamyl phosphate transferase, Biochemistry 7:3047 (1968).
16. W. H. Marsh, B. Fignehut and E. Kirsh, Determination of urea nitrogen with the diacetyl method and an automatic dialyzing apparatus, Am.J.Clin.Pathol. 28:681 (1957).
17. L. G. Ceriotti and A. Gazzaniga, A sensitive method for serum ornithine-carbamyl-transferase determination, Clin. Chim. Acta 14:57 (1966).

18. O. H. Lowry, N.H. Rosebrough, A.L. Farr and R.J. Randall, Protein measurement with the Folin phenol reagent. J.Biol.Chem. 193:265 (1951).
19. L. Hudson and F.C. Hay, "Practical Immunology", Blackwell Scientific Publication, Oxford (1976).
20. G. Mancini, A.O. Carbonara, J.F. Heremans, Immunochemical quantification of antigens by single radial immuno diffusion, Immunochemistry 2:235 (1965).
21. F. Kalousek, B. François and L. Rosenberg, Isolation and characterization of ornithine transcarbamylase from normal human liver, J.Biol.Chem. 253:3939 (1978).
22. J. G. Conboy, F. Kalousek and L.E. Rosenberg, In vitro synthesis of a putative precursor of mitochondrial ornithine transcarbamoylase, Proc. Natl. Acad. Sci. USA 76:5724 (1979).
23. M. Mori, S. Miura, M. Tatibana and P.P. Cohen, Characterization of a protease apparently involved in processing of pre-ornithine transcarbamylase of rat liver, Proc. Natl. Acad.Sci.USA 77:7044 (1980).
24. M. Mori, S. Miura, M. Tatibana and P.P. Cohe, Processing of a putative percursor of rat liver ornithine transcarbamylase, a mitochondrial matrix enzyme, J.Biochem. (Tokyo) 88:1829 (1980).

ENZYMATIC ANALYSIS OF CITRULLINEMIA (12 CASES) IN JAPAN

Takeyori Saheki*, Atsuko Ueda**, Masakazu Hosoya**,
Mariko Sase*, Kyoko Nakano*, Tsunehiko Katsunuma**

*Department of Biochemistry, School of Medicine,
Kagoshima University, Kagoshima, 890 and **Department of
Biochemistry, School of Medicine, Tokai University,
Isehara, 259-11 Japan

INTRODUCTION

Citrullinemia, first described by McMurrey et al[1], is considered a rare hereditary disorder of the urea cycle caused by a deficient activity of argininosuccinate synthetase(ASS). Shih[2] reviewed 12 cases in 1975. In Japan, however, more than 40 cases of citrullinemia have been reported. Most of them were characterized by higher age of onset and moderately high level of serum citrulline[3] in contrast to neonatal onset and extremely high concentration of serum citrulline of classical-or neonatal-type citrullinemia described by McMurrey et al. and others. These findings suggest that there may be some heterogeneities in citrullinemia. So we analyzed the properties of ASS in the liver and other organs of 12 cases of citrullinemia in Japan.

MATERIALS AND METHODS

Patients

The patients included in this report are twelve adults(18-48 years old) whose surgical or autopsy specimens of the liver and other organs were sent to our laboratory. All the patients suffered from disturbance of consciousness such as disorientation, restlessness and sleeplessness[3]. They were diagnosed as citrullinemia on the basis of laboratory findings of high serum citrulline(more than 100 nmol/ml as compared with control values of 20-40 nmol/ml) with normal levels of the other amino acids and hyperammonemia with almost normal liver function tests, as briefly summarized in Table 1. No one had a family history of hereditary disorder, although the mother and the son of patient No. 3 showed a slightly higher level of serum citrulline (52 nmol/ml). Histological examination of the liver specimens of all the patients revealed moderate fatty infiltration or mild portal fibrosis. Detailed clinical descriptions of patients No. 1[4], 2[5], 3[6], 6[7], and 7[8] have been reported and those of the other patients are in preparation for report.

Determination of Enzyme Activity

The liver, kidney and brain were homogenized in 9 vols. of 0.15 M KCl containing 0.05 M Tris-HCl, pH 7.5, with a Teflon homogenizer. The supernatant was prepared by centrifugation at 100,000 xg for 30 min. and used for the determination of ASS, argininosuccinase and arginase. For the determination of the activities of carbamylphosphate synthetase and ornithine transcarbamylase, the liver was homogenized in 9 vols. of 0.02 M Tris-HCl, pH 7.2, and 20% glycerol containing 2 mM dithiothreitol and 0.2% cetyltrimethylammonium bromide. Lysis of cultured fibroblasts was achieved by freeze-thawing two times in 50 mM Tris-HCl, pH 8.5. The enzyme assays were performed on the crude homogenate immediately after lysis of the cells. ASS activity in the brain and the cultured skin fibroblasts was determined by the radiochemical method of Schimke[9] at pH 8.5 and 37°C, and that in the liver and the kidney was determined by the modified method[3] at pH 7.5 and 25°C. The activities of the other urea cycle enzymes were determined essentially according to the method of Schimke[10].

RESULTS

Activities of Urea Cycle Enzymes in the Liver of the Patients

The ASS activities in the liver of the twelve citrullinemic patients were decreased to 2 to 50% of the control, while the other urea cycle enzymes were almost normal (Table 1). The liver of patient No. 7 had a considerably low activity of carbamylphosphate synthetase. This may be attributable to the storage of the liver sample for long time, since the high level of serum citrulline found in his clinical data seems to conflict with the deficiency of carbamylphosphate synthetase. In some cases, the activity of ornithine transcarbamylase was higher than the control.

These results indicate that the defect of ASS is characteristic to these citrullinemic patients.

Kinetic Analysis of ASS in the liver

The kinetic properties of hepatic ASS from all the patients except No. 3 and 6 were very similar to those of the control enzyme, and the Km values of the control enzyme, 0.049 mM, 0.033 mM and 0.24 mM for citrulline, aspartate and ATP, respectively, were in

Table 1. Laboratory Findings and Activities of Urea Cycle Enzymes in the liver of 12 Citrullinemic Patients.

	Sex	Blood Ammonia[a]	Serum Citrulline[a]	Serum Arginine[a]	Enzyme Activity Carbamyl-phosphate synthetase	Ornithine trans-carbamylase	Argininosuccinate synthetase	Argininosuccinase	Arginase
		μg/dl	nmol/ml		Unit/g liver (Unit/mg protein)				
Control[b]	—	<100	20 – 40	80 – 130	4.1±1.9 (0.059±0.035)	56±21 (0.56±0.16)	0.65±0.28 (0.0072±0.0029)	2.6±1.1 (0.030±0.013)	6.7±266 (7.1±2.9)
Patient									
1 (F.N.)	male	100-600	152-428	118	3.1 (0.027)	87 (0.76)	0.11 (0.0020)	2.2 (0.031)	551 (6.2)
2 (K.K.)	male	70-380	50-151	90	2.9 (0.031)	69 (0.65)	0.013 (0.00019)	1.2 (0.018)	298 (3.6)
3 (H.H.)	female	100-480	1994	43	3.9 (0.059)	70 (1.14)	0.040 (0.00080)	3.0 (0.080)	893 (18.7)
4 (S.W.)	male	40-870	373	358	4.5 (0.038)	180 (1.38)	0.12 (0.0014)	0.85 (0.011)	360 (3.6)
5 (S.N.)	male	160-440	370	391	*[c]	*	0.24 (0.0029)	*	617 (7.5)
6 (M.T.)	female	100-240	88-130	69	4.5 (0.089)	158 (1.19)	0.33 (0.0036)	2.8 (0.030)	576 (6.3)
7 (F.K.)	male	100-600	130-286	162±57	0.79 (0.0038)	72 (0.31)	0.29 (0.0029)	2.1 (0.023)	441 (4.7)
8 (J.A.)	male	600	669	*	1.5 (0.012)	131 (1.04)	0.048 (0.00065)	1.9 (0.025)	400 (5.6)
9 (K.N.)	male	140	173	68	5.1 (0.044)	141 (1.25)	0.096 (0.0016)	2.8 (0.045)	408 (6.5)
10 (H.M.)	male	400	751	213	6.3 (0.049)	124 (0.97)	0.043 (0.00080)	2.4 (0.044)	268 (5.0)
11 (M.Y.)	male	594	684	269	10.9 (0.048)	182 (0.80)	0.30 (0.0022)	7.8 (0.056)	1500 (10.7)
12 (K.I.)	male	600-700	290	254	10.8 (0.063)	99 (0.57)	0.16 (0.0015)	5.4 (0.047)	840 (7.2)

[a] The values of blood ammonia, serum citrulline and arginine represent ranges of several determinations or the values of single determination. [b] Enzyme activities of a control subjects represent mean ± S.D.
*[c], not determined

agreement with those of the purified human enzyme determined by O'Brien[11] and us (Table 2). Fig. 1-a shows a hyperbolic relationship between the enzyme activity and the concentration of citrulline for the ASS preparations from control and patient No. 7. Fig. 1-b shows peculiar property of the enzyme from patient No. 3. The saturation curve of activity versus concentration of citrulline was not hyperbolic, but of so-called negative cooperativity. A quite similar result was also obtained with aspartate. The concentration of both citrulline and aspartate showing the half-maximal reaction velocity was about 20 mM. The Km value for ATP was similar to that of the control (Table 2). Fig. 1 also illustrates the Lineweaver-Burk plots of a control and patients No. 3, 6 and 7. The straight-line plots of a control and patient No. 7 show a familiar type of kinetics, and the line of patient No. 3, which curves downwards, indicates again the negative cooperativity. Like in patient No. 3, the enzyme from patient No. 6 showed slightly negative-cooperative kinetics. These results suggest that the ASS in the liver of patients No. 3 and probably No. 6 is qualitatively abnormal.

Table 2. Kinetic constants of ASS in the liver of control and citrullinemic patients

	Km (mM)		
	Citrulline	Aspartate	ATP
Control	0.049 ± 0.010*	0.033 ± 0.014*	0.24 ± 0.14*
Patient			
1 (FN)	not increased	not increased	not determined
2 (KK)	0.038	0.016	not determined
3 (HH)	ca.20**	ca.20**	0.24
4 (SW)	0.032	0.032	0.18
5 (SN)	0.038	0.026	not determined
6 (MT)	0.046(0.13)**	0.071(0.18)**	0.17
7 (FK)	0.033	0.033	0.12
8 (JA)	0.042	0.051	0.17
9 (KN)	0.030	0.023	0.11
10 (HM)	0.049	0.033	0.18
11 (YM)	0.042	0.020	0.17
12 (KI)	0.020	0.029	0.079

* M ± SD ** Activity versus substrate concentration showed negative cooperativity.

ENZYMATIC ANALYSIS OF CITRULLINEMIA

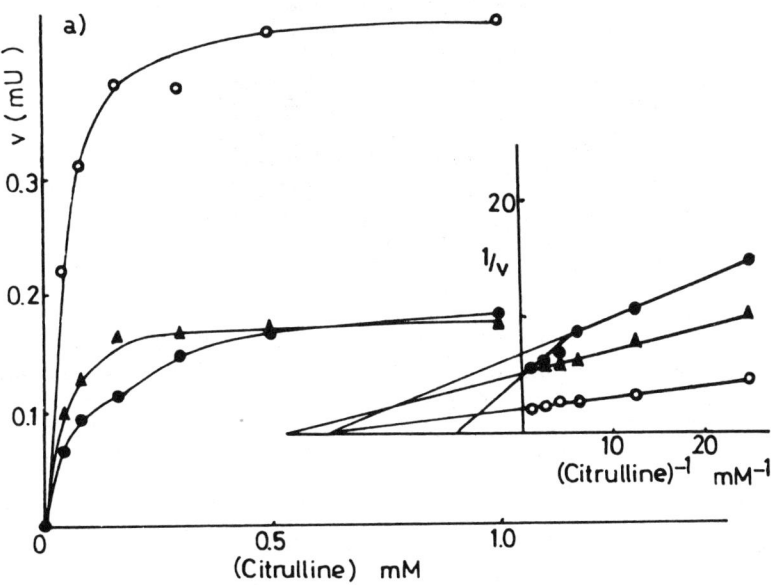

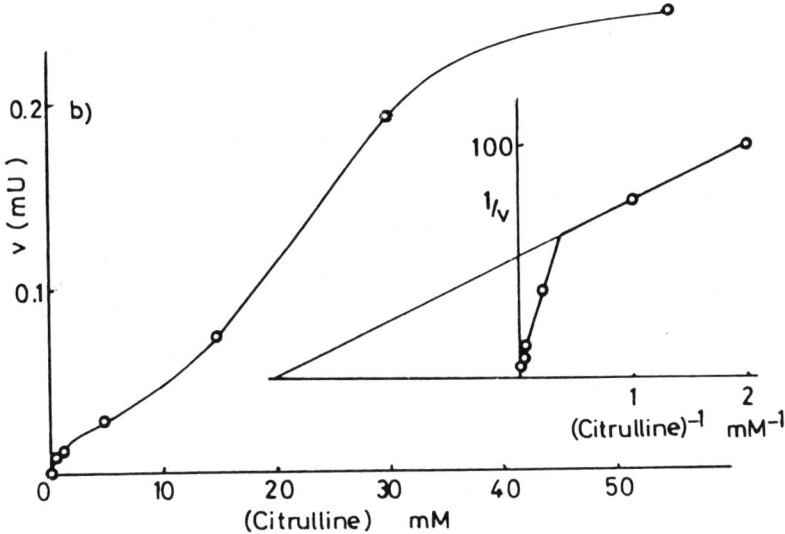

Fig. 1. Effect of the concentration of citrulline on ASS activity in the liver of control (O) and patients No. 6 (●) and No. 7 (▲) in fig. 1-a) and of patient No. 3 (O) in Fig. 1-b). See also reference No. 3.

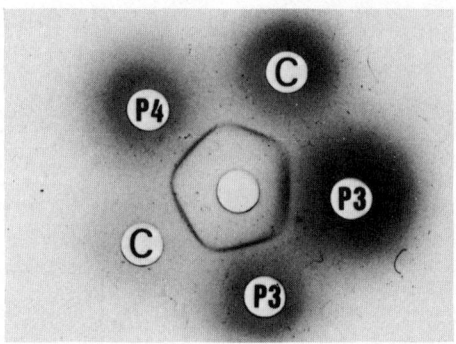

Fig. 2. Ouchterlony's double immunodiffusion analysis of antigenic relationship between ASS from a control person (c) and ASS from citrullinemic patients No. 3 (P3) and No. 4 (P4). The plate was stained with Coomassie brilliant blue, after extensive washing with 0.9 % NaCl.

Immunological Property of ASS in the Liver of the Patients

Immunological property of ASS in the liver of the patients was tested with anti-rat ASS antiserum. In the double immunodiffusion method of Ouchterlony, a purified rat liver ASS preparation, a crude rat liver sample, and a human liver sample formed a single precipitation line, and the lines formed with the samples from rat and human were fused. Similarly, the lines formed with the samples from control persons and the citrullinemic patients were fused. Fig. 2 shows a representative precipitation line formed with a control and patients No. 3 and 4. In the gel electrophoresis with sodium dodecylsulfate, the precipitate formed with the antiserum and the liver samples from control or the patients showed two protein bands derived from the light and heavy chains of immunoglobulin and an additional protein band having a molecular weight of the subunit of human ASS.

These results indicate that the antiserum raised against purified rat liver ASS can react with human ASS as well as rat ASS, and that the ASS from citrullinemic patients can not be immunologically distinguished from either.

The Amounts of ASS Protein and the Relative Specific Activities of ASS in the Liver of the Patients (fig. 3)

The amount of the enzyme protein in the liver samples was

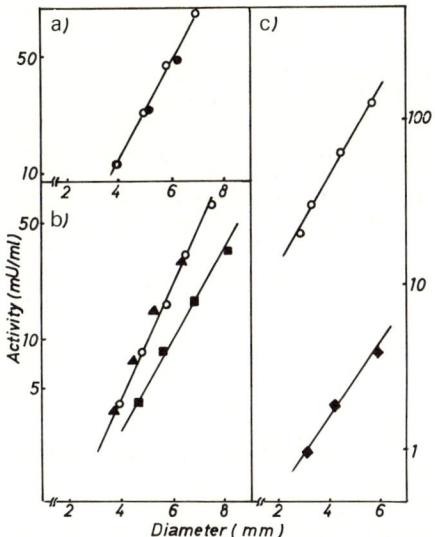

Fig. 3. Analysis by single radial immunodiffusion or ASS protein from control persons (○) and citrullinemic patients No. 5 (●) in Fig. 3-a), No. 6 (■) and No. 7 (▲) in Fig. 3-b) and No. 3 (◆) in Fig. 3-c).

determined by the single radial immunodiffusion method on the basis of the immunological properties described above and was plotted against the enzyme activity. The standard curves were constructed with the liver samples from control persons. The curves obtained with the samples from patients No. 5 and 7 were almost the same as the standard curves, indicating the similarity in the specific activity of ASS to that of control. In data not shown we have obtained similar results with patients No. 1, 2, 4, 8, 9, 10, 11 and 12 (not shown). By contrast, the curve obtained for patient No. 3 ran far below the standard curve, indicating a very low specific activity of the enzyme in the liver of this patient. Since the same amount of the liver samples was used for the control and patient No. 3, the similarity of the size of the precipitation ring suggests the presence of an ASS-like protein in the liver of patient No. 3 in an amount comparable to that of the normal liver.

These results indicate that the decreased ASS activity in the liver of patients No. 1, 2, 4, 5, 7, 8, 9, 10, 11 and 12 may be explained by the decrease in the amount of enzyme protein of the normal type, while the decreased activity in patient No. 3

Table 3. ASS activities in the kidney, brain and the cultured skin fibroblasts of control and citrullinemic patient

	ASS activity			
	Liver	Kidney	Brain	Fibroblast
	mU/g Tissue (mU/mg Protein)			
Control	650+280 (7.2+ 2.9)	160+40 (3.3+ 1.5)	4.3+1.5 (0.19+ 0.056)	– (0.34+0.14)
Patient				
2 (KK)	13(0.19)	220(6.7)	0.90(0.043)	*
3 (HH)	40(0.80)	*	*	(0.043)
5 (SN)	240(2.9)	*	*	(0.45)
7 (FK)	290(2.5)	*	0.52(0.025)	*
9 (KN)	96(1.6)	220(4.3)	2.8(0.17)	(0.30)
10 (HM)	43(0.80)	180(4.7)	5.2(0.17)	*
11 (YM)	300(2.2)	*	*	(0.31)

*: not determined.

represents a defectiveness of the enzyme protein rather than a decrease in amount of the enzyme protein present. The curve obtained for patient No. 6 was slightly off the standard curve, suggesting an abnormality analogous to that of patient No. 3.

<u>The ASS Activities in the Other Organs and the Cultured Skin Fibroblasts of the Patients (Table 3)</u>

We analyzed ASS activities in the kidney and the brain of three and four citrullinemic patients, respectively. These were all from the patients who had low activities of hepatic ASS caused by a decrease in the amount of the enzyme protein. All of the kidney samples and a half of the brain samples showed ASS

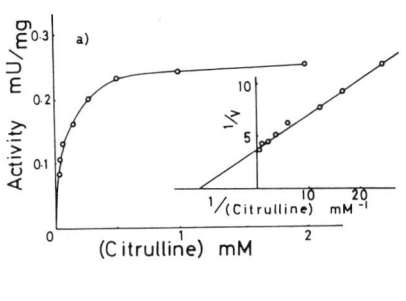
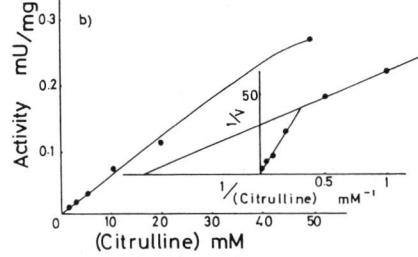

Fig. 4. Effect of the concentration of citrulline on ASS activity in the fibroblasts from a control (Fig. 4-a) and citrullinemic patients No. 3 (Fig.4-b). see also reference No. 12.

activities comparable to the controls. It is very interesting that the cultured skin fibroblasts of patients No. 5, 9 and 11 who had low activities of hepatic ASS caused by a decrease in the amount of the enzyme protein showed activities comparable to the controls, while those from patient No. 3 who had qualitatively abnormal ASS in the liver showed a low activity (13 % of the control)[12].

The Kinetic Properties of ASS in the Cultured Skin Fibroblasts

There has been no comparative study of the kinetic properties

Table 4. Kinetic constant of ASS in the cultured skin fibroblast of control and citrullinemic patients.

	Km value (mM)	
	Citrulline	Aspartate
Control	0.087	0.048
Patient		
3 (HH)	ca. 20	ca. 40
5 (SN)	0.087	0.056
11 (YM)	0.028	0.025

of ASS in liver and other organs or cells from one single patient[2].

Fig. 4-a shows a hyperbolic relationship between the enzyme activity and the concentration of citrulline for the fibroblast ASS preparation from a control. The kinetic properties of the fibroblast ASS from patients no. 5 and 11 were quite similar to those of the control fibroblast and also hepatic ASS (Table 4). Fig. 3-b shows peculiar properties, i.e. high Km value and negative cooperativity, of the fibroblast enzyme from patient No. 3 which were quite similar to those of the hepatic ASS from the same patient.

DISCUSSION

The kinetic and immunochemical analysis of hepatic ASS from twelve citrullinemic patients showed for the first time that there are two types of abnormalities of this enzyme. One type of defect is qualitative such as seen in ASS of patients No. 3 and probably No. 6 who showed abnormal kinetics but with no decrease in the amount of enzyme protein, suggesting a change in the structural gene. This view was supported by the fact that the cultured skin fibroblast from patient No. 3 also showed a low ASS activity with abnormal kinetic properties quite similar to those of the hepatic enzyme. This indicates that the fibroblast and hepatic ASS protein are products of the same structural gene.

By contrast, a second type of defect was quantitative where

the decrease in ASS activity in the liver of patients No. 1, 2, 4, 5, 7, 8, 9, 10, 11 and 12 was due to the decrease in the amount of enzyme protein with normal kinetic properties. The kidney, brain and the cultured skin fibroblasts of these patients possess quantitatively normal ASS. (Table 3). This fact is very important because the fibroblast and hepatic ASS are genetically of same origin, and suggests that the quantitative abnormality of the hepatic ASS of citrullinemic patients may be due to abnormal organ-specific gene expression, including enzyme synthesis and degradation. Some investigators [3,13,14] described that the enzyme deficiency in several citrullinemic patients was only present in the liver and not in the kidney or brain. The results described in this report may be one of the explanations for the difference in enzyme deficiency among various organs.

As described by Walser et al.[15] and Morrow et al.[16], so-called neonatal-type citrullinemia tends to be arginine deficient. It was shown by Funahashi et al.[17] that arginine is synthesized from citrulline by the pathway involving ASS and argininosuccinase mainly in the kidney. This suggests that citrullinemic patients with qualitatively abnormal ASS in the liver possess also defective ASS in the kidney, so that they can not synthesize arginine from citrulline. On the contrary, citrullinemic patients with quantitatively abnormal ASS in the liver, but with active ASS in the kidney may synthesize arginine in the kidney from citrulline excessively supplied from blood. This consideration may be supported by the fact showing normal or relatively higher concentration of arginine in the serum of the quantitative-type citrullinemia in contrast with lower arginine level in the serum of qualitative-type (patients No. 3 and 6), as described in Table 1.

We noticed sex difference of citrullinemic patients between qualitative and quantitative defect of ASS, especially all of the quantitative-type patients examined were male. This point must be further investigated. We consider that the type of quantitative abnormality of ASS constitutes the majority of citrullinemia in Japan, at least 10 out of 12 in our experiment, and may be the cause of the relatively high incidence in Japan.

SUMMARY

The enzymological and immunological analysis of argininosuccinate synthetase in the liver, kidney, brain and the cultured skin fibroblasts of twelve citrullinemic patients in Japan were performed.

Among the urea cycle enzymes in the liver, only the activity of argininosuccinate synthetase was specifically decreased from 2 to 50% of control.

The kinetic properties of argininosuccinate synthetase in the liver of two patients were quite different from control in terms of higher Km values and negative cooperativity, indicating abnormalities in the structural gene.

By contrast the argininosuccinate synthetase in the liver of the other ten patients showed normal kinetic properties indistinguishable from control and the decrease in the activity was explainable by the decrease in the amount of the enzyme protein determined immunologically with antisera raised against argininosuccinate synthetase.

This is the first report showing that there are thus two types of enzyme abnormalities in hepatic argininosuccinate synthetase of citrullinemia.

One type resulting in a quantitative deficiency of the enzyme in the liver, but not in cultured skin fibroblast, is responsible for the great majority of citrullinemia in Japan. At least 10 out of the 12 cases investigated were of this type.

The other type resulting in a qualitative deficiency of the enzyme where the cultured skin fibroblasts exhibited a decreased activity with abnormal kinetics similar to those of the hepatic enzyme.

From these results, we suggest that some of the quantitative abnormality of hepatic argininosuccinate synthetase of citrullinemic patient may be due to an abnormal organ-specific gene expression in the liver.

ACKNOWLEDGEMENT

This work was done in the cooperation with the following investigators ; Drs. T. Hirasawa, Y. Yajima, H. Suzuki, Y. Yamauchi, K. Fujisawa, N. Yamada, H. Sibata, K. Akamatsu, Y. Ohta, K. Kobayashi, K. Itahara, Y. Suzuki, S. Furuta, T. Karita, M. Takamizawa, A. Watanabe, I. Akaboshi, I. Matsuda, K. Takahashi, Y. Hara, T. Tahira and T. Kiryu.

We thank Drs. K. Tada, K. Omura, M. Ichihashi, K. Kitagawa, M. Owada and K. Iizima for preparing cultured skin fibroblasts, Dr. E.A. Khairallah for critical reading of the manuscript and Ms. M. Ogawa for technical assistance.

REFERENCES

1. W. C. McMurrey, F. Mohyuddin, R. J. Rossiter, C. Ratbun, G. H. Valentine, S. J. Koegler and D. E. Zargas, Citrullinemia, a new amino aciduria associated with mental retardation, Lancet 1:138 (1962).
2. V. E. Shih, Hereditary urea-cycle disorders, in: "The urea Cycle", S. Grisolia, R. Baguena and F. Mayor, eds., John-Wiley and Sons, New York (1976).
3. T. Saheki, A. Ueda, M. Hosoya, K. Kusumi, S. Takada, M. Tsuda and T. Katsunuma, Qualitative and Quantitative abnormalities of argininosuccinate synthetase in citrullinemia, Clin. Chim. Acta. 109:325 (1981).
4. Y. Yajima, T. Hirasawa and T. Saheki, Treatment of adult-type citrullinemia with oral administration of citrate, Acta Hepatl. Jap. 21:1682 (1980).
5. M. Yamauchi, T. Kitahara, K. Fujisawa, H. Kameda, S. Takasaki, R. Komori, T. Saheki, T. Katsunuma and N. Katunuma, An autopsied case of hypercitrullinemia in an adult caused by partial deficiency of liver argininosuccinate synthetase, Acta Hepatol. Jap 21:326 (1980).
6. N. Yamada, M. Fukui, K. Ishii, H. Shibata, H. Ohomiya, A. Matsunobu and M. Nishizima, A case of adult form hypertransaminasemia after delivery, Gastroenterologia. Jap. 77:1655 (1980)
7. Y. Suzuki, N. Yamamura, K. Nozawa, Y. Akahane, K. Kiyosawa, A. Nagata, S. Furuta and K. Chiba, A case of adult-type congenital citrullinemia, Acta Hepatol. Jap. , 21:1215 (1980)
8. M. Takamizawa, M. Toru, T. Kojima, A. Watanabe and K. Hirokawa, An autopsy case of juvenile hepato-cerebral degeneration (non-Wilsonian Inose-type) with mental retardation with special reference to amino acids metabolism, Psychiatr. Neurol. Jap. , 75:370 (1973).
9. R. T. Schimke, Enzymes of arginine metabolism in mammalian cell culture. 1. Repression of argininosuccinate synthetase and argininosuccinase. J. Biol. Chem. 239:136 (1964).
10. R. T. Schimke, Adaptive characteristics of urea cycle enzymes in the rat, J. Biol. Chem. 237:459 (1962).
11. W. E. O'Brien, Isolation and characterization of argininosuccinate synthetase from human liver, Biochemistry, 18:5353 (1979)
12. T. Saheki, A. Ueda, K. Iizima, N. Yamada, K. Kobayashi, K Kobayashi, K. Takahashi and T. Katsunuma, Argininosuccinate synthetase activity in cultured skin fibroblasts of citrullinemic patients, Clin. Chim. Acta, in press.
13. M. Vidailhet, B. Levin, M. Dautrevaux, P. Paysant, S. Gelot, Y. Badonnel, M Pierson, N. Neimann, Citrullinemie, Arch. Franc. Ped. 28:521 (1971).

14. F. H. Roerdink, W. L. M. Gouw, A. Okken, J. F. Van der Blij, G. Luitde Haan, F. A. Hommes and H. J. Huisjes, Citrullinemia, report of a case, with studies on antenatal diagnosis, Pediatr. Res. 7:863 (1973).
15. M. Walser, M. Batshaw, G. Sherwood, B. Robinson and S. Brusilow, Nitrogen metabolism in neonatal citrullinemia, Clin. Sci. Mol. Med. 53:173 (1977).
16. G. Morrow, L. A. Barness and M. L. Efron, Citrullinemia with defective urea production, Pediat. 50:565 (1967).
17. M. Funahashi, H. Kato, S. Shiosaka and H. Nakagawa, Formation of arginine and guanidinoacetic acid in the kidney in vivo. J. Biochem. 89:1347 (1981).

QUALITATIVE ABNORMALITY OF LIVER ARGININOSUCCINATE SYNTHETASE

IN A PATIENT WITH CITRULLINEMIA

Yoshiko Matsuda, Akihiko Tsuji and Nobuhiko Katunuma

Department of Enzyme Chemistry, Institute for Enzyme
Research, School of Medicine, Tokushima University
Tokushima 770, Japan

ABSTRACT

The enzymes involved in the urea cycle in normal liver and in the liver of a patient with citrullinemia were compared. The activities of carbamoyl-phosphate synthetase, ornithine-carbamoyl transferase, argininosuccinate synthetase, argininosuccinate lyase, and arginase in the patient's liver were normal under standard assay conditions. The properties of argininosuccinate synthetase of the patient were compared with those of the enzyme from normal liver. The enzyme from normal liver showed normal Michaelis-Menten kinetics and its Km values for ATP, L-aspartate, and L-citrulline were 1.8×10^{-4} M, 2.9×10^{-5} M, and 3×10^{-5} M, respectively. The enzyme from the patient with citrullinemia also gave hyperbolic curves with respect to the concentrations of ATP and L-aspartate and the Km values for ATP and L-aspartate were 1.8×10^{-4} M and 2.9×10^{-5} M, respectively. However, it gave a sigmoidal curve with respect to the concentration of citrulline, and Hill's coefficient was 1.66. The molecular weight of argininosuccinate synthetase from both normal liver and the patient's liver was estimated to be 185,000 by sucrose density gradient centrifugation both in the presence and absence of citrulline. Argininosuccinate stabilized the enzyme preparations from both livers against heat treatment, but the enzyme from the patient's liver was less stable than the normal enzyme against heat treatment without added argininosuccinate.

INTRODUCTION

Citrullinemia was first described at 1962 by McMurray et al.[1] and deficiency of the activity of liver[2,3] and fibroblast[4,5,6,7] argininosuccinate synthetase (EC 6.3.4.5) of the citrullinemia were reported. We experienced the patient who excreted the normal level of urea and represented high citrulline and slightly high ammonia in the plasma. We have studied the kinetic properties of the liver argininosuccinate synthetase of this patient and correlation of the metabolic abnormality with the specific properties of this enzyme.

CASE REPORT

The patient was a female child of the unrelated healthy japanese parents and born after an uncomplicated pregnancy of 40 weeks. However, four of her ten siblings suffered from vomiting and diarrhea and died at infancy. The patient was first brought to the hospital at the age of 18, because of frequent attack of vomiting and diarrhea. She has shown retarded physical and mental development since early childhood; her I.Q. is now that of a 3 year old child and she is a dwarf. She suffered from unconciousness with convulsion for two hours at 20 years and since then, the symptoms have become pronounced. Related laboratory data were: serum glutamic oxaloacetic transaminase (GOT) 101 IU, glutamic pyruvic transaminase (GPT) 55 IU and detoxication test of ammonia was markedly decreased with comparison to the control.

RESULTS

Concentrations of Serum Amino Acids

Table 1. Amino Acid Concentrations in the Serum

Amino acids	Patient (mM)	Control (mM)	Amino acids	Patient (mM)	Control (mM)
Citrulline	0.41	0.02±0.01	Serine	0.06	0.10±0.03
Urea	3.94	5.64±0.47	Histidine	0.04	0.06±0.01
Ammonia	0.13	0.07±0.03	Tyrosine	0.06	0.04±0.01
Taurine	0.05	0.06±0.01	Valine	0.16	0.19±0.03
Ornithine	0.06	0.05±0.01	Methionine	0.03	0.03±0.00
Arginine	0.16	0.08±0.03	Isoleucine	0.07	0.07±0.02
Glycine	0.10	0.17±0.07	Leucine	0.08	0.11±0.02
Alanine	0.12	0.31±0.08	Threonine	0.17	0.11±0.02
Proline	0.23	0.14±0.03	Lysine	0.21	0.14±0.04
Glutamic acid	0.06	0.05±0.02	Phenylalanine	0.07	0.04±0.01

Values of controls are means ±standard deviation from 5 samples.

The serum amino acid values were consistently abnormal and were high when protein intake and blood ammonia level were increased. The concentration of citrulline in the patients serum was 26 times higher than that in normal serum, as shown in Table 1. However, the concentration of ammonia was slightly higher than normal and the urea concentration was approximately 80% of the normal level. The other amino acids were almost the same level as normal level.

Activity of Urea Cycle Enzymes

The urea cycle enzymes were determined by the method reported previously[8]. Activities of urea cycle enzymes of the patient's liver were compared with those of the normal liver in Table 2. All enzyme activities measured were within their normal range under standard condition that contained the sufficient concentration of substrates and cofactors for each enzyme.

Kinetic Studies of Argininosuccinate Synthetase

In order to determine why citrulline accumulated in the patient's serum, in spite of normal range of argininosuccinate synthetase, the properties of argininosuccinate synthetase were studied. A linear relationship between enzyme concentration and the activity was obtained with the cytosol fractions of both normal liver and the patient's liver. This suggests that there is no

Table 2. Activity of Urea Cycle Enzymes in Liver

	Activity (μmol/h/g wet.wt)		Specific activity (μmol/h/mg protein)	
	Patient	Control(No.)	Patient	Control(No.)
Carbamoyl phosphate synthetase	101	108±42(5)	2.3	2.7±1.4(5)
Ornithine-carbamoyl transferase	5,570	4,310±121(3)	129	115.0±12.6(3)
Argininosuccinate synthetase	52	52±7(3)	0.6	0.9±0.1(3)
Argininosuccinate lyase	88	102±28(3)	1.8	2.7±0.9(3)
Arginase	8,270	4,990±52(3)	191	117.0±17.2(3)

Values of controls are means ±standard deviation from samples (No.).

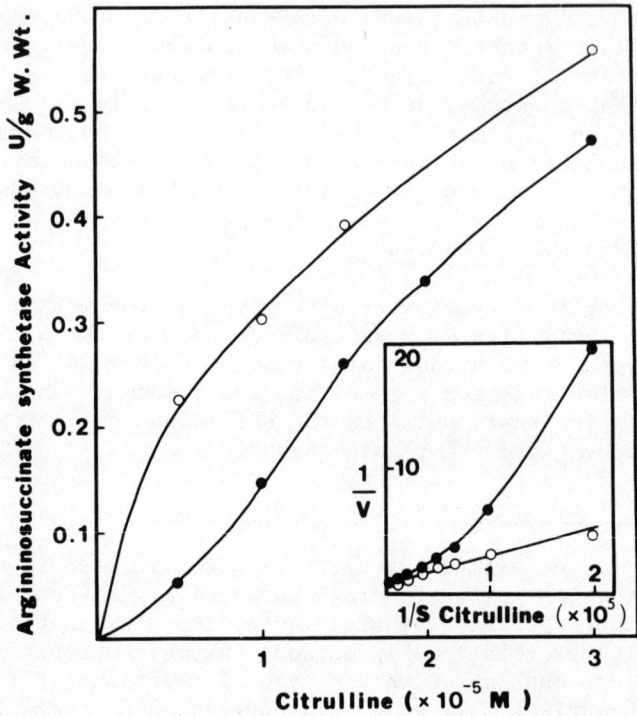

Fig. 1. Effect of L-citrulline on argininosuccinate synthetase activity. Assay was carried out with 0.98 mM ATP, 5 mM L-aspartate and the indicated concentration of citrulline. ○ Control; ● patient.

substance present that affects argininosuccinate synthetase activity in the cytosol fractions of these preparations. Lineweaver-Burk plots for ATP and aspartate of the two preparations with 5 mM citrulline were linear and the Km value of both preparations for ATP was calculated to be 1.8×10^{-4} M. The Km value for aspartate was calculated to be 2.9×10^{-5} M in the presence of 0.98 mM ATP and 5 mM citrulline. However, the preparation from normal liver gave a hyperbolic curve for velocity versus citrulline concentration and the Km value for citrulline was calculated to be 3.0×10^{-5} M, whereas the preparation from the patient's liver gave a sigmoidal curve for velocity versus citrulline concentration and the Hill coefficient was calculated to be 1.66, although Vmax was very similar to that of the normal enzyme, as shown in Fig. 1.

Sucrose Density Gradient Centrifugation

Since argininosuccinate synthetase from the patient's liver

exhibited a sigmoidal curve for velocity versus citrulline concentration and the plot of $1/v$ versus $1/s^2$ was almost linear, the possibility that the enzyme polymerized at high concentrations of citrulline was examined. The molecular weight of argininosuccinate synthetase in both preparations was calculated to be 185,000 in either the absence or presence of citrulline; in addition the activity of argininosuccinate synthetase was detected as a single peak in both cytosol preparations and citrulline did not change the pattern obtained by sucrose density gradient centrifugation. Therefore it is concluded that the enzyme does not polymerize in the presence of a saturating concentration of citrulline, but that the patient's enzyme has two binding sites for citrulline and that the first binding site interacts with the second site.

Heat Stability of Enzyme

Argininosuccinate synthetase in crude extracts of normal liver and the patient's liver was labile on heat treatment at 50°C. The enzyme from the patient seemed to be slightly less stable than the normal enzyme, but argininosuccinate stabilized both preparations.

DISCUSSION

The reaction catalyzed by argininosuccinate synthetase in the liver is believed to be a rate-limiting step in urea synthesis. The enzyme has been purified from rat[9], bovine[10] and human[11] liver, and some of their properties have been reported. Argininosuccinate synthetase is mainly present in the liver, kidney, and brain and most cases of citrullinemia have been reported to have an abnormality of the liver enzyme, although some cases have been found to have normal enzyme activity in the brain or kidney and to produce urea.

The present case showed marked citrullinemia, but the serum ammonia level was only slightly higher than normal. The serum urea concentration was also almost normal. It seems likely that in this patient, liver argininosuccinate synthetase is a variant type and catalyzes inefficiently the formation of argininosuccinate. In this patient the enzyme showed a typical sigmoidal curve for activity versus citrulline concentration, whereas the normal enzyme gave a hyperbolic curve. The results show that the activity of the patient's enzyme represents 0.14 Vmax and that of the normal enzyme is 0.35 Vmax at 10^{-5} M citrulline, which is the concentration of citrulline in normal human serum, and the activity of the patient's enzyme is 0.77 Vmax and that of normal enzyme is 0.80 Vmax at 4×10^{-4} M citrulline, which is the level of the patient's serum citrulline. Therefore, if the concentration of serum citrulline reflects its level in the liver, the activity of the patient's enzyme may be about 40% of that of normal liver under normal citrulline concentration, however, the enzyme of the patient liver is able to produce argininosuccinate when citrulline accumulates to a high level in the liver. Therefore, this patient can utilize ammonia to produce urea, suffering from citrullinemia.

ACKNOWLEDGMENTS

This work was supported in part by a Grant-in-Aid for Scientific Research from the Ministry of Education, Science and Culture of Japan. Authors are grateful to Ms Emiko Inai for her help preparing the manuscript.

REFERENCES

1. W. C. McMurray, F. Mohyuddin, R. J. Rossiter, C. Rathbun, G. H. Valentine, S. J. Koegler, and D. E. Zarfas, Citrullinemia, a new amino aciduria associated with mental retardation, Lancet 1:138 (1962).
2. I. Matsuda, M. Anakura, S. Arashima, Y. Saito, and Y. Oka, A variant form of citrullinemia, J. Ped. 88:824 (1976).
3. T. Saheki, A. Ueda, M. Hosoya, K. Kusumi, S. Takada, M. Tsuda and T. Katsunuma, Qualitative and quantitative abnormalities of argininosuccinate synthetase in citrullinemia, Clin. Chim. Acta 109:325 (1981).
4. T. A. Tedesco, and W. Mellman, Argininosuccinate synthetase activity and citrulline metabolism in cells cultured from a citrullinemic subject, Proc. Natl. Acad. Sci. U.S.A. 57:829 (1967).
5. D.T. Whelan, T. Brusso, and M. Spate, Citrullinemia: Phenotypic variations, Pediatrics 57:935 (1976).
6. N. R. M. Buist, N. G. Kennaway, C. A. Hepburn, J. J. Strandholm, and D. A. Ramberg, Citrullinemia : Investigation and treatment over four year period, J. Ped. 85:208 (1974).
7. N. G. Kennaway, R. J. Harwood, D. A. Ramberg, R. D. Koler, and R. M. Buist, Citrullinemia: Enzymatic evidence for genetic heterogeneity, Pediat. Res. 9:554 (1975).
8. Y. Matsuda, A. Tsuji, N. Katunuma, M. Hayashi, and Y. Takahashi, Studies on liver argininosuccinate synthetase in a patient with citrullinemia and in normal subjects, J. Biochem. 85:191 (1979).
9. T. Saheki, T. Kusumi, S. Takada, and T. Katsunuma, Studies of rat liver argininosuccinate synthetase. 1. Physicochemical, catalytic and immunochemical properties, J. Biochem. 81:687 (1977).
10. O. Rochovansky, H. Kodawaki, and S. Ratner, Biosynthesis of urea, molecular and regulatory properties of crystalline argininosuccinate synthetase, J. Biol. Chem. 252:5287 (1977).
11. W. E. O'Brien, Isolation and characterization of argininosuccinate synthetase from human liver, Biochemistry 18:5353 (1979).

ARGININOSUCCINIC ACIDURIA IN ADULT : A CLINICAL, ELECTROPHYSIOLOGICAL AND BIOCHEMICAL STUDY

T. Grisar

Institute of Medicine, Department of Neurology
University of Liège, Boulevard de la Constitution 66
B-4020 Liège, Belgium

INTRODUCTION

Argininosuccinic aciduria is an hereditary disorder of the urea cycle in which argininosuccinase (As ; EC : 4.3.2.1.) is defective or absent. The defect may be detected by the finding of the overexcretion of argininosuccinic acid (ASA) and of negligible levels of enzyme activity in erythrocyte lysates.

The clinical effect includes mental retardation, neurological signs (mostly seizure or lethargia) and hair or cutaneous abnormalities.

Since the initial paper of Allan et al. (1958), fourty four cases have been reported and recently reviewed by Formstecher (1978). In all cases, the diagnosis was established between the first hours (12 h) of life and the age of 20 years.

We have found argininosuccinic aciduria in a 63 year old patient presenting, together with seizure and mental retardation, neurogenic osteoarthropathies and hyperuricemia.

Biochemical investigations raised the hypothesis of a possible utilization of excessive ammonia for the synthesis of uric acid. This could explain the unusual survival of such a patient.

1. CASE REPORT

The patient (W.M.), a 63 year old woman was admitted to the hospital for a treatment of skin ulcerations.

She developed an Australia antigen positive hepatitis and then was transferred from the department of Dermatology to the department of Internal Medicine (infectious unit). She recovered of her hepatitis after three weeks. An advice of a neurologist was requested for adjustment of an antiepileptic treatment.

Indeed, since her infancy, the patient presented repetitive Grand Mal and absence type seizures, treated with Phenobarbital and Phenytoïne. Moreover, she exhibited a severe mental retardation and was rejected by her family when she was 34 because she became an unmarried mother. She worked as maidservant until 50 from then on she lived in a home for old people.

On admission, she was an elderly woman with an unusual sad and inexpressive facial appearance. She was drowsy and spoke with a cerebellar dysarthria. Cranial nerves were normal except for bilaterally impaired auditory acuity. The limbs were hypotonic and exhibited cerebellar static ataxia. Hands and feet presented a resting 3 to 5 per sec tremor. This sign was already noted in infancy. Tendon reflexes were depressed while power and sensation were normal. Icterus was present but disappeared after three weeks. Skin dystrophic lesions were noted in anterior parts of the legs while hairs and nails were known as fragile and crisp. One of the most striking observation was the existence of hand and feet osteoarticular deformations with shortened fingers and toe-joints. Again, those signs were present during the young age and led to difficulties of walking and gripping.

X-rays of patient's feet showed acro-osteolytic lesions of some toe-joints together with subluxations of others and osseous alterations indicated neurogenic and/or metabolic disorders of bone structures.

EEG demonstrated a subalpha-théta rhythm with some paroxysms and twice post-ictal alterations. A CT-Scan only showed an increased volume of cerebral ventricles without pictures or signs of severe atrophy.

The diagnosis was made by the detection of argininosuccinic acid in the urine (see below, biochemical findings). Dermatological examination confirmed the existence of trichorrhexis nodosa.

Two months after this first investigation, the patient developed a bronchopneumonia of the left inferior pulmonary lobe and one week later a complete palsy of the homolateral diaphragmatic cupola for which the investigations were negative.

Four years after the diagnosis was established, the patient died after a cardiac arrest following a long period of respiratory difficulties under hyperammoniemia.

2. FAMILY HISTORY

The detail of the genealogical study has been reported elsewhere (Husquinet et al., 1981).

Briefly, a medical report showed that a patient's sister (W.A. 1916-1941) suffered from convulsions during the first months of life and had mental retardation with severe seizures. She died when she was 25 year old after a prolonged coma attributed to a tuberculosis. The tracking down of heterozygotes was either estimated by genealogical deduction or proved by the measurement of the ASA-lyase enzyme level of erythrocyte lysates (Dr. S. Schoos-Barbette - Lab. of Genetic Biochemistry). The family tree of the patient is shown on Fig. 1. The patient's parents were healthy and, as shown, consanguineous since a detailed genealogical study found the existence of commun ancestor in 1580 (patronyme W.). Additional common and more recent ancestors also exist (patronymes M and S). In the mother's family no heterozygotes were found. Four negative results were detected in children from patient's brothers and sisters.

However, on the father's side, the grand-father (W.F. 1818-1893) certainly carried the abnormal gene. Indeed, he was married twice and six heterozygotes were proved in the lineage from both marriages. Only five descendants from the second marriage were tested and they were normal.

3. ELECTROPHYSIOLOGICAL FINDING

Regular electroencephalograms were always abnormal with a predominant electrical activity consisting of 6 to 7 per second theta waves rather than the 8 to 9 per second alpha activity. In addition, when ammonia levels were elevated (see biochemical section), there were bursts of 1 to 3 per second slow waves without spikes activity.

The contingent negative variation (CNV) was also recorded (Dr. M. Timsit-Berthier - Lab. of Electrophysiology) under both conditions : rest and during hyperammonemia following an enriched protein diet. Under basal conditions (i.e. 0.46 mg/l of ammoniemia), despite a good cooperation of the patient during the test, the reaction time (817.6444 + 13.6 msec) was exaggerated compared with normal values ($\pm$ 200 msec). The CNV ending is of type IV (i.e. low negative component, 6 μV; with slow back to normal, up to 1 sec after S_2).

The spectrum is abnormal, predominating on theta rhythms (32.6 %) with small increase during CNV. Under hyperammoniemia, the reaction time appeared surprisingly reduced while the spectrum

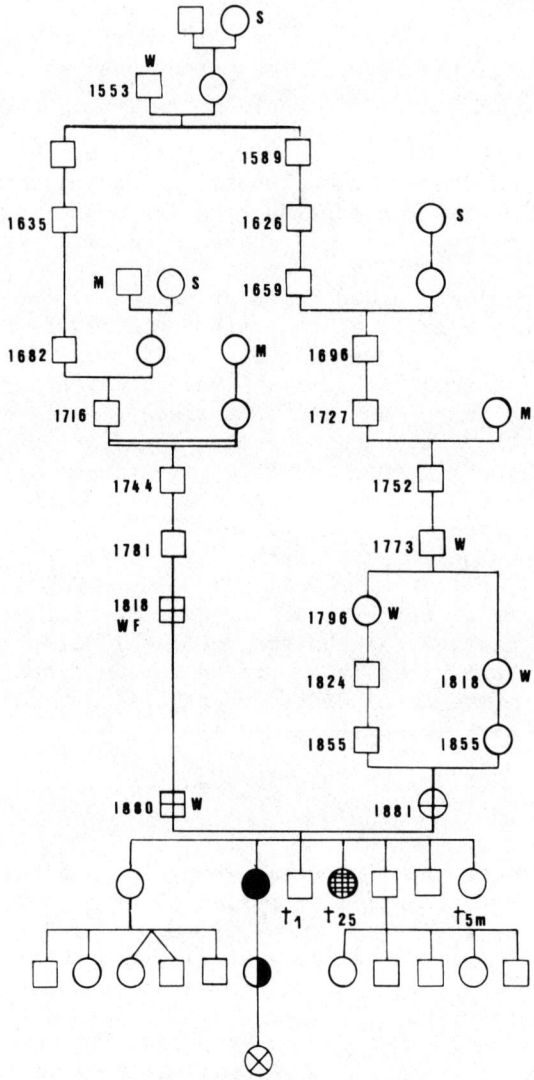

Fig. 1. The family tree. The patient is shown as black circle. The crosshatched circle refers to W.A. who is assumed to be homozygote from medical observations (see the text for details). The half-black symbols refer to patients thought to be heterozygote, on the basis of red cell argininosuccinase assay. The ⊞ ⊕ symbols refer to those who are assumed to be heterozygote by deduction whereas the ⊠ ⊗ symbols refer to patients who have been examined both clinically and biologically and are proved to be normal. The date indicates the birthday while the letters refer to the initial of the last name.

exhibited a decreased theta rhythm energy. The most striking observation however was the development of a positivity of the initial part of the CNV under elevated seric ammonia.

4. BIOCHEMICAL FINDINGS

At the time of admission, liver function tests were abnormal because of an Australia antigen positive hepatitis that completely recovered after a three week period.

Under an estimated 6 to 8 g protein daily intake (so-called "standard conditions") the blood urea nitrogen was normal whereas both veinous ammonia (1.3 $\pm$ 0.5 mg/l) and serum uric acid (0.90 $\pm$ 0.08 mg/l) were enhanced.

This hyperuricemia accompanied an increased urinary excretion of uric acid (1.08 $\pm$ 0.56 g/24 h) while the renal function was biologically normal.

Table I shows the urinary excretion of argininosuccinic acid (ASA) found in this patient. Under standard conditions, the patient's ASA daily elimination averaged 0.9 g/l, i.e. approximately 1.4/24 h. This value increased during the hepatitis episode to 2.4 g/l or 3.9 g/24 h.

Table II indicates that this overexcretion of ASA was due to a deficiency of ASA cleavage enzyme (ASA-lyase, EC 4.3.2.1.) both in erythrocytes lysates and liver tissue (liver biopsy). Approximately 13 % of residual activity remained in red blood cells, while only 1 % of residual activity was determined in liver cells.

Amino acid and guanidino derivatives in serum and urines were also determined under various conditions (fasting, "standard" conditions, 30 g protein intake in hyperammoniemia). These results will be presented elsewhere (Grisar et al. , in preparation).

In order to understand the significance of the unusual observed hyperuricemia, the variations of serum uric acid, ammonia and ASA and of urinary ASA were studied after a 30 g protein intake (Fig. 2).

Urea blood nitrogen remained normal during the test. As shown on Fig. 2, serum uric acid remained elevated but did not further increase under the intake of protein. In contrast, serum ammonia and ASA were enhanced together with an overexcretion of urinary ASA.

Figure 3 shows the influence of a 10 g intake of ammonia

Table I. Urinary argininosuccinic acid (ASA)

	Total μM/g creat.	ASA mg/l	ASA 1'ànhydr. μM/g creat.	ASA 2'anhydr. μM/g creat.
Standard conditions	58	649	357	1556
	1001	782	392	3890
	3122(1)	–	–	–
	1901(2)	–	103(2)	565(2)
	5453	850	–	–
	3505	1340	–	–
during hepatitis	9596	3280	–	–
	4614	1620	–	–

(1) from the Born-Bunge Institute Laboratories (Prof. A. Lowenthal UIA)

(2) from the Central Laboratory of Chemistry (Prof. J.P. Colombo Insespital Bern)

The other determinations are from the Laboratory of Genetic Biochemistry (Dr. Schoos-Barbette and Dodinval-Versie - University of Liège).

Table II. Argininosuccinase and arginase levels of erythrocytes lysates and liver tissue

Enzymes	Tissue	Patient	Controls	Units
Argininosuccinase	Erythrocytes	0.943(1) 0.960(2)	4.510–9.940	$\mu M \times gHb^{-1} \times h^{-1}$
	Liver	2.170(1)	128–294	$\mu M \times gHb^{-1} \times h^{-1}$
Arginase	Liver	20.915(1)	5.340–14.557	$\mu M \times gHb^{-1} \times h^{-1}$

(1) from the Central Laboratory of Chemistry (Prof. J.P. Colombo-Inselspital Bern)

(2) from the Laboratory of Genetic Biochemistry (Dr. Schoos-Barbette - University of Liège).

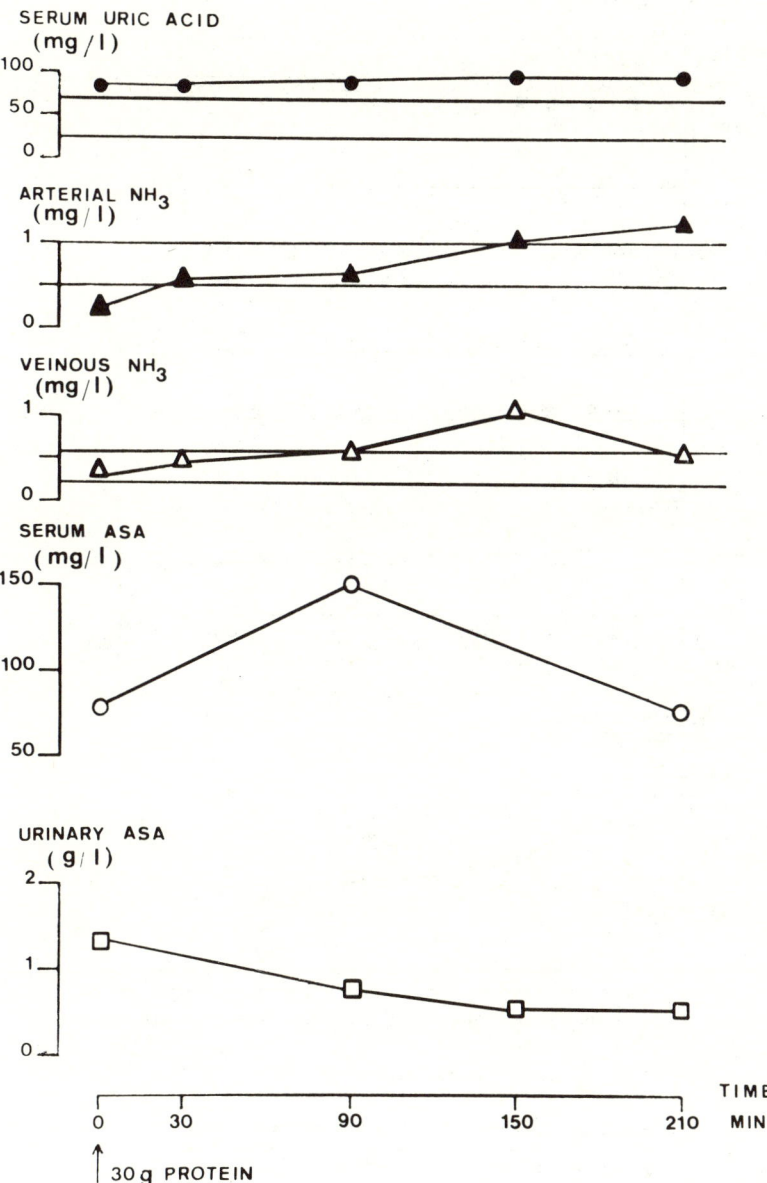

Fig. 2. Effect of a 30 g protein intake on serum uric acid, ammonia and ASA and on urinary ASA in a patient with argininosuccinic aciduria. Horizontal straight lines indicate the normal range.

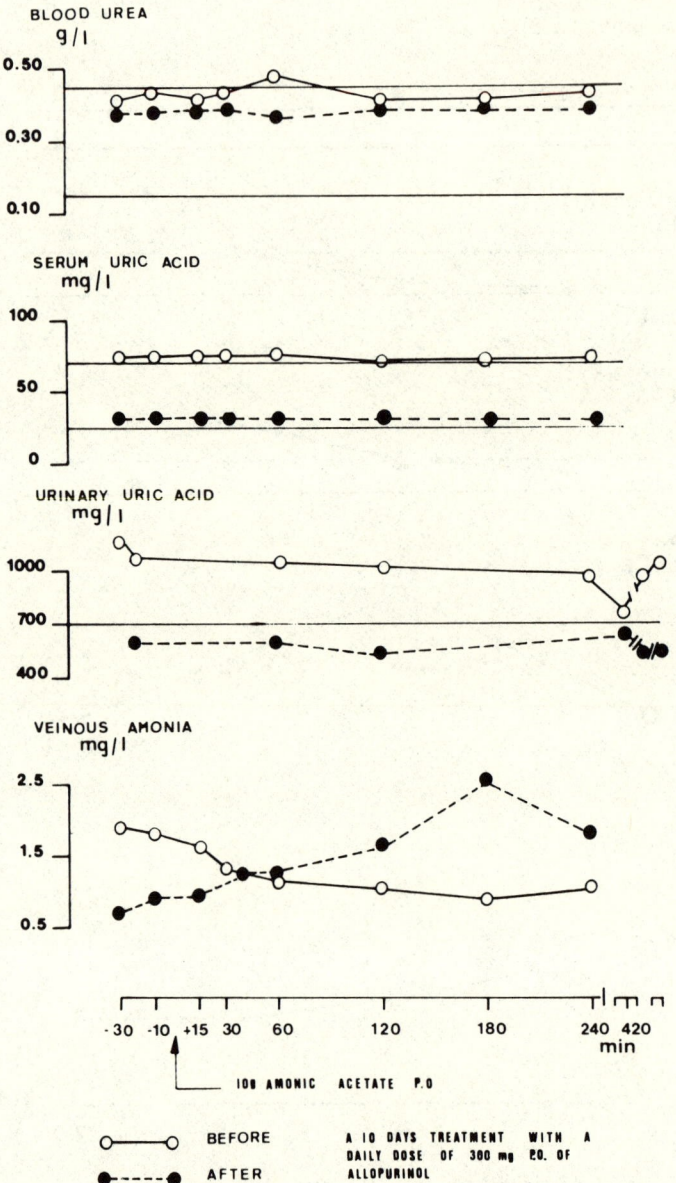

Fig. 3. Combined effects of a 10 g P.O. intake of ammonia acetate and a 10 days daily treatment with 300 mg of allopurinol in a patient with argininosuccinic aciduria. The results clearly show a decrease of the tolerance of the ammonia intake when blocking the in vivo synthesis of uric acid by allopurinol.

acetate. Neither blood urea, or serum ASA were influenced by such a diet. Surprisingly the veinous ammonia also remained unchanged.

However blocking the in vivo synthesis of uric acid by a daily dose of 300 mg of allopurinol, not only decreased serum and urinary uric acid elimination but led to a reduced tolerance of the ammonia charge since veinous ammonia increased after a same intake of ammonia acetate.

This experiment was reproduced twice and was clinically well supported by the patient.

COMMENTS

The clinical features of argininosuccinic aciduria are shown typically in the present case. Indeed, nearly all reports described the association of mental retardation, neurological signs like seizure, lethargia and tremor and cutaneous abnormalities (see Formstecher, 1978 for excellent review). Even the sad and inexpressive appearance of the face noted by Allan et al. (1958) and Schreier and Leuchte (1965) was also present.

Two clinical particularities however underlie our observation :

1) the late age of the diagnosis ;
2) the existence of osteoarticular rehandlings from neurogenic or metabolic origin.

The actual 44 cases were detected between the first days of life and 20. However, only few information is available concerning the age of the death. Hence, we do not have a clear idea about the exact prognosis of this "late and usual" form of the illness. We can only assume in front of previous reports that the short-dated prognosis of most of the cases was unfavourable. Thus we can assume, if not assert, that the present observation corresponds to a new "elderly" form of the argininosuccinic aciduria. Wether or not the osteoarticular abnormalities characterize this tardive form is of course unknown. Autosomal recessive inheritance is quite likely from the well documented family story (Husquinet et al., 1981, Fig. 1). Examination of the red blood cells for reduced ASA-lyase activity confirmed the Tolimson and Westall's suggestion (1964) that this enzyme determination might be an excellent test for detecting heterozygotes.

The interpretation of the positivity of the CNV is more difficult. Only few information is now available concerning this electrophysiological parameter in "organic" neurological diseases (i.e. non-psychiatric diseases). We recently found this paradoxal positivity in patients suffering from organic dementia (Timsit-Berthier et al., 1981) whereas it was not seen during hyperammoniemia in a patient presenting a porto-caval anastomosis syndrome (unpublished observation).

From the biochemical view-point, as reported by the majority of the authors, blood urea and the rate of urea production have been normal even elevated despite the well documented defect in urea cycle ASA lyase enzyme (Table I and II).

We had the opportunity to determine ASA metabolism during a slight liver deficiency during hepatitis. As shown (Table I), both ammoniemia and ASA elimination were enhanced. Also the protein diet (Fig. 2), as very often reported, increased both urinary ASA and serum ammonia.

The normal urea blood nitrogen level led to the hypothesis that alternative pathways exist not only for the synthesis of urea but also for the utilization of the excess of ammonia.

In this context, our principal finding in the present case was the existence of elevated serum and urinary uric acid (Fig. 2 and 3). The effects of an intake of ammonia salt has been investigated by some authors (Moser et al. , 1967 ; Tancredi et al., 1973), demonstrating either a subsequent hyperammoniemia (Moser et al., 1967) or, as in the present observation (Fig. 3) no change in veinous ammonia (Tancredi et al., 1973). Blocking the in vivo synthesis of uric acid by xanthine-oxydase inhibitor, allopurinol, led to the apparition of a progressive increase of serum ammonia, not seen under normal conditions. This might be an indirect but suggestive indication of a possible relationship between the urea cycle defect and elevated serum uric acid.

Thus, an alternative pathway could be the in vivo production of uric acid. Additional results on amino acid and guanidino derivatives metabolism are now available, attempting to support this hypothesis (Grisar et al. , in preparation).

Whether or not this presumed mechanism is responsible for the unusual survival of our patient remains an open question.

REFERENCES

Allan, J. D., Cusworth, D. C., Dent, C. E., and Wilson, V. K., 1958, A disease, probably hereditary, characterized by severe mental deficient constant gross abnormality of aminoacid metabolism, Lancet, 1:182.
Formstecher, P., 1978, L'argininosuccinylurie, in : "Le cycle

de l'urée et ses anomalies". J. P. Farriaux, ed., Doin, Paris.

Husquinet, H., Schoos-Barbette, S., Dodinval-Versie, J., Parent, M.Th., and Grisar, T., 1981, Argininosuccinylurie. Détection des hétérozygotes. J. Genet. Human, 29, 2, 286-295.

Moser, H. W., Efron, M. L., Brown, H., Diamond, R. and Neumann, G., 1967, Argininosuccinic aciduria : report of two new cases and demonstration tent elevation of blood ammonia. Amer. J. Med., 42:9.

Schreier, K. and Leuchte, G., 1965, Argininbernsteinsäure, Krankheit. Dtsch. Med. Wochenschr., 90:864.

Tancredi, F., Striano, S., Ragonese, G., Cedrola, G. and Guazzi, G.C., 1973, Argininosuccinico aciduria con a senza iperammoniemia. Studio di Minerva Pediatr. 25:280.

FIRST CASE OF ARGININOSUCCINIC ACIDURIA IN JAPAN :

CLINICAL OBSERVATIONS AND TREATMENT

T. Sakiyama, T. Suzuki, M. Owada and T. Kitagawa

Department of Pediatrics, Nihon University Hospital
1-8-13, Surugadai, Kanda, Tokyo, Japan

INTRODUCTION

Argininosuccinic aciduria is the most common disorder of the urea cycle in European countries and the U.S.A., but no case has ever been reported in Japan. It is caused by the deficiency of the enzyme L-argininosuccinic acid lyase (EC 4. 3. 2. 1., ASA lyse) with a resulting accumulation of large amounts of argininosuccinic acid (ASA) in body fluids. It is divided into three clinical types[1] : neonatal, subacute and late-onset form accompanied with mental retardation, hepatomegaly and friable hair, known as trichorrhexis nodosa. Genetic heterogeneity with the various clinical manifestations has been reported with the suggestion[2,3] of different biochemical forms in argininosuccinic aciduria. We report here the first case of argininosuccinic aciduria in Japan and the follow-up study under the treatment with l-arginine supplemented protein restricted diet for two and a half years.

CASE REPORT

A two-and-half-month-old Japanese boy, who was the first child of healthy unrelated parents, was hospitalized for his feeding difficulty, hepatomegaly and convulsion. He was born at 36 weeks gestation after uncomplicated pregnancy. Birth weight was 2980 g and no neonatal asphyxia was recorded. Family history revealed two unaccountable deaths during neonatal and infantile period respectively among the pedigree of a paternal uncle.

At the age of 4 days, he gradually manifested decreased sucking and feeding difficulty with weak physical movement, thin cry and depressed Moro reflex. He was studied through the routine laboratory examination of urine, blood and cerebrospinal fluid, but nothing abnormal was found. We looked again the fact retrospectively that the elevated level of methionine was found once at his first neonatal mass screening, but not detected at recall check. He was tube-fed, but was switched to bottles. At the age of one month, his body weight was 3840 g, but tremor of the extremities, hypertony, hepatomegaly about 3 cm below the right costal margin and convulsion were observed. At the age of 10 weeks, he was admitted to our hospital because of increasing frequencies of tremor and convulsion.

Physical examination at admission revealed his spastic and hypertonic figure with disappeared Moro reflex, frequent foot clonus and positive sun set phenomenon. Failure to thrive, a firm liver margin, palpable 4 cm below the right costal margin and light brown coarse and friable hair were noted. Height was 54.5 cm (-2 S.D.), weight 5240 g (-1 S.D.) and head circumference, 38.8 cm. Abnormal laboratory findings included serum glutamic-oxaloacetic transaminase 40 I.U. ; glutamic-pyruvic transaminase 48 I.U. ; creatine phosphofructokinase 123U ; acidemia pH 7.19 with bicarbonate 19 mEq/l and base excess -10 ; chloride 111 mEq/l ; calcium 11.1 mg/dl ; postprandial blood ammonia 298 μg/dl ; and lactate 38 mg/dl. Serum amino acid analysis by using Hitachi Amino Acid Autoanalyzer Model 835 revealed a marked elevation of glutamine 21.37 mg/dl ; citrulline 10.70 mg/dl ; ASA 5.86 mg/dl and a significant fall of arginine 0.45 mg/dl. Urinary ASA excretion was also remarkably increased with normal value of orotic acid. Electroencephalography showed a pattern of spike and wave complex and brain computed tomography revealed a moderate brain atrophy. He was diagnosed as having argininosuccinic aciduria and a protein restricted diet (1.5 g/kg/day) was started subsequently. After his dietary treatment, serum ammonia was normalized and abnormal neurological findings disappeared with improvement of some serum biochemical findings, which are listed in Table 1.

METHOD AND RESULTS

The activities of ASA lyase in red blood cells from the patient, mother and controls were assayed with using the method of Glick et al[2]. The activity in the patient's erythrocytes was reduced but maternal activity showed low normal value, which are listed in Table 1. After initiation of the therapy with l-arginine[4] supplemented (600 mg/day) protein restricted diet (1.5 d/kg/day), the neurological symptoms of the patient were markedly improved. Tremor and convulsion disappeared, and blood ammonia level became within normal range. The follow-up study for two and a half years, we changed the amount of l-arginine prescription

Table 1. Laboratory findings before and after dietary treatment, and ASA lyase activity in RBC.

		Before Therapy	After Therapy (protein 1.5g/kg)
Blood Gas	pH	7.19	7.27
	HCO_3	19 mEq/l	20
	BE	-10	-8
Pyruvate		0.48mg/dl	0.40
Lactate		38.0	28.5
Ammonia	1h after feed.	298 ug/dl	40
	2h after feed.	87	
BUN		10 mg/dl	6
GOT		40 I.U.	24
GPT		48 I.U.	23
Amino acid			
ASA	serum	5.86mg/dl	2.69
	spinal fluid		2.22
	urine	157.7	195.9

Orotic acid in urine : within normal limit
Organic acid in urine: not detectable

Argininosuccinate Lyase in RBC

    Control   100%
    Mother   low normal
    Patient   15.1% & 19.0% of control

from 600 mg/kg/day down to 400 mg/kg/day and up to 1000 mg/kg/day. But the levels of blood ammonia and serum arginine have maintained the normal range. During the period of l-arginine administered 400 mg/day, body weight gain seemed to be halted. ASA in the serum, however, was always detected and serum citrulline showed similar trend despite of different doses of arginine administration

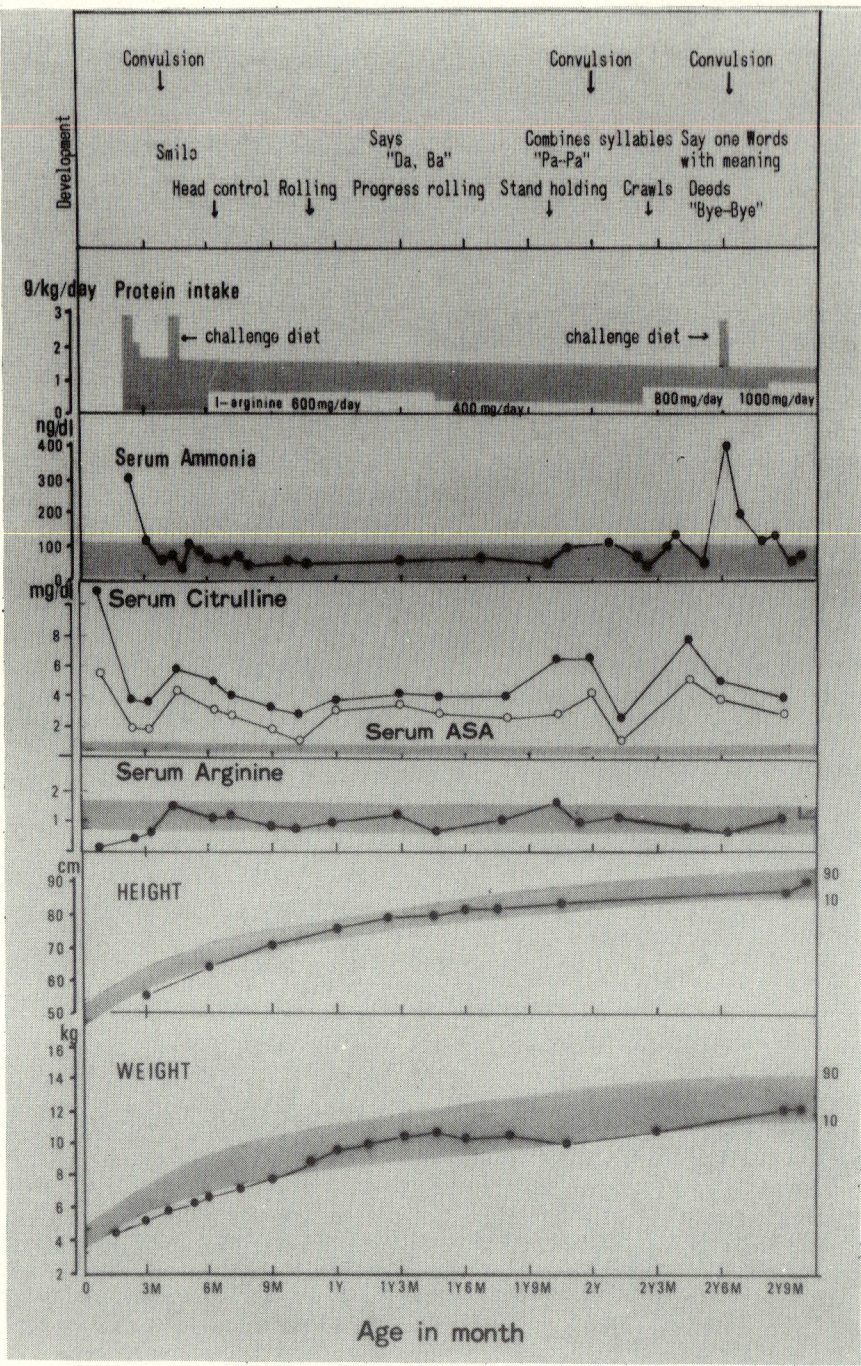

Fig. 1. Clinical course and the effect of treatment on blood ammonia and serum amino acids.

The value of ASA in cerebrospinal fluid was also elevated for several times, but did not exceed that in serum except once. But brain atrophy was markedly improved at the age of one. At the age of two and a half years, he was tried to have a challenge with free diet and he immediately manifested vomiting and lethargy with tremendous elevation in plasma ammonia, which is illustrated in Figure 1. This IQ level remained around 40 in spite of keeping this consistent ammonia and arginine level in the serum.

DISCUSSION

The incidence of argininosuccinic aciduria in Japan seems to be much lower than in Europe or in the U.S.A.[5]. It might be attributed to racial difference or might be overlooked because the clinical and biochemical manifestation of argininosuccinic aciduria is considerably variable. It might be, however, possible to find some other cases, if the determination of blood ammonia is used more frequently in suspected cases. The onset of our case was not as acute as in the neonatal type. But feeding difficulty, hepatomegaly, failure to thrive and convulsive attacks developed in early infancy. Clinical features are similar to the cases of Allan et al[6] and Levis B et al[7], which is classified as a subacute form or an infantile form with onset during the first year of life.

It has been established to make the diagnosis of argininosuccinic aciduria on the basis of the level of reduced enzyme activity in red blood cells[2,8]. Few cases, however, have been reported with residual erythrocyte enzyme activity over 10 % of control activity[2,5]. Our case also showed over 10 % residual ASA lyase activity in red blood cells, but it might be difficult to postulate ASA lyase activity in various organs. Because it has been explained that the absence of residual enzyme in erythrocytes does not always represent an acute or neonatal form. In addition to those heterogeneity, different enzyme assay systems should be considered.

The therapeutic administration of arginine was introduced to treat argininosuccinic aciduria by Brusilow and Batshaw. The treatment with arginine supplemented protein restricted diet was successful to control the blood ammonia and arginine level within normal range, but was not sufficient to reduce elevated ASA for our patient. Although the pathogenesis of neurologic abnormalities in this disease is not yet known, the cause of tremor, convulsion and other neurologically abnormal findings might be due to the accumulation of ASA and ammonia and the low level of arginine, because rapid improvement of these signs were observed after the initiation of this therapy. But a normal arginine content in the patients' brain has been found at four different laboratories[3]. This phenomenon suggests that ASA itself might play some important role to the mechanism of the neurotoxicity in this disease. Because first, the normal arginine content in brain might be expected and secondly the ASA in cerebrospinal fluid is also remarkably elevated. Under balanced distribution of amino acids in physiological fluids and in tissues due to the ASA lyase deficiency, such as high value of glutamine and glutamic acid may also affect brain function as the accumulation of nitrogen waste. Recently sodium benzoate and sodium phenylacetate were introduced to increase the excretion of nitrogen in compounds other than urea by Batshaw and Brusilow. Those compounds also would help for argininosuccinic aciduria to excrete the accumulated metabolites, or some specific substance would be expected for reduction of ASA in combination with dietary treatment.

REFERENCES

1. V. E. Shih, Urea cycle disorders, in "The Metabolic Basis of Inherited Disease", J. B. Stanbury, B. Wyngaarden and S. D. Fredrickson eds. McGraw-Hill, New York (1978).
2. N. R. Glick, P. J. Snodgrass, and I. A. Schafer, Neonatal Argininosuccinic Aciduria with Normal Brain and Kidney but Absent Liver Argininosuccinate Lyase Activity, Am. J. Hum. Genet. 28:22 (1976).
3. T. L. Perry, M. L. K. Wirtz, N. G. Kennaway, Y. E. Hsia, F. C. Atienza and H. S. Uemura, Amino acid and enzyme studies of brain and other tissues in an infant with argininosuccinic aciduria, Clin. Chim. Acta, 105:257 (1980).
4. S. Brusilow, and M. L. Batshaw, Arginine therapy of argininosuccinase deficiency, Lancet, 1:124 (1979).
5. I.A. Qureshi, J. Letarte, R. Ouellet, and B. Lemieux, Enzymologic and metabolic studies in two families affected by argininosuccinic aciduria, Pediat. Res. 12:256 (1978).
6. J. D. Allan, D. C. Cusworth, C. E. Dent, and V. K. Wilson, A disease, probably heriditary, characterized by severe mental deficiency and a constant gross abnormality of amino acid metabolism, Lancet, 1:182 (1958).
7. B. Levin, H. M. M. Mackay, and V. G. Oberholzer, Argininosuccinic aciduria. An inborn error of amino acid metabolism. Arch. Dis. Child., 36:622 (1961).
8. S. Tomlinson, and R. G. Westall, Argininosuccinic aciduria, Argininosuccinase and arginase in human blood cells, Clin. Sci., 26:261 (1964).
9. M. L. Batshaw, and S. W. Brusilow, Treatment of hyperammonemic coma caused by inborn errors of urea synthesis, J. Pediat. 97:893 (1980).

COMPLEMENTATION IN ARGININOSUCCINATE SYNTHETASE AND ARGININOSUC-

CINATE LYASE DEFICIENCIES IN HUMAN FIBROBLASTS

L. Cathelineau, D. Pham Dinh, P. Briand and P. Kamoun

Laboratoire de Biochimie Génétique, Hôpital Necker-
Enfants Malades, F-75730 Paris Cedex 15, France

INTRODUCTION

Gene complementation analysis of genetically defective cells is a useful tool in the knowledge of the heterogeneity of the inherited diseases. We have used the polyethylene glycol fusion on fibroblasts deficient either in argininosuccinate synthetase (ASS) or in argininosuccinate lyase (ASL) to detect a possible heterogeneity of citrullinemia or argininosuccinic aciduria. A slight modification of the method of Tedesco and Mellman[1] was used in the search of a positive complementation.

MATERIAL AND METHODS

Fibroblastic strains were established in our laboratory from six patients with citrullinemia and two unrelated patients with argininosuccinic aciduria. Enzymatic deficiencies are checked in liver biopsy by determination of argininosuccinate synthetase[1] or argininosuccinate lyase[2]. ASS was less than 1 p.c. of controls in citrullinemic patients and ASL was 30 p.c. of controls in argininosuccinic aciduria. Fibroblasts were also cultured from four strains of citrullinemic patients (obtained from the Human Genetic Cell Repository; strains 1679, 1684, 1044, and 0063) and from three strains of argininosuccinic aciduria (strains 2830, 533, and 525 ; the latter two are from siblings). All fibroblasts were cultured in RPMI 1640 medium supplemented with 10% fetal calf serum and antibiotics : penicillin (10^5 units/l) and streptomycin (250 mg/l). Mycoplasma contamination was checked by the method of Russel et al[3]. Incorporation of the radioactivity from [carbamoyl-^{14}C] citrulline into proteins was performed according to the method of Tedesco and

Mellman[1]. Cells were cultured in a standard medium in 25 cm^3 Falcon dishes. When confluent subcultures were obtained, cells were washed three times with a saline solution and then incubated at 37°C with arginine-free and serum-free medium. After 18h the medium was removed and 2 ml of arginine-free medium plus fetal calf serum (10%) containing 0.25 μCi of [carbamoyl-^{14}C] citrulline (50 μCi/mmol Amersham) and 2 μCi of [2,3-^{3}H] phenylalanine (32 Ci/mmol CEA France) were added. After 6h of incubation at 37°C the flasks were washed and the cells trypsinized and washed three times in cold saline solution. Proteins were precipitated with 5% cold trichloroacetic acid. After centrifugation 1 ml of 1 M NaOH was added to the pellet and the protein solution was added to 10 ml of Dimilume (Packard) for radioactivity counting. The tritium incorporation into proteins provides an estimation of the protein synthesis and the ratio ^{14}C/^{3}H indicates the ability of cells to transform ^{14}C-citrulline into ^{14}C-arginine which is then incorporated into proteins. This transformation needs the integrity of ASS and ASL activities.

Complementation was performed by mixing equal amounts of two different strains in a 25 cm^3 Falcon flask. After a few hours cells were washed three times. Polyethylene glycol (PEG 1000 Merck) 50% in RPMI was then added for 20 sec and after thorough washing the cells were incubated with RPMI 1640 and 20% fetal calf serum. Using a microscope with phase contrast we have checked that more than 50% of the cells are heterokaryons. After 24h ^{14}C-citrulline and ^{3}H-phenylalanine were added and the experiment performed as described above. Autoradiography was performed[4] on fibroblasts before and after complementation by use of incubation with [carbamoyl ^{14}C]-citrulline.

RESULTS

1) <u>Methodology</u>

In the method used the tritium incorporation provides an estimation of the protein synthesis and the ratio ^{14}C/^{3}H is related to the ability of cells to synthetize ^{14}C-arginine from ^{14}C-citrulline. This synthesis needs the integrity of ASS and ASL activities. The results obtained are unrelated to the amount of cells used. (Fig. 1).

Precision of the analysis may be defined as the random error expected of repeated measurements. It is, in other words, the reproducibility of the analysis under the prescribed conditions.

Precision is commonly expressed in terms of the standard deviation. Rather than running many replicate analyses of a single specimen it is usually more feasible to derive an estimate of standard deviation from many specimens analyzed in duplicate[5].

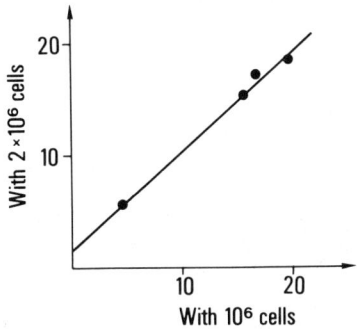

Figure 1 - $^{14}C/^3H$ ratios in fibroblasts after 6 hours of incubation with ^{14}C-citrulline and ^3H-phenylalanine : comparison of results with number of cells used.

$$(\text{standard deviation})^2 = \frac{(\text{differences between duplicates})^2}{\text{number of determinations}}$$

In our conditions the coefficient of variation ranged from 12.6 to 25.1 p.c. (Table 1).

2) <u>Citrullinemic and argininosuccinic aciduria strains</u>

The ratio $^{14}C/^3H$ radioactivities into proteins was very low (mean 4.1) in citrullinemic when expressed as percentage of the control strains. In contrast this ratio reaches 14.5 in argininosuccinic aciduria strains (Table II).

Table 1 - Precision of the $^{14}C/^{3}H$ ratio measurements

Strains	Mean	Standard deviation	Number of duplicates	Coefficient of variation
Citrullinemic				
without PEG	1.98	0.25	26	12.6
with PEG	1.83	0.46	28	25.1
Argininosuccinic aciduria				
without PEG	3.17	0.42	11	13.2
with PEG	4.53	0.62	11	13.8

Table II - $^{14}C/^{3}H$ ratio into proteins as percentage of control

Strains	Mean ± SEM	Number of experiments
Citrullinemic		
ROV	4.9 ± 1.1	7
MAR	2.7 ± 1.1	6
ROO	2.7 ± 0.3	3
SOU	3.8 ± 1.1	4
STE	6.0 ± 1.1	4
DAC	2.2 ± 0.7	5
0063	7.0 ± 3.1	3
1679	9.0 ± 4.0	2
1684	5.0	1
1044	2.3 ± 0.7	4
TOTAL	4.1 ± 0.5	39
Argininosuccinic aciduria		
DAM	8.3 ± 1.3	8
BOU	10.0 ± 1.7	8
2830	17.3 ± 2.9	8
525	31.3 ± 3.4	4
533	14.3 ± 1.4	8
TOTAL	14.5 ± 1.4	36

3) Complementation between citrullinemic strains.

The individual results of the tests are shown in figure 2 for 10 citrullinemic strains. The results are expressed by the ratio determined with or without polyethylene glycol. Cumulative results are shown in Table III. No complementation group can be defined in citrullinemia.

4) Complementation between argininosuccinic aciduria strains

Complementation by polyethylene glycol fusion is positive only between DAM and the four other strains (table IV). The cumulative

	0063	ROV	MAR	ROO	SOU	STE	DAC	1679	1684	1044
1044										1.2
1684									0.6	
1679								0.5		1.0
DAC							1.0			
STE						0.9	1.6	0.9		1.2
SOU					0.7	0.8	0.9	0.8	1.5	0.9
ROO				1.2	0.6	1.1		0.2		1.3
MAR			0.4	0.9	1.1	0.9				0.5
ROV		0.9	0.8	0.8	0.6	0.9		1.0	0.6	1.1
0063	1.1	0.7	0.8	1.6	1.1	1.1		0.7		0.8

Figure 2. Complementation tests between 10 strains from patients with citrullinemia. Results are expressed as the magnification of the ratio $^{14}C/^{3}H$ radioactivities incorporated into proteins after PEG treatment as compared to co-culture.

Table III - Complementation tests between 10 strains from patients with citrullinemia (cumulative results of figure 2). Results are expressed by ratio :

$$\frac{^{14}C/^{3}H \text{ with PEG}}{^{14}C/^{3}H \text{ without PEG}}$$

Strains	Mean ± SD	Number of strains used for complementation
0063	1.0 ± 0.4	7
ROV	0.8 ± 0.2	8
MAR	0.8 ± 0.2	6
ROO	0.9 ± 0.5	7
SOU	0.9 ± 0.3	9
STE	1.1 ± 0.3	8
DAC	1.3 ± 0.6	2
1679	0.8 ± 0.3	6
1684	1.1 ± 0.7	2
1044	0.9 ± 0.3	7

Table IV - Complementation tests between 5 strains from patients with argininosuccinic aciduria. Results are expressed by ratio :

$$\frac{^{14}C/^{3}H \text{ with PEG}}{^{14}C/^{3}H \text{ without PEG}}$$

Number of experiments are in parentheses.

	DAM	BOU	2830	525	533
DAM	0.8	6.4 ± 5.8 (5)	2.3 ± 1.5 (3)	1.7	1.5 ± 0.2 (3)
BOU		0.8	1.0	1.0	0.9
2830			1.1	0.8 ± 0.1 (2)	0.9 ± 0.1 (2)
525				0.9	0.6
533					1.0

results show a significant difference between DAM and the other strains (table V).

Autoradiographies of typical results are shown in figure 3 a positive complementation between DAM and other patients is clearly shown.

DISCUSSION

In citrullinemia four types are discernible according to their clinical manifestations : neonatal type, subacute type, mild type and atypical type[6]. Information on the enzyme defect also indicates heterogeneity. Activity of the ASS was not detectable in some patients. In other patients residual enzyme activity was 5 to 20 percent of the control values. Some patients have a deficiency of the enzyme in liver but not in other tissues. Kennaway et al[7] described kinetic abnormalities of ASS in cultured fibroblasts obtained from 3 patients. These biochemical findings suggest that citrullinemic patients comprise a heterogenous group. Liver ASS has been purified in rat[8] and human[9]. In all instances it is a tetramer formed of monomeric subunits of equal molecular weight. However, recently Takada et al[10] were able to separate three forms of ASS from rat liver. They assumed the enzyme has two kinds of subunits.

In regard of these results it was surprising that in ten strains of fibroblasts obtained from citrullinemic patients, complementation tests do not restore the enzyme activity as measured by ^{14}C incorporation into proteins from ^{14}C-citrulline.

Table V - Complementation tests between 5 strains from patients with argininosuccinic aciduria. Cumulative results of Table IV.

Complementation between	Number of experiments	Mean ± SD	t	p
[BOU; 2830; 525; 533] X [BOU; 2830; 525; 533]	7	0.9 ± 0.1	2.33	0.05
[BOU; 2830; 525; 533] X [DAM]	12	3.8 ± 4.3		

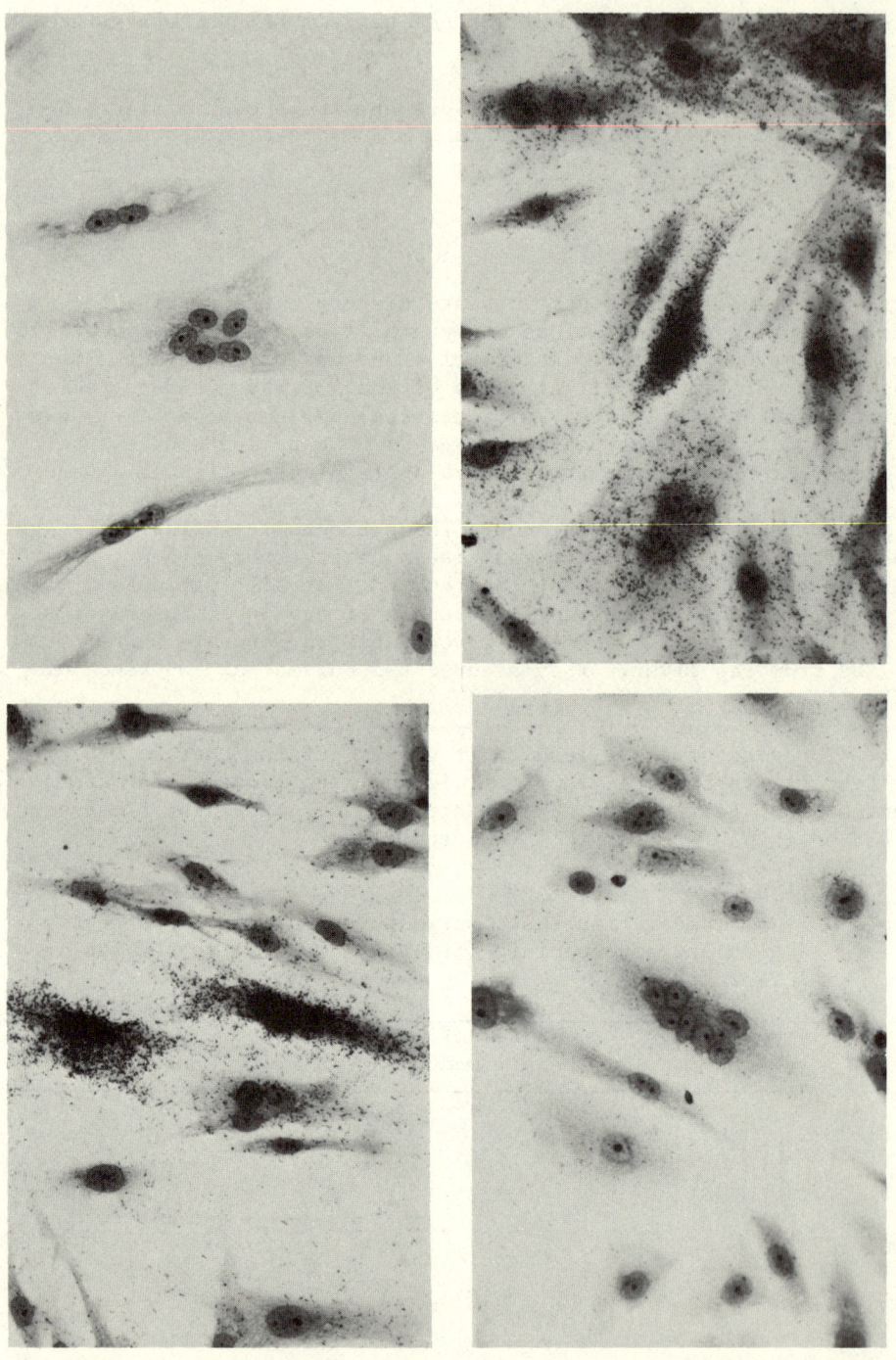

Fig 3 Autoradiography of
upper : right   control   ; left   DAM
lower : right   BOU x 533 ; left   BOU x DAM

CLINICAL AND BIOCHEMICAL FINDINGS IN ARGININEMIA

H.G. Terheggen, A. Lowenthal, J.P. Colombo

Municipal Children's Hospital, Cologne, Federal
Republic of Germany ; Born-Bunge Foundation, University
of Antwerp, Belgium ; Chemisches Zentrallabor,
University of Berne, Switzerland

I. CLINICAL OBSERVATIONS

In April 1968 two female siblings, 5 years and 2 months of age respectively, came under our observation. They were the offsprings of probably consanguinous parents. The elder sibling had a history of motor and mental retardation, seizures and spasticity predominantly in the lower extremities. The younger sibling presented us with major convulsive attacks, and showed a progressive loss of motor and mental skills together with increasingly more severe spasticity of both legs, during her first year of life In addition, she had intermittant episodes of vomiting, poor appetite and lethargy, features never present in the elder girl probably because of a self-selected low protein intake. In 1971 another female sibling was born who developed the same symptoms.

The similarity of the clinical findings in the two first mentioned sisters prompted us to do further biochemical investigations.

II. BIOCHEMICAL INVESTIGATIONS

Amino acid determinations were performed according to Efron (1). The determination of arginase activity has been described by Colombo (2).

Amino acid determination in the urine displayed a cystinuria pattern in both children. "Classic" cystinuria usually is not related to a neurological disease. Atypical cystinuria in our patients was characterized by the following features :

1. connection to a neurological disorder,
2. prominent argininuria,
3. normal or only slightly decreased reabsorption of cystine, lysine, ornithine and arginine (Table 1, 2).

Table 1. Urinary excretion of cystine, lysine, ornithine and arginine in the 3 affected children as compared to patients with "classic" cystinuria. The values are given in moles/gm of creatinine.

	A.W.	M.W.	I.W.	"Classic" cystinuria (n = 10)
Cystine	202	2960	352	2638
Lysine	2580	15600	471	5627
Ornithine	190	2540	21	2827
Arginine	1420	13700	108	3809

Normal values : cystine 83 ± 18, lysine 186 ± 77, ornithine 35 ± 16, arginine 33 ± 14

Table 2. Percentage of tubular reabsorption of cystine, lysine, ornithine and arginine in patients A.W. and M.W., patients with "classic" cystinuria (n = 10) and normal subjects.

	A.W.	M.W.	"Classic" cystinuria	Normal subjects
Cystine	68,4	80,4	7,4	97,4
Lysine	95,7	91,8	46,2	99,2
Ornithine	99,4	95,8	71,0	97,7
Arginine	99,4	97,9	42,5	99,8

The excretion of guanidino compounds in argininemia will be discussed by Dr. Marescau in a separate paper.

Determination of amino acids in serum showed a marked increase of the arginine concentration. Ornithine was normal or slightly

Analysis of the symptoms and the mode of onset of argininosuccinic aciduria shows that it may be divided into three types[6] : neonatal, subacute and late onset. A biochemical heterogeneity was also described in one case of argininosuccinic aciduria.[11]

In the five strains with ASL deficiency, we are able to define two groups of complementation : one strain complements with four others, but not to the same extent with each of the other strains. ASL from bovine liver[12], kidney[13], and brain[14] has been extensively studied. These enzymes all have the same molecular weight; all are tetramers formed of four identical subunits, and have the same catalytic, physical, and chemical properties. Regarding these biochemical results on purified ASL we did not think that we were able to find positive complementation with ASL deficient strains. However, recently McInnes et al[15] were able to restore the enzyme activity by complementation in this disease : three strains complement each other while a fourth strain failed to complement with the three others. Because in each experiment, restoration of activity was less than in the positive control which was ASL deficient strain fused with citrullinemic strain, they concluded at a probable intragenic complementation. We assumed also that our findings[16] are in favor of an intragenic complementation because the restoration of activity after fusion is also less than in the control. However, this hypothesis has to be supported by additional data including studies of the restoration of the enzyme activity in vitro as it has been performed for human B galactosidase by Hoeksema[17].

REFERENCES

1. Tedesco T.A. and Mellman W.J. (1967). Argininosuccinate synthetase activity and citrulline metabolism in cells cultured from a citrullinemic subject. Proc. Natl. Acad. Sci. USA 57 : 829.
2. Brown G.W. and Cohen P.P. (1959). Comparative biochemistry of urea synthesis. I. Methods for the quantitative assay of urea cycle enzymes in liver. J. Biol. Chem. 234 : 1769.
3. Russel W.C., Newman C. and Williams D.H. (1975). A simple cytochemical technique for demonstration of DNA in cells infected with mycoplasmas and viruses. Nature 253 : 461.
4. Stein G.H. and Yanishevsky R. (1979). Autoradiography in : "Methods in enzymology". Jakoby W.B., Pastan H. ed., Academic Press New York 58:279.
5. Henry R.J. and Dryer R.L. (1963). Some applications of statistics to clinical chemistry in : "Standard Methods of Clinical Chemistry". Seligson D. ed, Academic Press New York.
6. Shih V.E.(1978). Urea cycle disorders and other congenital hyperammonemic syndromes in : "The metabolic basis of inherited disease". Stanbury J.B., Wyngaarden J.B., Fredrickson D.S. ed, Mc Graw Hill New York.

7. Kennaway N.G., Harwood P.J., Ramberg D.A., Koler R.D. and Buist N.R.M. (1975). Citrullinemia : enzymatic evidence for genetic heterogeneity. Pediat. Res. 9 : 554.
8. Saheki T, Kusumi T., Takada S. and Katsunuma T. (1977). Studies of rat liver argininosuccinate synthetase. I. Physicochemical, catalytic and immunochemical properties. J. Biochem. 86 : 1353.
9. O'Brien W. (1979). Isolation and characterization of argininosuccinate synthetase from human liver. Biochem 18 : 5353.
10. Takada S., Kusumi T., Jaheki T., Tsuda M. and Katsunuma T. (1979). Studies of rat liver argininosuccinate synthetase. The presence of three forms and their physicochemical, catalytic, and immunochemical properties. J. Biochem. 86 : 1353
11. Qureshi I.A., Letarte J., Quellet R. and Lemieux B. (1978). Enzymologic and metabolic studies in two families affected by argininosuccinic aciduria. Pediat. Res. 12 : 256.
12. Lutsy C.J. and Ratner S. (1972). Biosynthesis of urea : XIV. The quaternary structure of argininosuccinase. J. Biol. Chem. 247 : 7010.
13. Bray R.C. and Ratner S. (1971). Argininosuccinase from bovine kidney. Comparison of catalytic, physical and chemical properties with the enzyme from bovine liver. Arch. Biochem. Biophys. 146 : 531.
14. Murakami-Murofushi K. and Ratner S. (1979). Argininosuccinase from bovine brain. Isolation and comparison of catalytic, physical, and chemical properties with the enzymes from liver and kidney. Anal. Biochem. 95 : 139.
15. McInnes R.R., Shih V. and Liunardo N. (1980). Intragenic complementation in argininosuccinic acid lyase (ASAL) deficiency. Pediatr. Res. 14 : 524 (abstract).
16. Cathelineau L., Pham Dinh D., Briand P. and Kamoun P. (1981). Studies on complementation in argininosuccinate synthetase and argininosuccinate lyase deficiencies in human fibroblasts. Human. Genet. 57 : 282.
17. Hoeksema H.L., Vandiggelen O.P. and Galjaard H. (1979). Intergenic complementation after fusion of fibroblasts from different patients with β-galactosidase deficiency. Biochim. Biophys. Acta. 566 : 72.

decreased. In the CSF arginine was also elevated (Table 3, 4).

Table 3. Amino Acid Concentration (μmoles/100 ml) in the Serum of the 3 Affected Children

Amino Acid	A.W.	M.W.	I.W.	Normal values (n = 25, X $\pm$ 2 s)
Ornithine	6.20	6.69	7.20	15.40 $\pm$ 10.0
Citrulline	3.90	5.27	4.27	3.58 $\pm$ 2.4
Arginine	64.00	99.60	157.92	9.16 $\pm$ 4.5

Table 4. Amino Acid Concentration (μmoles/100 ml) in the CSF of the 3 Affected Children

Amino acid	A.W.	M.W.	I.W.	Normal values (n = 13, X $\pm$ 2 s)
Ornithine	0.63	0.63	0.81	0.84 $\pm$ 0.46
Citrulline	0.17	0.25	1.21	0.21 $\pm$ 0.14
Arginine	5.33	9.48	11.32	1.42 $\pm$ 1.48

The assumption of dealing with an inborn error of the urea cycle was further supported by the finding of increased ammonia levels in all three siblings, the values ranging from 243 to 671 μg/100 ml on a normal protein intake and from 134 to 317 μg/100 ml on a low protein diet. It finally proved to be true by the demonstration of a decreased arginase activity in the patients' red blood cells (Table 5).

As parental consent for a liver biopsy was not obtained the site of the metabolic block was further investigated by loading tests. After an intravenous arginine load the arginine concentration of the serum was markedly increased, while the values for ornithine and citrulline were not influenced. The lack of any rise of the ornithine level points to a defective cleavage of arginine. The rise of the arginine concentration following an intravenous ornithine load, however, pleads for the functioning of the other metabolic steps of the urea cycle.

Table 5. Arginine Concentration in Serum and Red Blood Cell Arginase in the Family with Argininemia

	Serum arginine	Red blood cell arginase
Parents		
W.W.	16.11	743
I.W.	18.04	573
Affected siblings		
A.W.	64.00	absent
M.W.	99.60	absent
I.W.	157.92	absent
Healthy siblings		
M.W.	11.66	1294
I.W.	21.22	470
E.W.	14.98	616

Normal values : arginine 9.16 ± 4.5 µmoles/100 ml, arginase 710-1330 moles/hour/gm Hb

As red cell arginase activity accounts only for 2 % of total body arginase activity the results of the loading tests point to a defective liver enzyme.

Serum arginine concentration and red cell arginase activity was determined in the parents and their living children. Of the three clinically healthy children one had normal serum arginine values whereas the concentrations of the parents and the two other children exceeded the normal range. The red cell arginase activity corresponded in each case to the serum arginine level (Table 5). These findings point to a heterozygous gene state in the parents and in two other siblings and are compatible with an autosomal recessive inheritance.

After the diagnosis of an urea cycle disorder had been established, daily protein intake was restricted to 1.6 gm/kg of body weight. Although no clinical improvement was achieved by protein restriction, some important metabolic modifications occurred : the ammonia values in blood dropped to concentrations slightly above the normal range, the increased serum arginine went down, and concurrently the cystinuria pattern disappeared, indicating that cystinuria in argininemia is a secondary phenomenon due to the excretion of excess arginine. A trial of gene replacement using the Shope papilloma virus, although effective in the patients' cultured fibroblasts was not successful in vivo.

III. REVIEW OF THE LITERATURE

Until today 10 cases of argininemia have been published (3-11), 9 of which are symptomatic.

<u>Clinical features</u> are the result of the two major metabolic aberrations and may be ascribed either to ammonia intoxication or to increased arginine concentrations. Symptoms pointing to hyperammonemia are intermittent episodes of irritability, poor appetite and vomiting, lethargy eventually progressing to coma, and seizures. These features were found only in 7 of 9 symptomatic cases. They are neither persistent nor specific in argininemia, as they are also found in other inborn disorders of the urea cycle in which they are even more pronounced.

Typical features of argininemia common to all patients, except one treated effectively from birth, are the following : progressive loss of mental and motor skills, increasing spasticity predominantly in the lower extremities and hyperreflexia. Additional features not noted persistenly were abnormal spike and wave forms in the EEG, growth retardation, seizures, microcephaly, cerebral atrophy, ataxia, and athetosis. The severity of the spasticity and the absence of any improvement following introduction of a low protein diet distinguish arginase deficiency clinically from the other disorders of the urea cycle. Argininemia, therefore, may be considered to be an established disorder with specific characteristics.

The only <u>biochemical abnormality</u> present in all patients was an elevation of arginine in plasma and CSF. A decreased plasma ornithine concentration has infrequently been found (6, 9). The elevations of alanine, glutamine and glutamic acid levels observed in the hyperammonemia due to OTC deficiency were not present. In the patients reported by Cederbaum et al. (10) many other amino acids in the CSF were elevated including ornithine, citrulline, glutamine, asparagine, glycine, serine, tryptophan, and methionine. These authors, therefore, speculate that the pathogenesis of the neurological abnormalities in argininemic patients may in part be due to the perturbation of intracellular amino acid levels secondary to the elevated levels of arginine.

Urinary amino acid studies revealed increased excretion of arginine and to a lesser extent of the other dibasic amino acids and of cystine in 7 of 9 cases. In the case described by Michels and Beaudet (9) no atypical cystinuria pattern could be detected. In the patient published by Cederbaum et al. (7) urinary amino acid excretion was normal on most occasions, a modest dibasic aminoaciduria occurred only after a very high protein meal. Other abnormalities included an increased excretion of alanine, lysine and glutamine occurring after a single protein load (7) a transient homocitrullinuria (7), and an elevated excretion of citrulline, lysine and argininosuccinic acid (9) Urinary orotic acid excretion was greatly increased in four patients (8, 10). Excretion of uracil and uridine was increased in a similar fashion (10).

Hyperammonemia has neither been persistent nor striking in argininemic patients. Its detection may be difficult in children with a self-selected protein restriction as illustrated by the case reported by Michels and Beaudet (9) : of 30 ammonia levels, including many that were post-prandial, there were only two increases. Nevertheless it does occur in most patients at times. Conditions leading to hyperammonemia were febrile illnesses (10, 11), an increase in the daily protein intake (10), and loading tests with alanine, glycine or arginine (8). The connection of hyperammonemia to dietary protein intake is further demonstrated by the observation, that increased ammonia values in the three siblings reported by us were reduced following the introduction of a low protein diet (4). In opposition to these observations is the fact that one patient did not develop hyperammonemia in response to protein loading (7).

In 8 of 10 cases the demonstration of arginase deficiency was achieved using red blood cells as the sole diagnostic tissue (3-8, 10, 11). Arginase activity was either absent (3-6, 8, 11) or very low (3-6, 7, 10). The investigators of these cases had inferred that the liver enzyme must be deficient as well, an assumption supported by intravenous loading with arginine (12). The confirmation of this deficiency has been obtained in two cases (9, 10) and has been shown to include arginase in leukocytes (13) and the stratum corneum (9). Arginase activity in the fibroblasts from three argininemic patients was similar to that in controls (14).

The inconsistency of hyperammonemia and atypical cystinuria makes the diagnosis of argininemia more difficult. The abnormal amino acid excretion pattern may be construed as representing cystinuria coincidentally associated with spasticity and mental retardation. A normal urinary amino acid pattern found in patients on dietary (5) or self selected (7) low protein intake indicates this most frequently used criterion for the identification of inborn errors may be inadequate. Therefore all retarded or neurologically handicapped children with "cystinuria" and all patients with progressive spasticity and psychomotor retardation should have the amino acid levels in plasma investigated (10). With regard to this difficulty in diagnosis we agree with other authors (7, 10) that argininemia may not be rare.

Considering the pathogenesis of this unusual clinical picture, one might raise the question, whether it is simply the result of ammonia intoxication or whether it is due to the accumulation of

arginine. The increasing spasticity which has not been observed
in other forms of congenital hyperammonemia is suggestive of a
detrimental effect of high arginine levels. This view is supported
by the results of therapeutical trials : restriction of protein
intake, while successful in controlling hyperammonemia (4-6), has
reduced but not normalized the increased plasma arginine levels
(5)8, 10) and has not improved the clinical picture. The insensitivity of this chemical abnormality to a low protein diet demonstrates that protein restriction alone is not a suitable treatment
for this disorder.

Argininemia has now been reported in a total of 10 patients
originating from 6 families. Impaired arginase activity has been
demonstrated in 10 parents (in two families no paternal values
are given (9, 10), in 13 out of 19 healthy siblings, and in 4 of 6
grandparents investigated. The sex distribution and the half-
normal levels of arginase in parents, grandparents and siblings
establish the inheritance pattern of arginase deficiency as autosomal recessive.

Enzyme replacement constitutes the only causal treatment of
inborn errors of metabolism. In argininemia an attempt was made
to replace the deficient enzyme by administering packed red blood
cells (9). Although there was a small immediate decrease in serum
arginine, no significant changes occurred. The application of erythrocytes loaded with human liver arginase must still await in
vivo studies on animal models (15).

The Shope papilloma virus induces arginase in rabbits and
certain other animals including man (16). Skin fibroblasts derived
from an argininemic patient were cultured and infected with the
Shope virus. In these cells virus-induced arginase could be
demonstrated by measuring the amount of ^{14}C-ornithine liberated from
^{14}C-arginine. In addition virus-coded arginase could be shown
in the cells by fluorescent antibody studies (17). Following
injection of the Shope virus in three argininemic patients the serum
arginine values remained unchanged (18).

Symptomatic therapy : Lysine supplementation should be an effective form of therapy since it should interfere with tubular
reabsorption of arginine. One trial of lysine supplementation in
an argininemic patient had no effect on the plasma arginine level
(9). Moreover the supplementation of lysine may have adverse
effects, since arginase is inhibited by lysine (2, 19, 20). In
addition it has been found that a load of lysine increased the plasma
arginine level in two children with argininemia (11). This effect
in the argininemic patients is most likely due to inhibition of
residual arginase activity.

As mentioned above, on a low protein diet plasma arginine

levels remained high and there was no improvement of the clinical picture (5-8, 10).

Snyderman et al. (8), therefore, treated two argininemic children with an essential amino acid mixture simultaneously restricting the daily protein intake to 0.9 gm/kg. Treatment was successful in controlling the biochemical and clinical sequelae of the disease. The adequacy of the nitrogenous intake has been monitored by complete quantitative determination of the plasma amino acids at short intervals. None of the changes representative of early protein deficiency has been found (21). Recently Snyderman et al (11) reported on a neonate in whom the diagnosis of argininemia was established immediately after birth. A diet which provided the nitrogen moiety as a mixture of essential amino acids was started at 18 hours of age. At 32 months of age the child was physically, neurologically and mentally normal. Both height and weight were between the tenth and twenty-fifth percentiles.

Regarding the course of this child we can hopefully state, that argininemia may be another metabolic disorder, in which nutritional management is successful in controlling the detrimental effects of an inborn enzymatic deficiency.

IV. REFERENCES

1. M. L. Efron. Quantitative estimation of amino acids in physiological fluid using a Technicon amino acid analyser, in: "Automation in Analytical Chemistry", Technicon Symposia 1965, Mediad, New York (1965).
2. J. P. Colombo, W. Bürgi, R. Richterich, and E. Rossi. Congenital lysine intolerance with periodic ammonia intoxication, Metabolism 16:910 (1967).
3. H. G. Terheggen, A. Schwenk, A. Lowenthal, M. van Sande, and J. P. Colombo. Argininemia with arginase deficiency, Lancet 2:748 (1979).
4. H. G. Terheggen, A. Schwenk, A. Lowenthal, M. van Sande, and J. P. Colombo. Hyperargininämie mit Arginasedefekt. Eine neue familiäre Stoffwechselstörung, Z. Kinderheilk. 107: 298 and 313 (1970).
5. H. G. Terheggen, F. Lavinha, J. P. Colombo, M. van Sande, and A. Lowenthal. Familial hyperargininemia, J. Génét. hum. 20:69 (1972).
6. H. G. Terheggen, A. Lowenthal, F. Lavinha, and J.P. Colombo. Familial hyperargininaemia, Arch. Dis. Childh. 50:57 (1975).
7. S. D. Cederbaum, K. N. F. Shaw, and M. Valente. Hyperargininemia, J. Pediatr. 90:569 (1977).
8. S. E. Snyderman, C. Sansaricq, W. J. Chen, P. M. Norton, and S. V. Phansalkar. Argininemia, J. Pediatr. 90 : 563 (1977).
9. V. Michels and A. L. Beaudet. Arginase deficiency in multiple tissues in argininemia, Clin. Genet., 13:61 (1978)

10. S. D. Cederbaum, K. N. F. Shaw, E. B. Spector, M. A. Verity, P. J. Snodgrass, and G. I. Sugarman. Hyperargininemia with arginase deficiency, Pediat. Res., 13:827 (1979).
11. S. E. Snijderman, C. Sansaricq, P. M. Norton, and F. Goldstein. Argininemia treated from birth, J. Pediatr. 95:61 (1979).
12. T. Strauven, Y. Mardens, R. Clara, and H. G. Terheggen. Intravenous loading with arginine-hydrochloride and ornithine-aspartate in siblings of two families presenting a familial neurological syndrome associated with cystinuria, Biomedicine 24:191 (1976)
13. B. Marescau, J. Pintens, A. Lowenthal, H. G. Terheggen and K. Adriaenssens. Arginase and free amino acids in hyperargininemia: leukocyte arginine as a diagnostic parameter for heterozygotes, J. Clin. Chem. Clin. Biochem. 17:211 (1979).
14. A. F. van Elsen and J. G. Leroy. Human Hyperargininemia : a mutation not expressed in skin fibroblasts?, Am. J. Hum. Genet. 29:350 (1977).
15. K. Adriaenssens, D. Karcher, A. Lowenthal, and H. G. Terheggen. Use of enzyme-loaded erythrocytes in in-vitro correction of arginase-deficient erythrocytes in familial hyperargininemia, Clin. Chem. 22:323 (1976).
16. S. Rogers. Shope papilloma virus : a passenger in man and its significance to the potential control of the host genome, Nature (Lond.) 212:1220 (1966).
17. S. Rogers, A. Lowenthal, H. G. Terheggen, and J. P. Colombo. Induction of arginase activity with the Shope papilloma virus in tissue culture cells from an argininemic patient. J. Exper. Med. 137:1091 (1973).
18. H. G. Terheggen, A. Lowenthal, F. Lavinha, J. P. Colombo, and S. Rogers. Unsuccessful trial of gene replacement in arginase deficiency, Z. Kinderheilk. 119:1 (1975).
19. M. Statler and A. Russell. Competitive interrelationships between lysine and arginine in rat liver under normal conditions and in experimental hyperammonemia, Life Sci. 22:2097 (1978).
20. H. Sogawa, K. Oyanagi, and T. Nakao. Periodic hyperammonemia, hyperlysinemia, and homocitrullinuria associated with decreased argininosuccinate synthetase and arginase activities, Pediat. Res. 11:949 (1977.
21. S. E. Snyderman, L. E. Holt Jr., P. M. Norton, E. Roitman, and S. V. Phansalkar. The plasma aminogram. I. Influence of the level of protein intake and a comparison of whole protein and amino acid diets. Pediat. Res. 2:13 (1963)

ARGININEMIA: REPORT OF A NEW CASE AND MECHANISMS OF OROTIC ACIDURIA
AND HYPERAMMONEMIA

Makoto Yoshino, Kaoru Kubota, Ichiro Yoshida, Tatsuo Murakami and Fumio Yamashita

Department of Pediatrics, Kurume University School of Medicine, Kurume, Japan

Argininemia is a rare inborn error of ureagenesis due to a deficiency of arginase activity (Terheggen et al., 1969; Terheggen, et al., 1970a; Terheggen et al., 1970b; Terheggen et al., 1975; Cederbaum et al., 1976; Snyderman et al., 1977; Cederbaum et al., 1977; Michels and Beaudet, 1978; Snyderman et al., 1979). Despite the apparent defect in enzyme activity, hyperammonemia is only intermittently observed in this disease, unlike other enzymopathies of the urea cycle. Increase in orotic acid excretion is also another biochemical characteristic of this disease (Snyderman et al., 1977; Bachmann and Colombo, 1980). However, the mechanisms of hyperammonemia and orotic aciduria have not been thoroughly explained. The purpose of this communication is to describe clinical features at 2 hours of testing. Concentration of arginine was elevated also in cerebrospinal fluid (1.91 mg/dl, normal range; 0.15 - 0.55 mg/dl), orotic aciduria.

CASE REPORT

A 4 years and 2 month-old neurologically under developed girl was admitted to hospital because of pneumonia. She was born to nonconsanguineous, healthy parents after uncomplicated pregnancy and delivery. She had been well until the 18th day of life, when she began vomiting after feeding. At the age of 23 days, she developed fever, complicated by clonic convulsions, after which she was first brought to medical attention. Her extremities were spastic, with opisthotonic posture, and primitive reflexes were absent. Convulsive episodes have occurred since then despite treatment with anticonvulsants. Her mother noticed that the baby appeared "drowsy" after feeding at 4 months of age. Her neurological development was

totally arrested since early infancy; head control, rolling over
and language development have not been attained. Computerized axial
tomography of the skull at 12 months revealed dilatation of the lat-
eral ventricles and subdural collection of fluid.

At the age of 4 years and 2 months, she was admitted to hospital
because of pneumonia. Physical examination on admission revealed
a markedly retarded girl with microcephaly, spastic tetraplegia and
intermittent convulsion. The body weight was 9.5 kg (M-3.6 SD) and
height, 82 cm (M-4.7 SD). On the 13th day of hospitalization (Table
1), she was found to have hyperammonemia (960 µg/dl) and marked
increase in orotic acid excretion (6,900 µg/mg creatinine), measured as previously described (Adachi et al., 1963). Serum arginine
and ornithine levels were 3.1 and 0.56 mg/dl, respectively. After
reintroduction of protein feeding, serum arginine level began to
rise gradually to reach a maximum value of 19.4 mg/dl, whereas
excretion of orotic acid in urine declined to normal or slightly
above normal values soon after hyperammonemia resolved. Serum
citrulline rose in parallel with elevation of serum ornithine.

A diagnosis of argininemia was made in view of the elevated
serum arginine concentrations and practically nil activity of arginase of red cells (16 and 19 µmol/hr-g Hb, control; 3,716 ± 2,236),
measured according to Shih et al. (1978). An arginine loading test
(100 mg/kg, per orally) revealed a marked delay in clearance of
arginine from plasma, with a maximum concentration of 18.4 mg/dl
at 2 hours of testing. Concentration of arginine was elevated also
in cerebrospinal fluid (1.91 mg/dl, normal range; 0.15 - 0.55 mg/dl),
with concurrent serum value of 13.1 mg/dl at the age of 4 years and
7 months. She was fed with natural foods containing 0.5 to 1.0 g/
kg/day of protein and 0.5 g/kg/day of the mixture of essential amino
acids (Snyderman et al., 1977). The serum arginine level ranged
from 10.9 to 12.3 mg/dl throughout this period, whereas plasma ammonia remained within normal limits or only slightly elevated, and
urea nitrogen concentrations were 2.1 to 5.9 mg/dl.

DISCUSSION

Hyperammonemia is only intermittently observed in patients
with arginase deficiency, unlike in the other enzymopathies of the
urea cycle. This would imply that accumulation of ammonia in body
fluids is not the sole etiology of brain damage in argininemia
Prevention or improvement of neurological signs can be expected by
reducing serum arginine level (Snyderman et al., 1977; Snyderman
et al., 1979). Elevated arginine concentration in cerebrospinal
fluid has been found in previous patients (Terheggen et al., 1970;
Terheggen et al., 1975; Cederbaum et al., 1977) as well as in the
present patient. These observations may indicate that argininemia,

Table 1. Changes in blood and urine chemistries during clinical course, including increase in dietary protein intake.

	normal range	Date										
		6/30/81[a]	7/1	7/6	7/10	7/12	7/15	7/17	7/22	7/25	7/28	7/30	8/5
protein intake (g/kg-day)	—	0	0	0.2	0.5	0.5	1.0	1.0	1.5	1.5	1.5	2.0	2.0
plasma ammonia (μg/dl)	20 – 60	960[b]	450[b]	—	43	43	41	77	70	50	79	111	65
blood urea nitrogen (mg/dl)	7 – 20	—	—	1.4	2.6	—	1.6	—	3.3	—	—	—	3.5
serum amino acids[c] (mg/dl)													
arginine	1.68 ± 0.70	3.1	—	—	6.3	—	6.8	—	9.1	—	9.6	—	19.4
ornithine	1.37 ± 0.59	0.56	—	—	0.85	—	1.1	—	1.2	—	1.2	—	0.62
citrulline	0.39 ± 0.23	0.45	—	—	0.31	—	0.31	—	0.40	—	0.57	—	0.78
orotic acid (μg/mg Cr)	5 – 9	6,900	—	—	13	9	—	8	—	17	—	2	—

[a] 13th day of hospitalization
[b] The initial two values were determined by the indophenol method (normal range, 70–120 μg/dl)
[c] in our laboratory

either through arginine or its metabolite(s), exerts toxic effects on the brain.

Pathogenesis of hyperammonemia in this disease has not been fully understood. It is apparent that hyperammonemia is not due to accumulation of precursors of arginine, because neither serum citrulline nor argininosuccinate in urine were increased even in the presence of hyperammonemia.

Orotic aciduria has been reported in three previous patients with this disease (Snyderman et al., 1977; Bachmann and Colombo, 1980). The following has been suggested by Bachmann and Colombo (1980) to explain orotic aciduria in argininemia: first, enhancement of carbamylphosphate synthesis via the N-acetylglutamate-arginine system (Tatibana and Shigesada, 1976), and second, insufficient supply of ornithine due to a defect in arginase activity and competitive inhibition by arginine of ornithine resorption through the renal tubules. In our patient, however, serum arginine was only slightly elevated when excretion of orotic acid in urine was prominently elevated in the acute phase of pneumonia (Table 1), and orotic acid excretion declined to normal or slightly above normal values when serum arginine rose significantly in the later period. This observation would indicate that stimulation of carbamylphosphate biosynthesis by accumulated arginine was not the primary cause of the orotic aciduria, at least in the acute phase of pneumonia, although it is likely that such mechanism may have been responsible for the slight elevations of orotic acid excretion in the later period.

Serum ornithine levels in normal individuals (0.10 ± 0.04 mM, in our laboratory) are less than 1/10 of apparent Km for ornithine (1.5 to 1.7 mM at pH 7.2) of human liver ornithine transcarbamylase (Mori et al., 1980). Despite this fact, blood ammonia level does not increase significantly in normal individuals when nitrogen load is increased. This may imply that the intramitochondrial ornithine pool available for citrullinogenesis increases to meet the requirement for the substrate, in response to increase in nitrogen load. Tatibana and Shigesada (1976) have shown in the mouse that the hepatic level of ornithine rises as the content of dietary protein increases. In individuals with intact arginase activity, enhancement of the nitrogen load due to, for example, infection, is well tolerated, and is probably associated with elevation of hepatic intramitochondrial ornithine without producing hyperammonemia or orotic aciduria. It appears likely that an increase in hepatic (intramitochondrial) ornithine level does not result in patients with arginase deficiency. In fact, the serum ornithine concentration was significantly reduced (0.04 mM) when the patient developed hyperammonemia as well as orotic aciduria in the acute phase of pneumonia.

REFERENCES

Adachi, T., Tanimura, A. and Asahina, M., 1963, A colorimetric determination of orotic acid, J. Vitaminol., 9:217-226.
Bachmann, S. D. and Colombo, J. P., 1980, Diagnostic value of orotic acid excretion in heritable disorders of the urea cycle and in hyperammonemia due to organic acidurias, Eur. J. Pediat., 134: 109-113.
Cederbaum, S. D., Shaw, K. N. F., Spector, E. B. and Snodgrass, P. J., 1976, Hyperargininemia with arginase deficiency in two siblings, Fifth International Congress of Human Genetics, Mexico City, Excerpta Med. Int. Congr. Ser., No. 397, p. 27, ab.
Cederbaum, S. D., Shaw, K. N. F. and Valente, M., 1977, Hyperargininemia, J. Pediat., 90:569-573.
Michels, V. V. and Beaudet, A. L., 1978, Arginase deficiency in multiple tissues in argininemia, Clin. Genet., 13:61-67.
Mori, M., Uchiyama, C., Miura, S., Tatibana, M. and Nagayama, E., 1980, Ornithine transcarbamylase deficiency: coexistence of active and inactive forms of enzyme, Clin. Chim. Acta, 104: 291- 291-299.
Shih, V. E., Jones, T. C., Levy, H. L. and Madigan, P. M., 1972, Arginase deficiency in *Macaca fascicularis*. I. arginase activity and arginine concentration in erythrocytes and in liver, Pediat. Res., 6:548-551.
Snyderman, S. E., Sansaricq, C., Chen, W. J., Norton, P. M. and Phansalkar, S. V., 1977, Argininemia, J. Pediat., 90:563-568.
Snyderman, S. E., Sansaricq, C., Norton, P. M. and Goldstein, F., 1979, Argininemia treated from birth, J. Pediat., 95:61-63.
Tatibana, M. and Shigesada, K., 1976, Regulation of urea biosynthesis by the acetylglutamate-arginine system, in: "The Urea Cycle", S. Grisolia, R. Báguena and F. Mayor, eds., pp 301-313, John Wiley & Sons, New York.
Terheggen, H. G., Schwenk, A., Lowenthal, A., Van Sande, M. and Colombo, J. P., 1969, Argininemia with arginase deficiency, Lancet, 2:748-749.
Terheggen, H. G., Schwenk, A., Lowenthal, A., Van Sande, M. and Colombo, J. P., 1970a, Hyperargininämie mit Arginasedefekt. Eine neue familiäre Stoffwechselstörung. I. Klinische Befunde, Z. Kinderheilk., 107:298-312.
Terheggen, H. G., Schwenk, A., Lowenthal, A., Van Sande, M. and Colombo, J. P., 1970b, Hyperargininämie mit Arginasedefekt. Eine neue familiäre Stoffwechselstörung. II. Biochemische Untersuchungen, Z. Kinderheilk., 107:313-323.
Terheggen, H. G., Lowenthal, A., Lavinha, F. and Colombo, J. P., 1975, Familial hyperargininemia, Arch. Dis. Childh., 50:57-62.

THERAPY OF NEONATAL ONSET UREA CYCLE ENZYMOPATHIES (UCE)

M. Batshaw and S. Brusilow

Johns Hopkins Medical Institute, Baltimore, Maryland U.S.A.

SUMMARY

We treated 18 children with neonatal onset UCE: CPS 2, OTC 3, AS 6, AL 6. CPS and OTC were treated with protein restriction (PR) 0.5-0.75 g/kg/d + essential amino acids (EAA) 1 g/kg/d + arginine 1 mmol/kg/d + benzoate (B) 1.75 mmol/kg/d. AS was treated with PR + EAA + Arg (3-4 mmol/kg/d) + B. AL was treated with PR + Arg (3-4 mmol/kg/d).

All patients are alive (mean age 12 mo., range 1-35 mo.). Plasma NH_4 levels were normal ($<35\mu M$) or near normal except when dietary therapy was interrupted by illness or non-compliance. Then hyperammonemia (150 - 380 µM) responded to intravenous Arg and/or B within 5 hours.

Weight gain is normal, linear growth delayed. Intellectual development has been normal in 8, mildly delayed in 8 and severely delayed in 2. Fasting plasma levels on therapy are : Arg 50 - 150 µM ; Gly 120 - 300 µM, B 1-5 mg %, hippurate 1-5 mg %. Partition of urinary true waste nitrogen (TWN) (TWN = Total N-urea N-1/2 ASA N-2/3 Cit N) revealed ; in OTC 42 % of TWN was found in hippurate (H), in AS 32 % of TWN was found in Cit and H, and in AL 45 % of TWN was found in ASA. There was a transient increase in SGOT in one case each of OTC and AS. Two AL patients developed hyperlipemia while receiving Arg. One patient, inadvertently given 6 mmol/kg B, developed vomiting and irritability with plasma Gly of 64 µM and benzoate of 124 mg %. SGOT was normal and the child recovered in 12 hours. Thus reduction of the requirement for waste nitrogen excretion (WNE) and promiting WNE as hippurate, Cit or argininosuccinic acid has been effective in permitting survival in these previously fatal diseases.

DIAGNOSTIC, CLINICAL, PATHOLOGICAL AND
BIOCHEMICAL ASPECTS OF THE DISEASES

3) <u>Secondary and Transient Hyperammonemia</u>

INTRODUCTION TO THE SESSION ON SECONDARY AND TRANSIENT

HYPERAMMONEMIA

H. G. Terheggen

Municipal Children's Hospital Cologne, Federal
Republic of Germany

It has been accepted that hyperammonemia is not only restricted to inborn disorders of the urea cycle, but is found in a great variety of different clinical conditions.

Secondary hyperammonemia is found in association with inborn errors of metabolism not involving urea synthesis. It is frequently connected to liver dysfunction of various origin or to portosystemic shunts. It may be due to the lack of urea cycle intermediates, to enzyme inhibition or to increased ammonia uptake from endogenous or exogenous sources. In some conditions the cause of hyperammonemia has not been discovered.

Inborn disorders of metabolism associated to elevated ammonia levels are above all organic acidemias: methylmalonic aciduria due to methylmalonyl CoA racemase or mutase deficiency, alpha-methylacetoacetyl CoA thiolase deficiency, propionyl CoA carboxylase deficiency, isovaleric acidemia, and lactic acidosis due to pyruvate carboxylase deficiency. Dr. Saudubray and co-workers are going to report on hyperammonemia in organic acidurias. A case of isovaleric acidemia associated with hyperammonemia will be presented by Dr. Yoshita. Dr Coudé and collegues detected increased ammonia values in 6 neonatal cases of pyruvate carboxylase deficiency and will discuss their results.

Hyperammonemia accompanying liver dysfunction may of course be related to hepatic failure in fulminant hepatitis or severe intoxication. Mental deterioration in chronic obstructive lung disease or congestive heart failure may not be the result of hypercapnia or hypoxia, but may be due to hepatic congestion and hyperammonemia resulting from the failure of the right ventricle.

Reduced activity of ornithine transcarbamylase may be the cause of hyperammonemia in Reye's syndrome and will be discussed by Dr. Arashima and co-workers. It has recently been observed that newborns with severe asphyxia or hyaline membrane disease have elevated blood ammonia levels originating from hypoxic liver damage. Dr. Sakaguchi and associates will report on the relation between hypoxia and hyperammonemia found in 82 neonates. It has been suggested that in hypovolemic shock the urea synthesizing systems would be impaired. This could result from anoxia of the liver due to stasis. Long term high dose aspirin therapy widely used in juvenile arthritis may be complicated by hyperammonemia and coma resulting from aspirin-induced liver injury. Biopsies indicate that the hyperammonemia found in fetal erythroblastosis also results from liver disease. Finally it is suggested that elevated ammonia levels found eventually in preterm infants may in part be due to liver function-limitation.

Porto-systemic shunt due to liver cirrhosis deserves no further comment. Very recently Belgian authors reported on angiographic studies in four preterm newborn infants with transient hyperammonemia. At the time of hyperammonemia low portal hepatic perfusion was demonstrated in all infants and shunting through the ductus venosus in three. The portal hepatic hypoperfusion was transient, as shown by control angiography.

Hyperammonemia observed in a patient with renal insufficiency treated with essential amino acids was thought to be due to the lack of urea cycle intermediates i.e. of arginine.

An increase of the plasma ammonia level in human volunteers on restricted zinc intake and in zinc-deficient rats has been reported. Since zinc deficiency has been observed in sickle cell anemia, Prasad et al. measured plasma ammonia levels in such individuals. In their report they document hyperammonemia in sickle cell anemia patients that was corrected with zinc therapy. As demonstrated by animal experiments hyperammonemia in zinc deficiency is probably due to OTC deficiency.

Hyperammonemia in epileptic patients taking the anticonvulsive drug valproic acid is a newly recognized problem. There was no evidence of liver disease in the hyperammonemic children. Coulter and Allen relate the increased ammonia levels in these cases to an inhibition of CPS I activity by valproate metabolites. Dr. Coude and co-workers will present further studies on valproate-induced hyperammonemia.

Hyperammonemia as the result of increased ammonia uptake has been found in newborn infants fed intravenously with amino acid solutions. In these cases hyperammonemia was due to the contami-

INTRODUCTION

nation of the solutions with great amounts of ammonia. Hyperammonemia with coma, tachypnea and respiratory alkalosis developed in a 3-year-old boy with prune-belly-syndrome during a urinary tract infection with Proteus mirabilis. Hyperammonemia is thought to have resulted from the production within the massively dilated urinary tract of excessive amounts of ammonia due to bacterial urease and its subsequent reabsorption into the systemic circulation.

In some patients the cause of elevated ammonia levels remains unclarified. Examples for this group are some of the newborn infants eventually developing hyperammonemia and patients with cerebral arteriovenous malformation.

HYPERAMMONEMIA SECONDARY TO HEREDITARY ORGANIC ACIDURIAS : A STUDY OF 29 CASES

J.M. Saudubray, F.X. Coudé, H. Ogier, L. Cathelineau, P. Briand and C. Charpentier

INSERM U 12, Laboratoire de Biochimie Génétique,
Hôpital Necker, 149 rue de Sèvres, 75730 Paris Cedex 15
FRANCE

Hyperammonia has been reported in several disorders of branched chain amino acids metabolism including propionic, isovaleric, methylmalonic acidemia and B ketothiolase deficiency [1]. Nevertheless the true incidence of hyperammonemia and its variation during the course of these diseases are not yet well known. The purpose of this study was to compare the blood ammonia concentrations and the concomitant serum organic acid accumulation in patients with propionic, isovaleric and methylmalonic acidemia.

MATERIAL AND METHODS

Seven patients with propionic acidemia have been studied. The diagnosis was made usually during the first episode of the disease by gas chromatography. It was always confirmed by the direct measurement of propionyl CoA carboxylase activity in fibroblasts [2]. All patients had a total defect (0 to 2% of controls). Four of them had a neonatal onset of the disease and three of them a later one.

Five patients with isovaleric acidemia have been followed. All of them had a neonatal onset of the disease. Three of them were sibs of the same family. The diagnosis has been made by gas chromatography. All patients but one received a glycine supplementation and they were growing well with a low protein diet. Unfortunately the last one died during the first days of his life in another hospital before starting the glycine supplementation.

Seventeen patients with methylmalonic acidemia were followed. Ten of them had a severe form of the disease with a neonatal onset of symptoms; all had a total methylmalonyl CoA mutase apoenzyme deficiency except one with an impaired cobalamin metabolism typed as cbl B (performed by L.E. Rosenberg, Yale University, New Haven, Conn.). Seven patients had less severe methylmalonic acidemia with a later onset of symptoms; four of them had impaired cobalamin metabolism (3 cbl A, 1 cbl B), one had methylmalonyl CoA mutase apoenzyme deficiency, and two patients could not be investigated.

Blood ammonia concentrations were measured by the method of Miller and Rice : normal is considered to be less than 90 μM. Serum methylmalonic, propionic and isovaleric acids were concomitantly determined by gas chromatography.

RESULTS

In propionic acidemia (see fig.1) there was a perfect correlation between blood ammonia and serum propionic acid whatever the kind of onset of the disease. We observed approximatively an ammonia increase of 160 μmoles for mmole of serum propionic acid. Usually ammonia concentrations were higher in the neonatal period (up to 1mM) but this corresponds also to a greater accumulation of propionic acid, probably related to the high protein catabolism of the neonatal period.

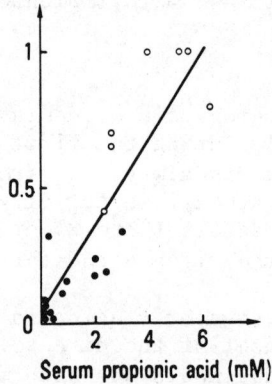

Fig. 1. Relationship between blood ammonia and serum propionic acid by linear regression analysis is demonstrated in 7 patients with propionic acidemia (neonatal (o) or late onset (●) of the disease). Blood ammonia (nmoles/l) = 159.9 (serum propionic acid in nmoles/l) - 42.4, r=0.893, n=20, (p<0.001).

In isovaleric acidemia (see fig. 2) there was also a good correlation between blood ammonia and serum isovaleric acid. Nevertheless this was less striking than in propionic acidemia and the ammonia increase was less important (approximatively 70 μmoles per nmole of serum isovaleric acid). After starting the glycine supplementation we never observed hyperammonemia.

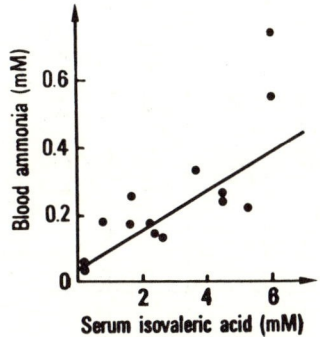

Fig. 2. Relationship between blood ammonia and serum isovaleric acid by linear regression analysis is demonstrated in 5 patients with isovaleric acidemia. Blood ammonia (nmoles/l) = 72.08 (serum isovaleric acid in nmoles/l) - 33.7, r=0.766, n=15, (p<0.001).

In eight neonatal methylmalonic acidemia (see the table 1), blood ammonia concentration was constantly and significantly elevated at the first examination, and in six of them reached values usually encountered in enzymatic defects of the urea cycle (over 300 μM). In two neonatal cases the ammonia was not determined during the neonatal period. Serum methylmalonic acid concentration was always increased above 1 mM and, when measured, serum propionic acid concentration was in the same range as that for methylmalonic acid. No correlation between serum methylmalonic acid and blood ammonia concentrations was observed. Later, during attacks of ketoacidosis

Table 1

Onset of symptoms	Assays	Results during		
		Initial neonatal period	Later keto acidosis attacks	Remissions
Neonatal 10 patients	Blood ammonia μM	447 ± 282 (8)	125 ± 32 (14)	108 ± 55 (9)
	Serum MMA mM	1.91 ± 0.97 (8)	1.21 ± 0.74 (10)	0.20 ± 0.04 (9)
Later in life 7 patients	Blood ammonia μM		112 ± 44 (9)	80 - 80 (2)
	Serum MMA mM		1.32 ± 0.92 (9)	0.18 ± 0.04 (9)

Blood ammonia and serum methylmalonic acid concentration in 17 patients with methylmalonic acidemia (mean ± SD).

serum methylmalonic acid reached approximately the same values as during the neonatal period. Nevertheless, concomitant blood ammonia concentrations were not significantly different from normal values. In the seven cases with later onset, there was no hyperammonemia during attacks of ketoacidosis in spite of a dramatic increase of the serum methylmalonic acid concentration (up to 3mM).

DISCUSSION

The lack of orotic acid elevation and the normal aminoacid pattern in serum and urine in our patients support the hypothesis of improved carbamyl phosphate synthesis. In vitro studies in rat have shown that N acetyl-glutamate synthetase activity was inhibited by propionyl and isovaleryl CoA but not by methylmalonyl CoA [3,4,5,6,7].

Hyperammonemia has been well documented in propionic adidemia (for review see ref. 8), Wolf et al. have followed the blood ammonia and propionic acid during isoleucine load in two patients and they observed a correlation between this two parameters. But to our knowledge no extensive study including numerous cases has yet been published. The significant correlation between propionic acid and ammonia which we observed in patients agree well with the in vitro studies in rats which have suggested that a metabolite of propionate inhibits ureagenesis [7]. Therefore serum propionic acid seems to be a direct reflect of the propionyl CoA accumulated into mitochondria.

In isovaleric acidemia hyperammonemia has been seldom reported. Nevertheless it was observed in all our patients and sometimes at concentration as high as 0.75 mM. Clearly ammonia levels were less high in isovaleric acidemia than in propionic acidemia and this may be explained probably by the fact that isovaleryl CoA is more rapidly metabolised by glycine N acylase than propionyl CoA. Indeed large amounts of isovaleryl glycine are excreted by patients with isovaleric acidemia whereas only small amounts of propionyl glycine are excreted by patients with propionic acidemia.

In methylmalonic acidemia hyperammonemia was exclusively observed in the neonatal period and was not correlated with serum methylmalonic acid values. Undoubtedly hyperammonemia is related to the accumulation of organic acids. But serum methylmalonic acid concentration may not be a true reflection of the intramitochondrial level of propionyl CoA, and this may account for the lack of correlation between methylmalonic acid and ammonia.

In all three defects hyperammonemia was always striking in the neonatal period but usually absent in subsequent attacks and this may be explained either by the high protein catabolism of the neonatal period, or by a particular susceptibility of the mitochondria to propionyl CoA, possibly related to physiologic enzymatic maturation. Enzymes of the urea cycle are known to be almost mature in the human newborn infants[9] but our knowledge of the maturation of the mitochondrial effectors of ureagenesis at this time is poor, especially with regard to acetylglutamate biosynthesis which seems to be an important regulator of ureagenesis[7].

REFERENCES

1. C. Bachmann, "Mental disorders of amino acid metabolism", W.L. Nyhan, ed., John Wiley and Sons, New York, page 361 (1974).
2. M. Saunders, L. Sweetman, B. Robinson, K. Roth, R. Cohn and R.A. Gravel, Biotin-response organic aciduria. Multiple carboxylase defects and complementation studies with propionic aciduria in cultured fibroblasts, J. Clin. Invest. 64 :1695 (1979).
3. L. Cathelineau, F. Petit, F.X. Coudé and P. Kamoun, Effect of propionate and pyruvate on citrulline synthesis and ATP content in rat liver mitochondria, Biochem. Biophys. Res. Commun. 90 :327 (1979).
4. D. Rabier, L. Cathelineau, P. Briand and P. Kamoun, Propionate and succinate effects on acetylglutamate biosynthesis by rat liver mitochondria, Biochem. Biophys. Res. Commun. 91 :456 (1979).
5. J.A. Gruskay and L.E. Rosenberg, Inhibition of hepatic mitochondrial carbamyl phosphate synthetase (CPSI) by acyl CoA esters. Possible Mechanism of hyperammonemia in the organic acidemias, Pediatr. Res. 13 :475 (1979).
6. F.X. Coudé, L. Sweetman and W.L. Nyhan, Inhibition by propionyl coenzyme A of N acetylglutamate synthetase of rat liver mitochondria. A possible explanation for hyperammonemia in propionic and methylmalonic acidemia, J. Clin. Invest. 64 :1544 (1979).
7. P.M. Stewart and M. Walser, Failure of the normal ureagenic response to amino acids in organic acid-loaded rats. Proposed mechanism for the hyperammonemia of propionic acid methylmalonic acidemia, J. Clin. Invest. 66 :484 (1980).
8. B. Wolf, Y.E. Hsia, K. Tanaka and L.E. Rosenberg, Correlation between serum propionate and blood ammonia concentrations in propionic aciduria, J. Pediatr. 93 :471 (1978).
9. N.C.R. Raiha, Developmental changes of urea cycle enzymes in mammalion liver, in "Urea cycle", Grisolia A., Baguena R. and Mayor F., Eds., John Wiley and Sons, Inc., p.261, New York (1976).

NEONATAL ISOVALERIC ACIDEMIA ASSOCIATED WITH HYPERAMMONEMIA

M. Yoshino, I. Yoshida, F. Yamashita, M. Mori[*], C. Uchiyama[*] and M. Tatibana[*]

Department of Pediatrics, Kurume University, Kurume Japan, Department of Biochemistry, Chiba University[*] Chiba, Japan

Isovaleric acidemia (IVA) is an inborn error of leucine metabolism due to a congenital deficiency of isovaleryl-CoA dehydrogenase (Rhead and Tanaka, 1980). Association of significant hyperammonemia with the disease has not been previously described. The present paper reports on a neonate with IVA who developed hyperammonemia. This observation may indicate that secondary derangement of ammonia metabolism can take place in IVA and that IVA should be included in the differential diagnosis of hyperammonemia in the neonatal period.

CASE REPORT

Y.Y. (#79-3809) was a 2,600 g female, product of a full-term pregnancy and a normal vaginal delivery. The pregnancy was complicated by maternal hypertension. The neonate's mother is the father's first cousin, once removed. She did well until the third day of life when her sucking became poor. At the age of four days, traction response and Moro reflex were felt incomplete and an offensive body odour was noted. On the same day, she was transferred to a neonatal intensive care unit because of progressive decrease in neonatal reflexes, poor sucking and loss of voluntary movements. Physical examination on admission revealed an inactive, moderately dehydrated, jaundiced newborn, responding poorly to stimuli. Tremor of fingers and staring gaze were occasionally observed and muscle tonus was moderately increased. Liver and spleen were not palpable. Laboratory data on admission included : blood glucose, 110 mg/dl; plasma bilirubin, 20.4 mg/dl; hemoglobin, 20.4 g/dl; leukocyte count, 6,600/cmm and pH of capillary blood, 7.34 with base deficit of 13 mEq/L. At the age of five days, blood ammonia, determined according to Tada et al. (1979), was 400 µg/dl (normal range, 50-100 µg/dl). The neonate was treated by phototherapy and albumin infusion for hyperbilirubinemia.

The combination of metabolic acidosis and the unusual body odour led to a presumptive diagnosis of IVA. The diagnosis was comfirmed by the demonstration of massive excretion of isovalerylglycine by gas chromatography and gas chromatography-mass spectrometry (Kuhara et al., 1979). Other organic acids characteristic of glutaric aciduria type II were not detected. At the age of seven days, plasma ammonia was 232 µg/dl (glutamate dehydrogenase method, normal range, 20-60 µg/dl). A complete blood count revealed: hemoglobin, 14.5 g/dl; leukocyte count, 3,600/cmm with a differential count of 12% band forms, 34% polymorphs, 50% lymphocytes, and 2% monocytes and platelet count, 25,000/cmm. Intermittent hypoglycemia and hypocalcemia, the latter ranging from 5.7 to 7.8 mg/dl, were found despite administration of calcium gluconate. Determination of amino acids in blood done on this day revealed moderate elevations of glycine and alanine (4.2 and 6.8 mg/dl, respectively, normal ranges, 1.9 ± 0.5 mg/dl for glycine and 3.3 ± 0.7 mg/dl for alanine). She was found to be hypotonic and deep tendon reflexes were depressed.

At the age of eight days, feeding was begun with a formula free of leucine, isoleucine and valine (prepared for maple syrup urine disease), later changed to a leucine free diet. Glycine administration (250 mg/kg-day in three divided doses, p.o.) was initiated on the same day. Since glycine therapy did not appear to improve her condition, an exchange transfusion was done at the age of ten days, after which blood ammonia concentration fell to 109 µg/dl. She died on the eleventh day of life, before further exchange transfusions were initiated.

MATERIALS AND METHODS

Liver tissue was obtained six hours after the death of the patient and control liver specimens were isolated from neonates who died without known liver disease six to twelve hours after death. All tissues were kept at -80°C until use and assays were done within three weeks. Urine specimens were stored at -20°C until analysis. Activities of urea cycle enzymes were measured according to Brown and Cohen (1959). For the study on kinetic properties of ornithine transcarbamylase (OTC), colorimetric and radiochemical methods were employed. For the measurement of apparent Km for ornithine, incubation mixture contained 50 mM KPO_4 buffer, pH 7.2, 10 mM carbamylphosphate, enzyme and variable concentrations of ornithine in the final volume of 200 µl, and citrulline formed was determined colorimetrically. The radiochemical method (Marshall and Cohen, 1972) was used to measure apparent Km for carbamylphosphate.

Table 1. Activities of urea cycle enzymes (μmol/hr/g)

Enzymes	Patient	Concurrent control[a]	Control (N=4)
Carbamylphosphate synthetase I	143	114	107 ± 7
Ornithine transcarbamylase			
pH 8.5	2,070	4,060	4,350 ± 760
pH 7.0	1,090	2,110	2,080 ± 110
Argininosuccinate synthetase	1.6	5.3	5.7 ± 2.3
Argininosuccinate lyase	81	103	125 ± 31
Arginase	18,700	15,800	3,380 ± 60[b]

[a] Control assays done concurrently with assays for the patient's specimen

[b] Arginase activities were measured without preincubation with manganese in these previous control assays, which gave a lower mean value of the activity than those of patient's and concurrent control tissues treated with manganese

Incubation mixture contained 50 mM diethanolamine-acetate buffer, pH 8.5, 5 mM ornithine, variable concentrations of ^{14}C-carbamylphosphate (specific activity, 1000 cpm/nmol) and enzyme in the final volume of 400 μl. For dialysis experiments, liver homogenates were dialyzed against 20 mM Hepes buffer, pH 7.2, containing 0.5 mM ornithine and 0.5 mM dithiothreitol. The content of N-acetylglutamate (N-AGA) in the liver was determined according to Shigesada and Tatibana (1971). Orotic acid in urine was measured by the colorimetric method of Adachi et al. (1963).

RESULTS

Moderate decreases in the activities of OTC and of argininosuccinate synthetase were found in the patient's liver (Table 1). The activities of the other enzymes were similar to control values. Apparent Km's of OTC for ornithine (1.25 mM) and for carbamylphosphate (0.07 mM) did not significantly differ from those of control

(1.33 and 0.07 mM, respectively). The reduction in the activity of patient's OTC was not restored by dialysis and the homogenate of the patient's liver did not inhibit OTC activity of control when the homogenates were mixed in various ratios and assayed for OTC activity. These results would rule out presence of a substance inhibitory to OTC activity.

The content of N-AGA, a key modulator of ureagenesis, was 2.18 nmol/g in the patient's liver, whereas those of two control specimens were 2.51 and 2.43 nmol/g, respectively.

Concentrations of orotic acid in urine (μg/mg creatinine) were 6.1 at five days of age, 9.1 and 27.1 (two specimens) at eight days of age and 3.3 at 10 days of age (normal range, 5-9).

DISCUSSION

Significant hyperammonemia, as observed in the present patient, has not been reported in previous patients with IVA. Association of hyperammonemia with IVA in our patient may suggest that hyperammonemia is one of the biochemical characteristics of IVA and that IVA should be included in diseases causing hyperammonemia in neonatal period, as with propionic acidemia and methylmalonic acidemia (Shih, 1978).

It has been postulated that under normal carbamylphosphate production, impairment of citrullinogenesis leads to enhanced influx of carbamylphosphate in the pyrimidine pathway. In our patient, urinary excretion of orotic acid, a known intermediate of pyrimidine synthesis, was not increased or only slightly elevated despite the presence of hyperammonemia and the reduction in OTC activity. This would indicate that impairment of carbamylphosphate synthesis was involved in the etiology of the hyperammonemia. The apocarbamylphosphate synthetase I was apparently intact (Table 1). The biosynthesis of carbamylphosphate is critically dependent on availability of N-AGA, an obligatory activator of carbamylphosphate synthetase I (Grisolia and Cohen, 1953; Shigesada and Tatibana, 1971; Tatibana and Shigesada, 1976; Shigesada et al., 1978). The content of N-AGA in the patient's liver did not appear to be depressed significantly. However, it has been shown that isovaleryl-CoA can also serve as substrate for N-AGA synthetase (Coude et al., 1979). It is possible, therefore, that accumulation of isovaleryl-CoA leads to formation of N-isovalerylglutamate and this acyl-glutamate inhibited activation of carbamylphosphate synthetase I by N-AGA. Occurrence of isovalerylglutamate has been demonstrated in a patient with IVA (Lehnert and Niederhoff, 1981). Isovaleric acid decreases oxidation rate of pyruvate and 2-oxo-glutarate (Gregersen, 1979). This may reduce generation of ATP in mitochondria. Such a possible reduction in ATP supply may also affect carbamylphosphate synthesis, since it is dependent on ATP.

The decreases in activities of OTC and argininosuccinate synthetase in the liver of the patient do not appear to have played a significant role in the pathogenesis of hyperammonemia.

Hyperglycinemia, though moderate in our patient, has also been found in previous patients with IVA (Ando et al., 1971; Guibaud et al., 1973; Saudubray et al., 1976). These observations may indicate that secondary derangement of glycine metabolism may take place in IVA, as in propionic acidemia or methylmalonic acidemia.

The authors are greatly indebted to Drs. T. Shinka, T. Kuhara and I. Matsumoto for organic acid analysis.

REFERENCES

Adachi, T., Tanimura, A. and Asahina, M., 1963, A colorimetric determination of orotic acid, J. Vitaminol., 9:217-226.
Ando, T., Klingberg, W. G., Ward, A. N., Rasmussen, K. and Nyhan, W. L., 1971, Isovaleric acidemia presenting with altered metabolism of glycine, Pediatr. Res., 5:478-486.
Brown, G. W. and Cohen, P. P., 1959, Comparative biochemistry of urea synthesis. I. Methods for the quantitative assay of urea cycle enzymes in liver, J. Biol. Chem., 234:1769-1774.
Coude, F. X., Sweetman, L. and Nyhan, W. L., 1979, Inhibition by propionyl-coenzyme A of N-acetylglutamate synthetase in rat liver mitochondria, J. Clin. Invest., 64:1544-1551.
Gregersen, N., 1979, Studies on the effects of saturated and unsaturated short-chain monocarboxylic acids on the energy metabolism of rat liver mitochondria, Pediatr. Res., 13:1227-1230.
Grisolia, S. and Cohen, P. P., 1953, Catalytic role of glutamate derivatives in citrulline biosynthesis, J. Biol. Chem., 204: 753-757.
Guibaud, P., Divry, P., Dubois, Y., Collombel, C., Larbre, F., 1973, Une observation d'acidémie isovalérique, Arch. Franc. Ped., 30:633-645.
Kuhara, T., Shinka, T. and Matsumoto, I., 1979, Gas chromatography-mass spectrometric studies on the urinary organic acids of patients with methylmalonic acidemia, Proc. Jap. Soc. Med. Mass Spectom., 4:315-320.
Lehnert, W. and Niederhoff, G., 1981, A series of new metabolites in isovaleric acidemia: 4-hydroxyisovaleric, 3-hydroxyisoheptanoic and N-isovalerylglutamic acid, Eur. J. Ped., 135: 332, abs.
Marshall, M. and Cohen, P. P., 1972, Ornithine transcarbamylase from Streptococcus faecalis and bovine liver II. Multiple binding sites for carbamyl-P and L-norvalline, correlation with steady state kinetics, J. Biol. Chem., 247:1654-1668.
Rhead, W. J. and Tanaka, K., 1980, Demonstration of a specific mito-

chondrial isovaleryl-CoA dehydrogenase deficiency in fibroblasts from patients with isovaleric acidemia, Proc. Nat. Acad. Sci. USA., 77:580-583.

Saudubray, J. M., Sorin M., Depondt, E., Herouin, C., Charpantier, C. and Pousset, J. L., 1976, Acidémie isovalérigue étude et traitement chez trois frères, Arch. Franc. Ped., 33:795-808.

Shigesada, K. and Tatibana, M., 1971, Role of acetylglutamate in ureotelism I. Occurrence and biosynthesis of acetylglutamate in mouse and rat tissues, J. Biol. Chem., 246:5588-5595.

Shigesada, K., Aoyagi, K. and Tatibana, M., 1978, Role of acetylglutamate in ureotelism. Variations in acetylglutamate level and its possible significance in control of urea synthesis in mammalian liver, Eur. J. Biochem., 85:385-391.

Shih, V., 1978, Urea cycle disorders and other congenital hyperammonemic syndromes, in :"The Metabolic Basis of Inherited Disease", J. B. Stanbury, J. B. Wyngaarden and D. S. Fredrichson, eds., pp 362-386, McGraw-Hill, New York.

Tatibana, M. and Shigesada, K., 1976, Regulation of urea biosynthesis by the acetylglutamate-arginine system, in:"The Urea Cycle", S. Grisolia, R. Baguena and F. Mayor, eds., pp 301-313, John Wiley & Sons, New York.

HYPERAMMONEMIA IN THE NEONATE WITH HYPOXIA

Yusuke Sakaguchi, Ken Yuge, Makoto Yoshino,
Fumio Yamashita and Takeo Hashimoto[*]

Department of Pediatrics, Kurume University School of
Medicine, and St. Mary's Hospital[*], Kurume, Japan

INTRODUCTION

Inherited defects in enzymes of the urea cycle and those in metabolism of several organic acids and some amino acids can cause neonatal hyperammonemia (Shih, 1978). Ballard et al. (1978) described five newborns with respiratory distress who developed hyperammonemia, but exhibited normal activities of urea cycle enzymes. Similarly significant hyperammonemia in neonates has been reported (Pollack et al., 1978; Le Guennec et al., 1980). Goldberg et al. (1979) have suggested that hyperammonemia may be associated with perinatal asphyxia. These observations would indicate that such hypoxic states as respiratory distress or neonatal asphyxia can cause nonspecific transient hyperammonemia. However, there have been no studies on the relationship between hypoxia and transient hyperammonemia. The relationship between levels of blood ammonia and blood gases was studied to elucidate the role of hypoxia in the pathogenesis of transient hyperammonemia in neonates.

PATIENTS AND METHODS

Among newborn infants admitted to the High-Risk Newborn Center of St. Mary's Hospital during a 5-month period, 82 newborns were studied. Of the 82 neonates studied, 71 were measured for blood ammonia from birth to 24 hours of age, 5 at one day of age, 4 at two days of age, one at three days and one at four days of age.

The blood ammonia level was measured according to Tada et al. (1979), a semiquantitative method using specimens obtained by heel stick. Blood gas values were measured concomitantly by the Astrup method (Corning Model 175) on the same specimens. Values of ammonia above 150 µg/dl (normal range for newborns; 50-100 µg/dl) have been defined as hyperammonemia.

Neonatal asphyxia was diagnosed when Apgar scores at 1 or 5 min were 6 or less. The relationship between the incidence of hyperammonemia and neonatal asphyxia was studied in newborns younger than 24 hours of age to evaluate effects of neonatal asphyxia on blood ammonia values. Newborns exhibiting retraction scores (Silverman and Anderson) of 3 or more at the time of blood sampling were classified as having respiratory distress.

The study group was divided into four. All newborns of 1 to 4 days of age were regarded as being without neonatal asphyxia. Group A (a1, a2) included neonates with both neonatal asphyxia and respiratory distress, group B (b1, b2), newborns with neonatal asphyxia alone, and group C (c1, c2), newborns with respiratory distress alone. Group D (d1, d2) consisted of neonates with neither neonatal asphyxia nor respiratory distress. These newborn infants were further divided according to the presence (a1 through d1) or absence (a2 through d2) of hyperammonemia.

RESULTS

In 82 neonates studied, 19 (23%) were found to have hyperammonemia. The incidence of hyperammonemia in neonates with blood pH < 7.2, base deficit (BD) $\leq$ -11 mEq/L, pO_2 < 40 mmHg and $pCO_2 \geq$ 60 mmHg, was significantly higher ($p < .001$) than for neonates with blood pH $\geq$ 7.2, BD $\geq$ -10 mEq/L, $pO_2 \geq$ 40 mmHg and pCO_2 < 60mmHg, respectively. The ranges of blood gas values (mean ± SD) in 19 neonates with hyperammonemia were: blood pH, 7.0 ± 0.16; BD, -17.2 ± 7.4 mEq/L, pO_2, 38.3 ± 14.5 mmHg; and pCO_2, 69.4 ± 19.3 mmHg (Fig. 1). Each patient in this group was given oxygen (FiO_2, 25-100%), and 17 of them required assisted ventilation (FiO_2, 60-100%). In 63 neonates with normal levels of blood ammonia, 35 were administered oxygen (FiO_2, 26-100%) and 12 of the 35 newborns were treated by assisted ventilation (FiO_2, 50-100%). Blood gas values in these 63 neonates were as follows: blood pH, 7.30 ± 0.09; BD, -3.2 ± 4.4 mEq/L; pO_2, 45.4 ± 8.2 mmHg; and pCO_2, 40.3 ± 10.1 mmHg (Fig. 1). The differences were all statistically significant.

The incidence of hyperammonemia in newborns with and without neonatal asphyxia is presented in Table 1, and the incidence of hyperammonemia in the groups with or without respiratory distress is shown in Table 2. In groups A through D, 13 cases (40%) in group A, one case (33%) in group B and 5 cases (22%) in group C had hyperammonemia, and in group D, there was no case of hyperammonemia (Table 3). The incidence of hyperammonemia was significantly higher in groups A, B and C ($p < .001$, $p < .01$, $p < .05$, respectively) than

HYPERAMMONEMIA IN THE NEONATE

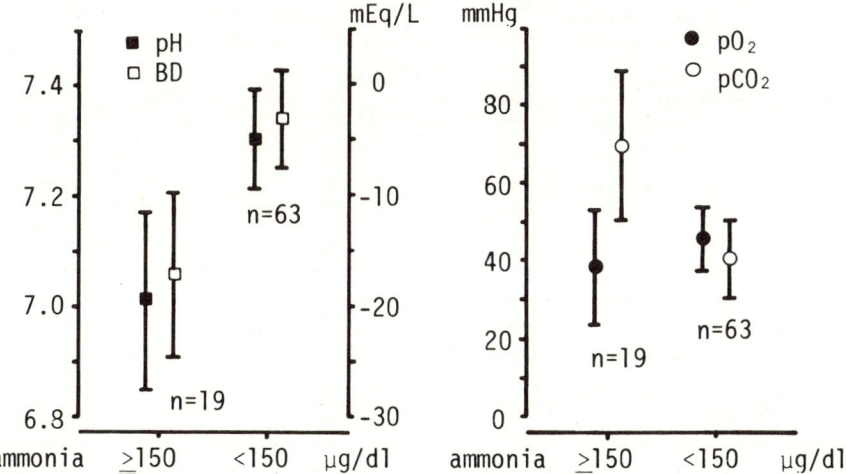

Fig. 1. Levels of blood gas values in neonates with hyperammonemia

In all of blood gas values, there are significant differences (blood pH, $p < .001$; BD, $p < .001$; pO_2, $p < .01$; and pCO_2, $p < .001$) between hyperammonemic and non-hyperammonemic neonates.

Table 1. The incidences of hyperammonia in newborns with and without neonatal asphyxia

Apgar score		1 min		5 min		total
		≤ 6	≥ 7	≤ 6	≥ 7	
blood ammonia	≥ 150	14	3	10	7	17
(μg/dl)	< 150	21	33	12	42	54

There is a significant difference between neonates in Apgar score (1 min and/or 5 min) ≤ 6 and neonates in Apgar score (1 min and/or 5 min) ≥ 7 ($p < .01$).

Table 2. The incidences of hyperammonemia in neonates with and without respiratory distress

retraction score		≥ 3	≤ 2	total
blood ammonia	≥ 150	18	1	19
(μg/dl)	< 150	37	26	63

There is a significant difference between newborns with retraction scores ≥ 3 and newborns with retraction scores ≤ 2 ($p < .001$).

Table 3. The incidences of hyperammonemia in groups A through D

group		blood ammonia (µg/dl)	number	total
A	a1	≥ 150	13	32
	a2	< 150	19	
B	b1	≥ 150	1	3
	b2	< 150	2	
C	c1	≥ 150	5	23
	c2	< 150	18	
D	d1	≥ 150	0	24
	d2	< 150	24	

A: newborns with both neonatal asphyxia and respiratory distress
B: newborns with neonatal asphyxia alone
C: newborns with respiratory distress alone
D: newborns with neither neonatal asphyxia nor respiratory distress

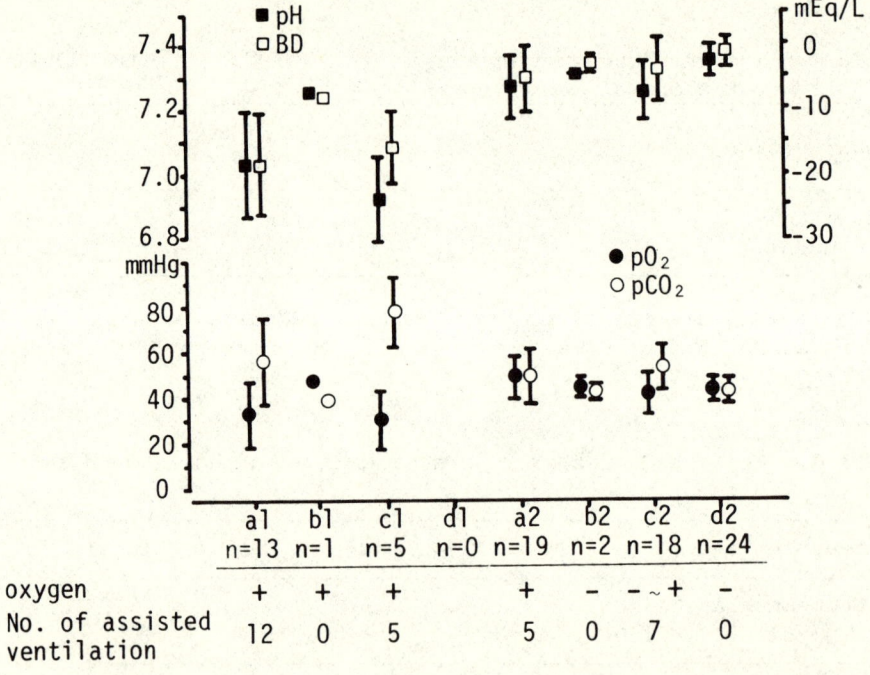

Fig. 2 Levels of blood gas values in groups with hyperammonemia (groups a1 through d1) and in groups with normal blood ammonia concentrations (groups a2 through d2)

in Group D. Fig. 2 depicts the ranges of blood gas values in groups a1 through d1 and groups a2 through d2. Groups b1, b2 and d1 were excluded from statistical analysis because the number of patients in these three groups were very small or zero. Blood pH and BD in group a1 were significantly lower ($p < .001$) than those in group a2, and there were similar differences in blood pH and BD ($p < .001$) between groups c1 and c2. The pO_2 level was significantly lower in group a1 ($p < .01$) than group a2, and there was no significant difference between groups c1 and c2. The values of pCO_2 in groups a1 and c1 were significantly higher ($p < .01$, $p < .001$, respectively) than in groups a2 and c2. There were no significant differences in blood gas values (blood pH, BD, pO_2 and pCO_2) between groups a1 and c1. There were no significant differences between groups with or without hyperammonemia and either birthweight or gestational age.

DISCUSSION

The pathophysiologic process common to previously described neonates with transient hyperammonemia is hypoxia (Ballard et al., 1978; Pollack et al., 1978; Le Guennec et al., 1980; Goldberg et al., 1978). The hypoxic state, associated with either neonatal asphyxia or respiratory distress, may cause hyperammonemia. However, little work has been done on the relationship between levels of blood ammonia and blood gas values in neonates.

From the results of this study, we believe hyperammonemia can occur in newborns with neonatal asphyxia and/or respiratory distress with hypoxia severe enough to cause metabolic acidosis even after oxygen is given. It is presumed that hypoxia impairs ureagenesis by affecting hepatic energy metabolism, upon which ureagenesis is critically dependent. A decrease in portal blood flow through the liver due to heart failure in respiratory distress may also reduce ammonia clearance from blood.

The 23% incidence of hyperammonemia in our series of newborns would be presumably higher than in the general population of newborn infants, because neonates in this study were at high-risk. Batshaw et al.(1978) found higher blood ammonia levels in low birthweight infants than in mature infants. However, it may be evidenced that the hyperammonemia observed in our neonates is not due to low birthweight but rather to hypoxia. We found no significant difference in body weight between hyperammonemic and normoammonemic newborns. Beddis et al (1980) also found no correlation between blood ammonia concentrations and birthweights in neonates admitted to their intensive care baby unit.

Association of hyperammonemia with neonatal hypoxia was demonstrated in this study. However, it is possible that inborn errors of ureagenesis or organic acidemias may mimic transient hyperammonemia due to hypoxia. We have identified two patients with methylmalonic acidemia and one with isovaleric acidemia (not included in the present series) among the neonates with metabolic acidosis and hyperammonemia. Such disorders should be vigorously excluded before diagnosis of transient hyperammonemia in the newborn infants is made.

REFERENCES

Ballard, R. A., Vinocur, B., Reynolds, J. W., Wennberg, R. P., Merritt, A., Sweetman, L. and Nyhan, W. L., 1978, Transient hyperammonemia of the preterm infant, N. Engl. J. Med., 299: 920-925.

Batshaw, M. L., and Brusilow, S. W., 1978, Asymptomatic hyperammonemia in low birthweight infants, Pediatr. Res., 12:221-224.

Beddis, I. R., Hughes, E. A., Rosser, E. and Fenton, J. C. B., 1980 Plasma ammonia levels in newborn infants admitted to an intensive care baby unit, Arch. Dis. Chilh., 55, 516-520.

Goldberg, R. N., Cabal, L. A., Sinatra, F. R., Plajstek, C. E. and Hodgman, J. E., 1979, Hyperammonemia associated with perinatal asphyxia, Pediatrics, 64:336-341.

Le Guennec, J. C., Qureshi, I. A., Bard, H., Siriez, J. Y. and Letarte, J. E., 1980, Transient hyperammonemia in an early preterm infant, J. Pediatr., 96:470-472.

Pollack, L., Hansen, T., Adams, J., Jr. and Beaudet, A., 1978, Transient hyperammonemia in term and preterm infants, Pediatr. Res., 12:532, abs.

Shih, V. E., 1978, Urea cycle disorders and other congenital hyperammonemic syndromes, in:"Metabolic Basis of Inherited Disease", J. B. Stanbury, J. B. Wyngaarden, D. S. Fredrickson, eds., McGraw-Hill Book Co., New York.

Tada, K., Okuda, K., Watanabe, K., Imura, Y. and Yamada, S., 1979, A new method for screening for hyperammonemia, Eur. J. Pediatr., 130:105-110.

A MECHANISM FOR VALPROATE-INDUCED HYPERAMMONEMIA

F. X. Coudé, D. Rabier, L. Cathelineau, G. Grimber,
P. Parvy and P. Kamoun

Laboratoire de Biochimie Génétique
Hôpital des Enfants Malades, 149, rue de Sèvres
75730 Paris Cedex 15, France

Sodium salt of valproic acid (VPA) is an anticonvulsant which has been successfully used in the treatment of several types of epilepsy, particularly in petit mal[1]. It has been shown recently that VPA induced hyperammonemia in children [2,3]. In the search of the mechanism by which VPA inhibits ureagenesis and because of the analogy between some other side effects of VPA and the ketotic hyperglycinemia syndrome (hyperglycinuria [4,5,6], leucopenia [7] and thrombocytopenia [8]) we have studied the in vivo and in vitro metabolic effects of VPA on the mitochondrial steps of ureagenesis in rat.

MATERIAL AND METHODS

Male Wistar rats (200-300 g) were fed a standard diet provided by Usine d'Alimentation Rationelle France, AO_2 diet 19 % protein. Rats were killed 2 hours after intraperitoneal injection of sodium valproate (100 mg/kg in 2 ml of water). Liver mitochondria were isolated according to the method of Hogeboom[9] and suspended in 250 mM mannitol, 2mM Tris-HCl pH 7.4. To study N-acetylglutamate biosynthesis mitochondria (about 5 mg of protein) were incubated for 15 minutes at 25° C in the following medium : 75 mM Tris-HCl pH 7.4, 15 mM KCL, 1 mM EDTA, 5 mM KH_2PO_4, 16.6 mM $KHCO_3$, 10 mM ornithine, 10 mM NH_4Cl, 3 mM ATP, 10 mM mannitol. Glutamate, succinate, valproate, propionate and 2 oxobutyrate were added as potassium salts. Rotenone (10 µg/ml) was always added with succinate. At the end of the incubation, N-acetylglutamate was assayed by its activating effect on carbamyl phosphate synthetase I as described by Meijer et al[10]. Acetylglutamate (0 to 15 nmoles) was added as internal standard.

To measure citrullinogenesis, mitochondria (about 5 mg protein) were incubated in 2 ml of reaction mixture containing : 50 mM Tris-HCl pH 7.4, 35 mM KCl, 1 mM EDTA, 5 mM KH_2PO_4, 16.6 mM $KHCO_3$, 10 mM ornithine, 10 mM HN_4Cl, 1 mM $MgCl_2$, 6.75 mM sucrose. The pH was adjusted to 7.4. After 10 minutes incubations were stopped by 1 ml of 1 M perchloric acid. The citrulline was assayed in the supernatant obtained by centrifugation at 11,000 g for

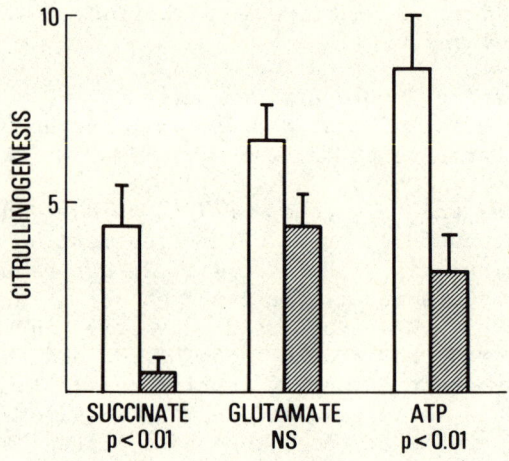

Fig. 1. Citrullinogenesis (nmol/min/mg protein) of isolated mitochondria from rat liver of 3 controls (open bars) and 3 valproate-treated (hatched bars) rats. Concentrations of succinate and glutamate are 16 mM. ATP was 4 mM with 10 μg/ml oligomycine, 0.04 mM 2,4-dinitrophenol. Results are expressed as mean ± SD.

2 minutes and determined according to a modified procedure of Ceriotti's method[11]. N-acetylglutamate synthetase, carbamylphosphate synthetase I and ornithine transcarbamylase activities were measured in isolated mitochondria as previously described[13,14]. Protein determination were made by the biuret method[15] L-[1 C] glutamic acid and [^{14}C] - HCO_3Na were purchased from Amersham England.

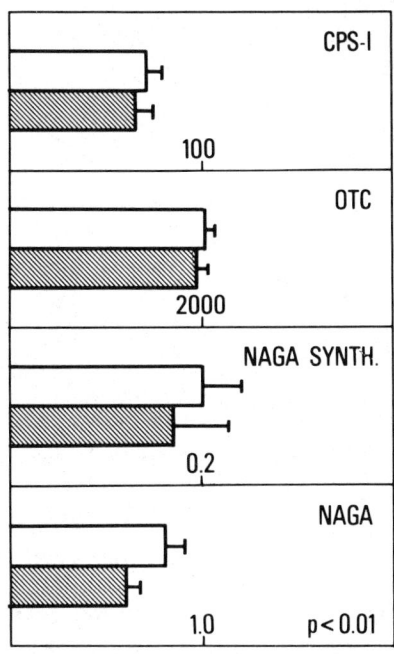

Fig. 2. NAGA concentrations ; NAGA synthetase, CPS-I and OTC activities in rat liver of 3 controls (open bars) and 3 valproate-treated (hatched bars) rats. Results of the enzymatic activities are expressed in compounds formed (nmol/min/mg of protein). For NAGA concentrations are expressed in nmol/mg protein.

RESULTS

In vivo experiments

In mitochondria from valproate-treated rats we observed a highly significant inhibition of citrullinogenesis in the presence of 16 mM succinate but not in the presence of 16 mM glutamate (Fig. 1). In the presence of ATP, oligomycine and 2,4-dinitro-

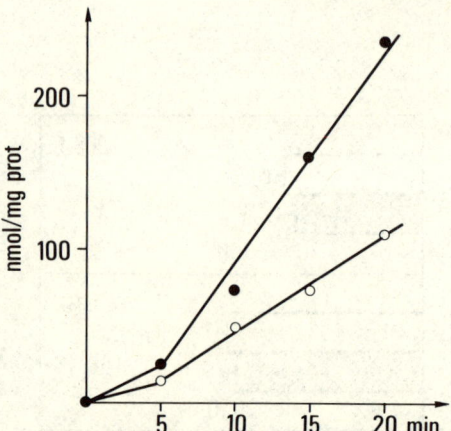

Fig. 3. Effect of 2mM valproate on citrullinogenesis of isolated mitochondria of rat liver.

Table 1. Comparative effects of 8 mM propionate, 8 mM 2-oxobutyrate and 8 mM valproate on NAGA synthesis and citrullinogenesis from rat liver mitochondria.

Added compounds	NAGA Synthesis (%)	Citrullinogenesis (%)
Glutamate 16 mM	100	100
+ propionate	14	69
+ 2-oxobutyrate	14	77
+ valproate	1	54

phenol the citrullinogenesis was increased but there was always a significant inhibition in valproate-treated rats. Concomitantly there was a significant decrease of mitochondrial N-acetylglutamate (NAGA) level in valproate-treated rats (Fig. 2) NAGA synthetase,

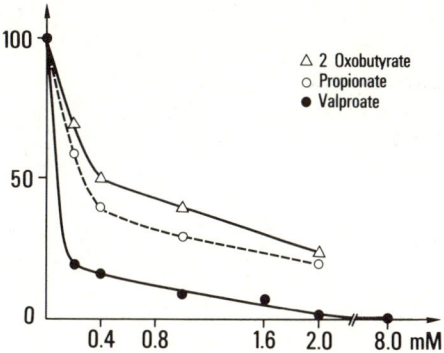

Fig. 4. Effect of valproate, 2-oxobutyrate and propionate on NAGA synthesis in isolated mitochondria of rat liver.

carbamylphosphate synthetase and ornithine transcarbamylase activities were in the same range in both groups (Fig. 2).

In vitro assays

Potassium valproate inhibits significantly citrullinogenesis in the presence of 16 mM glutamate (Fig. 3). There was a stricking concomitant inhibition of NAGA synthesis (Fig.4 and Table 1).

The same although less important effects were disclosed in the presence of 2-oxobutyrate and propionate. Valproate has no effect on NAGA synthetase in disrupted mitochondria (data not shown).

DISCUSSION

These results show that valproate or one of its metabolites is a potent inhibitor of the mitochondrial steps of ureagenesis. A direct enzymatic inhibition may be excluded because there was no change in the CPS-I, ornithine transcarbamylase and NAGA synthetase activities in valproate-treated rats. The normalization of the citrullinogenesis in the presence of glutamate which resulted in a

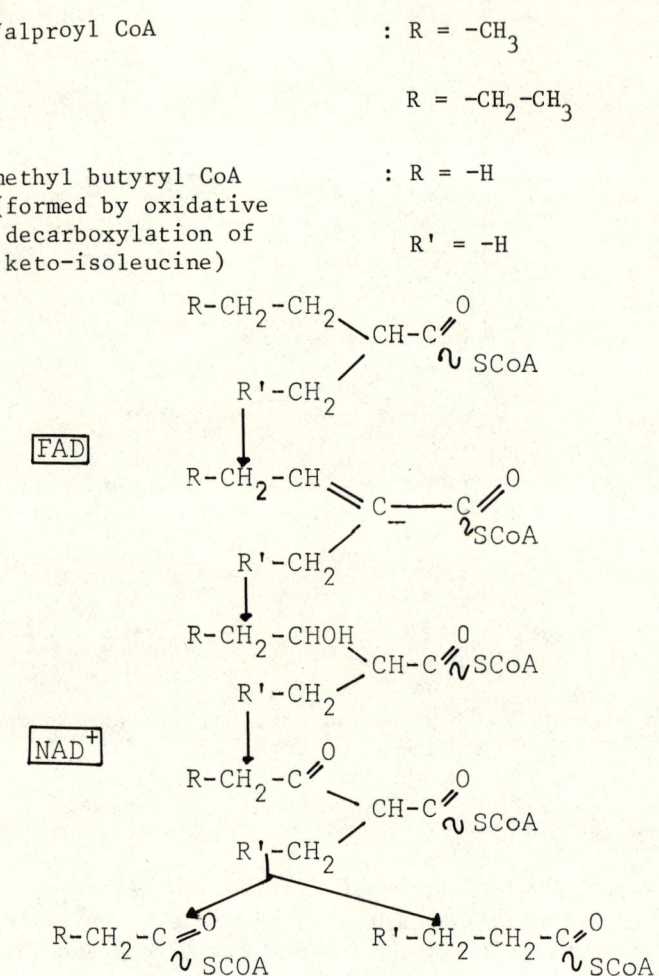

Fig. 5. Similarity between β-oxidation of valproic acid and catabolism of isoleucine.

de novo synthesis of NAGA in intact mitochondria as we have shown previously[16] indicates that NAGA must be the key factor. The addition of ATP increases the citrullinogenesis in both groups, but it remains significantly lower in valproate-treated rats. This implies that when ATP freely enters the mitochondria and is not anymore a rate-limiting factor, the citrullinogenesis is limited by the NAGA level. In vitro assays in normal mitochondria confirm these data. These valproate effects are analogous to the effect of propionate and 2-oxobutyrate which are metabolized to propionyl-CoA into mitochondria[17]. We have shown that the intraperitoneal injection of propionate at the same levels in rat does not decrease the citrullinogenesis. Therefore propionyl-CoA does not seem to be responsible for the valproate inhibition and valproyl-CoA or some closely related compounds can be the effective inhibitor. Indeed VPA is a eight-carbon branched-chain fatty acid which is structurally related to α methylbutyryl-CoA, a metabolite intermediate of the branched-chain aminoacid isoleucine. Moreover the β oxidation steps for valproate are suggestive of the oxidation steps for isoleucine (Fig. 5) and it has been shown that isoleucine interferes competitively with the β oxidation of the drug[18]. Therefore it is not so surprising that some side effects of valproate are very similar to the biological pattern of ketotic hyperglycinuria syndrome.

Elevated concentration of valproyl-CoA or one of its metabolites in mitochondria of the liver of patients with valproate treatment may therefore significantly inhibit the synthesis of N-acetylglutamate. Indeed seizure control needs usually a plasma concentration ranging from 0.3 to 0.6 mM[19]. A decreased level of N-acetylglutamate would decrease the activation of carbamyl phosphate synthetase and then clearance of ammonia by the urea cycle[20]. It may be possible to prevent acute episodes of hyperammonemia during valproate therapy by administering an activator of carbamylphosphate synthetase to compensate for decreased levels of N-acetylglutamate. Analogues of N-acetylglutamate such as N-carbamylglutamate have been shown to activate carbamylphosphate synthetase[21]. Therefore N-carbamylglutamate which seems to cross freely the biological membrane and activates CPS-I[22] might be useful in the management of these patients. Although valproyl-CoA levels in mitochondria are not known, a stricking inhibition of NAGA synthesis was present in our experiments with valproate concentration as low as 0.2 mM. These data agree very well with valproate concentration usually encountered in treated patients (range 0.3 to 0.6 mM).

REFERENCES

1. R. M. Pinder, R. N. Brodgen, T. M. Speight, and G. S. Avery, Sodium valproate : A review of its pharmacological properties and therapeutic efficacy in epilepsy, Drugs 13:81 (1977).
2. D. Coulter and R. J. Allen, Secondary hyperammonemia : a possible mechanism for valproate encephalopathy, Lancet i : 1310 (1980).
3. J. A. Sius, R. H. Trefor Jones, and W. H. Taylor, Valproate, hyperammonemia and hyperglycinemia, Lancet ii:260 (1980).
4. P. Kamoun, Ph. Parvy, and P. Debray-Ritzen, Hyperglycinurie induite par le N-dipropylacetate : modèle possible de l'acidemie propionique, Nouv. Pres. Med. 6:2162 (1977).
5. J. Jacken, L. Corbeel, P. Casaer, H. Carchon, L. Eggermond, and R. Eeckeli, Dipropylacetate (valproate) and glycine

metabolism, Lancet ii : 8038 (1977).
6. P. Kamoun, and Ph. Parvy, Effet du N-dipropylacetate sur l'élimination urinaire des acides aminés, Helv. Paediatr. Acta 33:373 (1978).
7. L. Boutillier, L. de Lumley, R. Saura and J. Boulesteix, Aplasie medullaire transitoire au cours d'un traitement par le dipropylacetate de sodium, Nouv. Press. Med. 8:611 (1979).
8. D. A. Winfield, P. Benton and M. C. Espir, Sodium valproate and trombocytopenia, Br. Med. J. 2:981 (1976).
9. G. H. Hogeboom, Fractionation of cell components of animal tissues in:Methods in Enzymology, S. P. Colowick, N. Kaplan eds. : Academic Press, New York (1955).
10. A. J. Meijer and M. Van Voerkom, Control of rate of citrulline synthesis by short term changes in N-acetylglutamate levels in isolated rat liver mitochondria, Febs. Lett. 86:117 (1978).
11. G. Ceriotti and L. Spandrio, Catalytic acceleration of the urea-diacetylmonoxime phenazone-reaction and its application to automatic analysis, Clin. Chem. Acta 11:519 (1965).
12. Fx. Coude, C. Sweetman and W.L. Nyhan, The inhibition by propionyl-CoA of acetylglutamate synthetase in rat liver mitochondria : a possible explanation for hyperammonemia in propionic and methylmalonic acidemia, J. Clin. Invest. 64: 1544 (1979).
13. L. Cathelineau, F. Petit, Fx. Coude, and P. Kamoun, Effect of propionate and pyruvate on citrulline synthesis and ATP content in rat liver mitochondria, Biochem. Biophys. Res. Commun. 90:327 (1979).
14. L. Cathelineau, F. Petit, Ph. Parvy, C. Charpentier, Fx. Coude, J. M. Saudebray and P. Kamoun, Vitamin B_{12} deficiency in rats and ammonia metabolism in Hommes FA. Ed. Models for the study of Inborn Errors of Metabolism, Elsevier, Amsterdam, (1979).
15. E. Layne, Spectrophotometric and turbidimetric methods for measuring proteins in:Methods in Enzymology, S. P. Colowick and N. O. Kaplan eds. Academic Press, New York (1957).
16. D. Rabier, L. Cathelineau, P. Briand, and P. Kamoun, Propionate and succinate effects on acetylglutamate biosynthesis by rat liver mitochondria, Biochem. Biophys. Res. Commun. 91:456 (1979).
17. T. Bohmer, The information of propionyl-carnitine in isolated rat liver mitochondria, Biochem. Biophys. Acta, 164:157 (1968).
18. I. Matsumoto, R. Kuhara, and M. Yoshino, Metabolism of branched medium chain length fatty acid. II. β oxidation of sodium dipropylacetate in rats, Biomed. Mass. Spectrom. 3:235 (1976). (1976).
19. J. Koch-Weser and T. R. Browne, Valproic acid, N. Engl. J. Med. 302:661 (1980).
20. M. Tatibana and K. Shigesada, Regulation of urea biosynthesis

by the acetylglutamate-arginine system, in: "The Urea Cycle", S. Grisolia, R. Baguena and F. Mayors eds. J. Wiley and Sons, Inc., New York (1976).
21. S. Grisolia and P. P. Cohen, Catalytic role of glutamate derivatives in citrulline biosynthesis, J. Biol. Chem. 204:753, (1953).
22. S. W. Kim, K. Paik and P. P. Cohen, Ammonia intoxication in rats : protection by N-carbamoyl-L glutamate plus L-Arginine, Proc. Natl. Acad. Sci. USA 69:3530 (1972).
23. Fx. Coude, D. Rabier, L. Cathelineau, G. Grimber, Ph. Parvy and P. Kamoun, Letter to the Editor : A mechanism for valproate-induced hyperammonemia, Pediat. Res. 15:974 (1981).

REDUCED ACTIVITY OF OTC IN THE LIVER OF A PATIENT WITH REYE'S SYNDROME

Shinichiro Arashima, Yasuro Takekoshi,* Mishiya Anakura Anakura, Haruo Nanbu* and Ichiro Matsuda

Hokkaido University, Sapporo, Japan
* Nakanoshima Chuo Hospital, Sapporo, Japan
** Kumamoto University, Kumamoto, Japan

In Reye's syndrome, it is said that all of the activities of the mitochondrial enzymes are lowered.

This is not only seen in the liver but is also seen in the brain and muscles[1].

Yamashita et al[2] conducted a nation wide survey and has confirmed the presence of Reye's syndrome in Japan[2]. However the number of cases are limited.

Recently we encountered a case of Reye's syndrome in a 1 year and 8 month old male child. We conducted a liver biopsy on the 8th day of the disease, and under an electronmicroscope aberrations of the mitochondria were recognized. On the 15th day after the onset of the disease the liver samples were obtained at autopsy. In the enzymes of the urea cycle, especially in OTC, we had the opportunity to investigate the enzyme kinetics. The results are reported in this paper.

CASE REPORT

The afflicted child was born to a healthy Japanese mother at term of primagravida the body weight 2580 gr. Because of incompatibility of the Rh blood type on the 5th day after birth exchange transfusion was conducted.

The infant gained head control at 5 months after birth, at 1 year and 3 months the child crawled and the first steps were

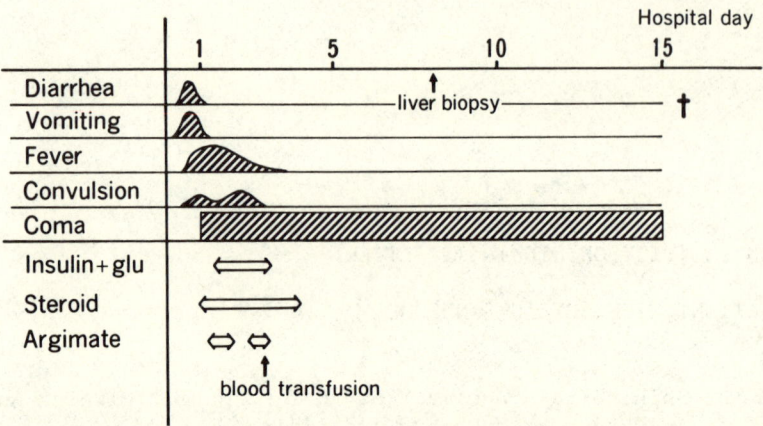

Fig. 1. Clinical Course of Reye's Syndrome.

taken at 1 year 7 months (after birth). However a slight mental retardation was recognized. At 1 year and 8 months one afternoon, the onset of watery diarrhea was seen (Fig. 1). At several hours after the child vomited and pyrexia was seen. Early in the following morning toxic seizure was recognized and the patient lapsed into a coma. The patient was admitted at the Nakanoshima Chuo Hospital in the morning of the same day. Laboratory studies run at the time of admission were blood pH 7.39, BE-9.5 mEq/L, blood ammonia 214/100 ml (normal under 50) SGOT 132 and LDH 2000 units (Fig. 2). Abnormal increases in alanine, phenylalanine were found in the plasma. The cell count in the cerebrospinal fluid was normal but glucose was lowered to 21 mg/100 ml.

In spite of the exchange transfusion and administration of steroids combined with glucose solution infusion no improvement of the symptoms were seen. On the 8th day after the onset of the disease liver biopsy was conducted and electronmicroscopic samples were prepared. On the 10th day after onset, the coma continued and while the blood ammonia and SGOT showed a tendency of lowering, LDH showed abnormally high values. The EEG on the 7th day after the onset was flat. And the child succumbed on the 15th day after the onset of the disease. The autopsy was done 3 hours after death and liver tissue was obtained. The liver tissue was stored for 2 months at -70°C after which biochemical analysis was conducted. Mitochondrial GOT was 24 units on the 2nd day after the onset and was 16 units (normal under 4 units) on the 5th day after onset.

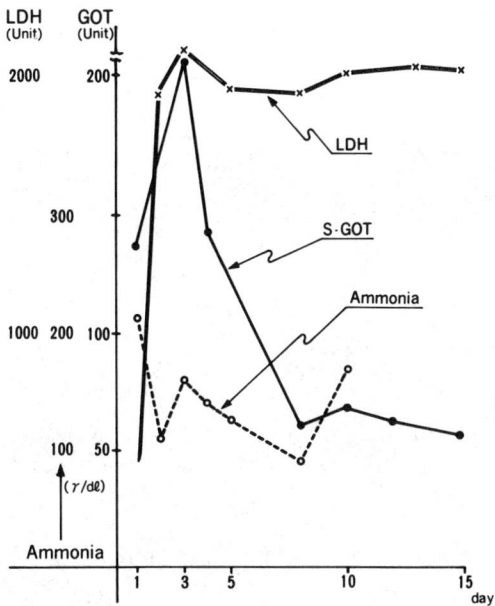

Fig. 2. Laboratory Data of Reye's Syndrome.

METHODS

Liver biopsy specimen obtained with (1.4 x 70 mm) Menghini needle was immediately placed in 5 ml of ice-cold 1 % osmium tetroxide in S-collidine buffer, where it remained for two hrs. Each was then dehydrated in graded alcohol and embedded in Epon 812. One-micron sections were cut and stained with toluidine blue. Ultrathin sections were cut by an LKB 111 ultratome and stained with lead citrate. The specimens were examined with Hitachi HU-12A electron microscope.

The urea cycle enzymes were determined by method of Brown and Cohen[3] after 2 months of storage. Liver specimens were homogenized with cethyltrimethylammonium bromide solution and OTC activity assayed in several different concentrations of either ornithine or carbamylphosphate. The rate of production of citrulline was expressed in μmole/min/gm wet weight tissue and the results plotted by the Lineweaver-Burk method. The apparent Km and Vmax of the enzyme for ornithine and carbamylphosphate were determined graphically.

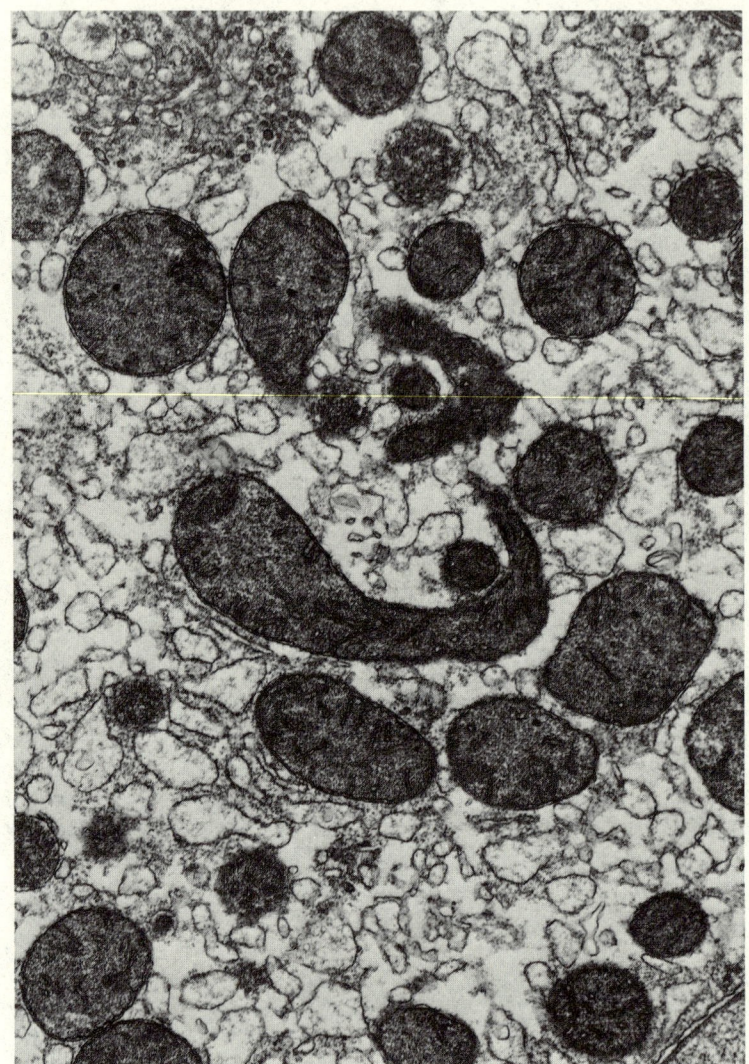

Fig. 3. Electron micrograph of hepatocyte showing various shape mitochondria (M) with distorted cristae. x 27,000

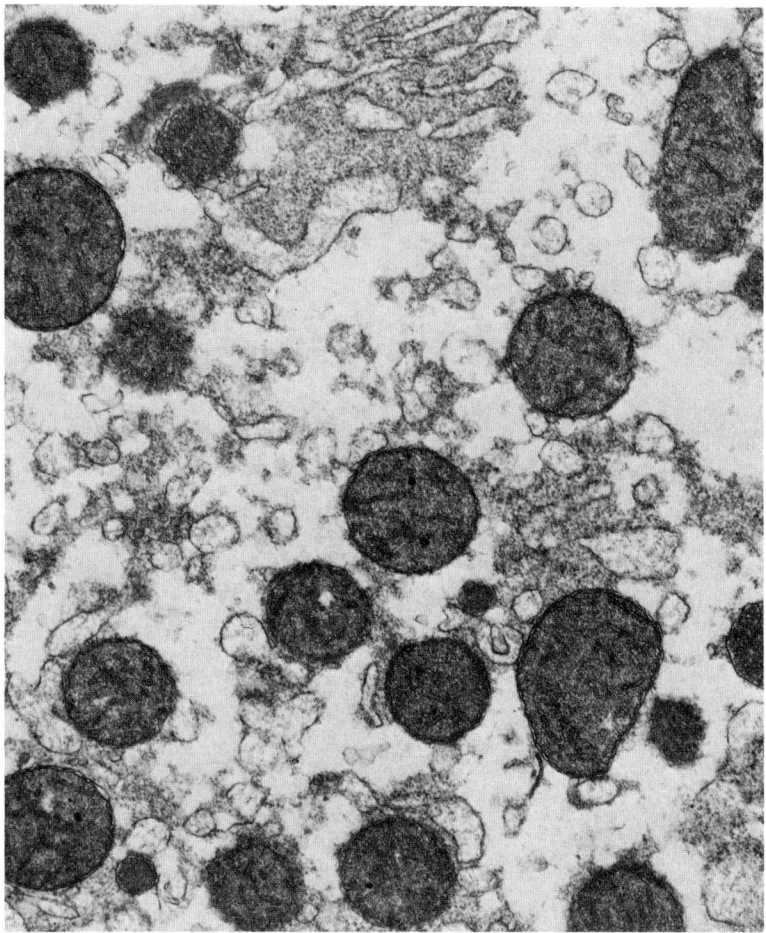

Fig. 4. Electron micrograph of hepatocyte showing various shape mitochondria (M) with distorted cristae. x27,000

Table 1. Urea cycle enzymes.

	Patient	Control (Necropsy)
Carbamyl phosphate Synthetase	130	100
Ornithine transcarbanylase	2700	6850
Arginin synthetese System	75	84
Argininosuccinase	410	386
Arginase	5800	6300

Unit : μ mole/gm wet weight/hr

RESULTS

In liver biopsy electron microscopy samples mitochondria were small and many of them had a round appearance otherwise there were various shapes such as V shaped, Y shaped or rod shaped (Fig. 3). Matrix substance appeared as rough granules or wool yarn like and the density was seen to increase. The matrix space was expanded and in addition an increase of microbodies was seen. Glycogen granules were remarkably decreased in number (Fig. 4).

In liver tissue obtained at autopsy urea cyle enzymes were analysed. It was shown that the OTC activity was lowered to 40 % of the normal liver (Table 1). Carbamylphosphate synthetase activity was normal. The urea enzymes of other than mitochondria were all normal.

Regarding OTC, Km for ornithine and Km for carbamylphosphate did not show any difference from normal values (Table 2).

DISCUSSION

From the pathological changes as seen electronmicroscopically of the liver mitochondria it was surmised that an all over dysfunction of the mitochondria prevailed. In our case since we conducted a liver biopsy on the 8th day after the onset, we surmised that there were stronger pathological changes before the biopsy. Three

Table 2. Apparent Km of hepatic OTC in Reye's syndrome.

	Patient	Control
For ornithine		
Km (mM)	1.45	1.47
Vmax	70	154
(μmole/min/g tissue)		
For carbamylphosphate		
Km (mM)	1.73	1.67
Vmax	75	160
(μmoles/min/g tissue)		

days after the onset, the damage was at its severest, and since there after it is said that it takes its recovery course, in our case in spite of the aggravation of the clinical manifestation, it may be that the liver cell had entered the recovery period. Routine liver function tests showed a normalization at that period and it provides proof of the recovery. Recently Shapiro et al[4] reported that abnormalities of mitochondria was seen in OTC deficiency. However, we did not find similar evidence.

In enzyme analysis while it was shown that the OTC activity was lowered, CPS was normal. We conducted autopsy 15 days after the onset of the disease and samples taken showed that liver samples were in a state of recovery ; hence we can surmise that the liver in general would be in a recovery stage. It may be considered that CPS was lowered in the early stage of the disease. Hitherto, it has been considered that in one week after the onset of the disease both OTC and CPS activities would be normalized, however in our findings even at 15 days after onset the disease OTC was still lowered. Sinatora et al[5] reported four cases in which even 15 days after the onset of the disease, the Km of OTC for carbamylphosphate was still lowered and they also stated that Vmax was also lowered.

In our case Km for ornithine and carbamylphosphate were normal and Vmax alone was lowered. We consider that the degree of abnormalities of mitochondrial enzymes varies with each individual case.

SUMMARY

The liver urea cycle enzymes were measured in a 20 months old male with Reye syndrome. OTC activity was lowered to 40 % of normal values in the liver necropsy sample after 15 days from onset. CPS and apparent Km of OTC for ornithine and CPS were normal.

In the biopsy specimen obtained on the 8th day after the onset ultrastructural change was seen in hepatocyte mitochondria.

REFERENCES

1. B. H. Robinson, J. Taylar, E. Cutz and D. G. Gall, Reye's syndrome : Preservation of mitochondrial enzymes in brain and muscle compared with liver, Pediat. Res. 12:1045 (1978).
2. F. Yamashita, M. Yamamoto, S. Okada, I. Yoshida and M. Yoshino, Reye's syndrome in Japan : Epidemiology, clinical features and indicators for mortality and the sequellae, in "Reye's Syndrome II" J. F. S., Crocker, ed., Grune & Stratton, New York (1979).
3. G. W. Brown and P. P. Cohen, Comparative biochemistry of urea synthesis I. Method for the quantitative assay of urea cycle enzyme in liver, J. Biol. Chem. 234:1769 (1959).
4. J. M. Shapiro, F. Schaffner, H. H. Tallan and G. E. Gaull, Mitochondrial abnormalities of liver in primary ornithine transcarbamylase deficiency, Pediatr. Res. 14:735 (1980).
5. F. Sinatora, T. Yoshida, M. Applebaum, W. Mason, N. J Hoogeuraad and P. Sunshine, Abnormalities of carbamyl phosphate synthetase and ornithine transcarbamylase in liver of patients with Reye's syndrome, Pediatr. Res. 14:735 (1980).

DIAGNOSTIC, CLINICAL, PATHOLOGICAL AND
BIOCHEMICAL ASPECTS OF THE DISEASES

4) <u>Animal Models</u>

SPONTANEOUS ANIMAL MODELS OF ORNITHINE TRANSCARBAMYLASE

DEFICIENCY: STUDIES ON SERUM AND URINARY NITROGENOUS METABOLITES*

I.A. Qureshi, J. Letarte and R. Ouellet

Laboratoire de Nutrition, Centre de Recherche Pédiatrique
Hôpital Sainte-Justine and Département de Pédiatrie
Université de Montréal, Canada

SUMMARY

Various groups of spf (sparse fur) and spf^{ash} (allele with abnormal skin and hair) mice, were compared in respect of their liver ornithine transcarbamylase (OTC) activity, urinary orotate and concentrations of serum NH_3 and glutamine. While liver OTC activity was comparable in both the strains, the excretion of urinary orotate was lower in spf^{ash} mice. In contrast to spf females, the $spf^{ash}/+$ heterozygotes cannot be distinguished from normal +/+ females, as urinary orotate excretion has no significant correlation with liver OTC activity. However, serum glutamine is significantly correlated with liver OTC deficiency in both the strains. Administration of 1% sodium benzoate to spf/Y males resulted in a pattern of excretion of hippurate and urea N similar to clinical trials in humans. Urinary orotate was significantly reduced, while ad lib dietary intake increased almost 3 fold. Both the spontaneous animal models have considerable potential to be used in the investigation of the etiology, specific pathology and nutritional therapy of congenital hyperammonemias.

---

*This work was supported by grants from the Medical Research Council of Canada (MA-7394), Quebec Network of Genetic Medicine and Fondation Justine Lacoste-Beaubien, Montréal. We acknowledge assistance provided by the Ross Laboratories and the Upjohn Company for participation in the Symposium. IAQ is also being supported by Conseil des Clubs Sociaux, Montréal.

INTRODUCTION

X-chromosomal "sparse-fur" (spf) mutation originated spontaneously in the progeny of an irradiated mouse at Oakridge (1). An abnormal form of ornithine transcarbamylase (E.C.: 2.1.3.3) (OTC) was discovered in spf-bearing mice by Demars et al. (2). It was observed that the spf and abnormal OTC phenotypes were closely associated, and were transmitted in the same manner as the OTC deficiency hyperammonemia in humans. The report included a study of a second abnormal form of OTC in mice having the "abnormal skin and hair" (ash) mutation (3). Complementation and linkage tests conducted by Doolittle et al. (4) had indicated that spf and ash genes were allelic. They suggested the name spf^{ash} for the new allele.

The suitability of spf mouse as animal model of human disease, in respect of its enzyme defect and variations in orotic acid excretion, has already been reported by us (5). The present study aims at a comparison of spf and spf^{ash} mice in respect of certain serum and urinary nitrogenous metabolites e.g. NH_3, glutamine, orotate, etc. It reports changes in urinary nitrogenous metabolites, as a result of preliminary studies done with sodium benzoate, using spf mice selectively bred at our laboratory (6).

MATERIALS AND METHODS

Animals and their management

Spf mice used in these studies were descendants of the parent stock supplied by Dr. L.B. Russell, of the Oakridge National Laboratory (2). Since our earlier study (5), the viability and fur characteristics of the spf/Y males have been improved by transferring the gene from 22-A strain to the CD-1 outbred albino (Canadian Breeding Farms and Laboratories, Ltd, St. Constant, Que.). Details of selective breeding have been reported separately (6).

Spf^{ash} mice were given to us by Dr. D.P. Doolittle of Purdue University (4). He supplied us with 15 heterozygous females, along with 15 normal males from the same stock. These animals had been selected on the basis of the morphological phenotype and breeding results. It was found later in our laboratories, on the basis of enzyme activity, that some of the normal males were actually mutants (spf^{ash}/Y); and two of the females were probably homozygous for the mutant gene (spf^{ash}/spf^{ash}). The original stock as well as the first generation progeny obtained by crossing +/Y males with spf^{ash}/+ females were used for the present studies.

All the experimental animals were fed on "Purina Mouse Chow" (Ralston Purina, St. Louis, MO, USA), containing not less than 22%

protein. They were kept in a temperature, humidity and lighting controlled environment.

Collection of urine and serum samples

For the collection of urine samples, the mice were kept individually in metabolic cages (Maryland Plastics Inc., Federalsburg, MD, USA 21632). During this period, diet and drinking water were provided ad lib. After 24 hours (09:00 AM to 09:00 PM), the urine collecting system was rinsed with 1 ml of distilled water, and the total volume collected was recorded. The samples were frozen until analyzed. All laboratory determinations on urine samples were expressed on the basis of a 24 h excretion.

Soon after the urine collection was completed, the mice were killed by decapitation; and the blood immediately collected in a "vacutainer" tube with the help of a funnel. Serum was separated by centrifugation, and frozen until analyzed. Liver was removed within two minutes of death and kept on aluminium foil placed on dry ice. It was stored at $-70°C$ until analyzed for OTC activity.

Sodium Benzoate administration

Only spf mice were used for the study of the effects of 1% sodium benzoate on dietary intake and excretion of urinary metabolites. The animals were preliminarily separated into normals or mutants on the basis of orotic acid excretion. Ten males, including 8 spf/Y hemizygotes and 2 normals (+/Y), were given the regular diet with or without 1% sodium benzoate. The diet was mixed with sodium benzoate and pelleted. Laboratory grade sodium benzoate (Fisher Scientific Co. Ltd., Montreal, Que., H4P 2L3) was used. After acclimatization to metabolic cages, urine samples were collected for 1 day on diet without benzoate and for 2 consecutive days with added 1% sodium benzoate. Intake was recorded each day. At the end of the experiment, the animals were killed and their livers removed, to verify the OTC activity. Spf/Y males had an average of 13% of enzyme activity as compared to normals.

Laboratory analyses

Urinary orotate was measured by the method of Adachi et al. (7), with a non-brominated blank for each sample tube (8). Urea N was measured by the technique of Ceriotti (8), and hippurate by the method of Tomokuni and Ogata (10). Results of urinary determinations were expressed as quantities excreted/24 h.

Serum NH_3 was determined by a miniaturization of the ion exchange resin technique (11) as adopted for the "Hyland" blood ammonia kit (Travenol Laboratories, Cosa Mesa, CA, USA). Serum glutamine was determined by the technique described by Welbourne et al. (12), in which glutamine is decomposed by glutaminase. The released NH_3 was trapped onto the ion exchange resin. Results of serum NH_3 and glutamine were expressed as μmol/litre.

Activity of liver OTC was measured by the colorimetric technique of Ceriotti (13) as adapted for liver tissue (14). The activity was expressed as μmol of citrulline/mg liver protein/h. Liver protein was measured by the method of Lowry et al. (15).

Statistical analyses

Data between groups were compared by the unpaired "t" test; and between treatments by the paired test. Statistical tables of Snedecor et al. (16) were used to determine the significance of "t" values and correlation coefficients.

RESULTS

Excretion of Urinary Orotate

Table 1 gives the results of urinary orotate excretion by various groups of spf and spf^{ash} mice. Animals were classified into each group by known breeding plans and by assays of liver OTC activity.

Although the activity of liver OTC in various groups of spf^{ash} mice is comparable to spf group, the excretion of orotate in all groups of spf^{ash} mice is lower than the corresponding groups of spf strain. Within spf strain, the orotate excretion in hemizygous affected males and heterozygous females is significantly higher ($p < 0.001$) as compared to the normal groups of the same sex. In spf^{ash} mice, there is a lower level of significance ($p < 0.05$) between hemizygous affected males and normal males, while $spf^{ash}/+$ females do not show any significantly different excretion as compared to normal females.

Since orotate excretion has been used by us as a biochemical criterion for separating spf/+ females from normal litter mates, the lack of difference in the excretion of orotate between $spf^{ash}/+$ and normal females was considered as a handicap, if these animals were to be bred for nutritional experimentation. These animals were, therefore, further tested by a study of correlation between the liver OTC activity and urinary orotate. The results are represented in Fig. 1. Whereas there is a significant negative correlation ($p < 0.01$) between these variables in spf/+ heterozygous females, no significance exists for the correlation between the level of liver OTC activity and orotate excretion in $spf^{ash}/+$ mice. There is a considerable spread of liver OTC activity amongst individual $spf^{ash}/+$ females. However, the excretion of orotate seems to stay within a minimum threshold level.

Table 1. Liver OTC activity and urinary orotate

Groups	Spf		Spf^{ash}	
	OTC^a activity	Orotate^b excretion	OTC^a activity	Orotate^b excretion
Males, normal	73.7±4.8^c (10)	1.80±0.18 (10)	91.7±8.7 (9)	1.13±0.21 (9)
Males, hemizygous	11.3±1.4*** (15)	11.1 ±1.86*** (15)	3.8±0.5*** (3)	6.65±2.46* (3)
Females, normal	82.4±5.8 (9)	1.81±0.32 (9)	96.6±5.7 (12)	0.80±0.08 (12)
Females, heterozygous	36.2±3.4*** (18)	3.39±0.38** (18)	37.8±3.8*** (20)	0.70±0.07 NS (20)
Females, homozygous	8.8 (2)	42.5 (2)	3.4 (2)	11.0 (2)

^{a} µmol citrulline produced/mg liver protein/h
^{b} µg orotate excreted/g body wt/24 h
^{c} X ± SEM (no. of animals in parenthesis)
* $p < 0.05$; ** $p < 0.01$; *** $p < 0.001$ (compared to the normal group of the same sex). NS: non significant.

Serum NH$_3$ and glutamine concentrations

Table 2 represents a study of serum NH$_3$ and glutamine levels in spf and spf^{ash} female mice, separated into various groups on the basis of liver OTC activity and breeding data.

While serum NH$_3$ levels between the heterozygous females and normal females are not different in both spf and spf^{ash} mice, the serum glutamine levels have a tendency to be higher in heterozygous females of both the strains. However, the values of serum glutamine are significantly different in spf female groups only. A study of correlation between liver OTC activity and serum glutamine represented in Fig. 2 indicates that a significant relationship exists between these variables in both the mutant strains.

Effect of sodium benzoate on urinary nitrogenous metabolites

Figure 3 represents the effects of administration of 1% sodium benzoate on feed consumption, and excretion of orotate, hippurate and urea N. Although feed consumption increases almost 3 fold, the excretion of urinary orotate declines significantly. The concomitant

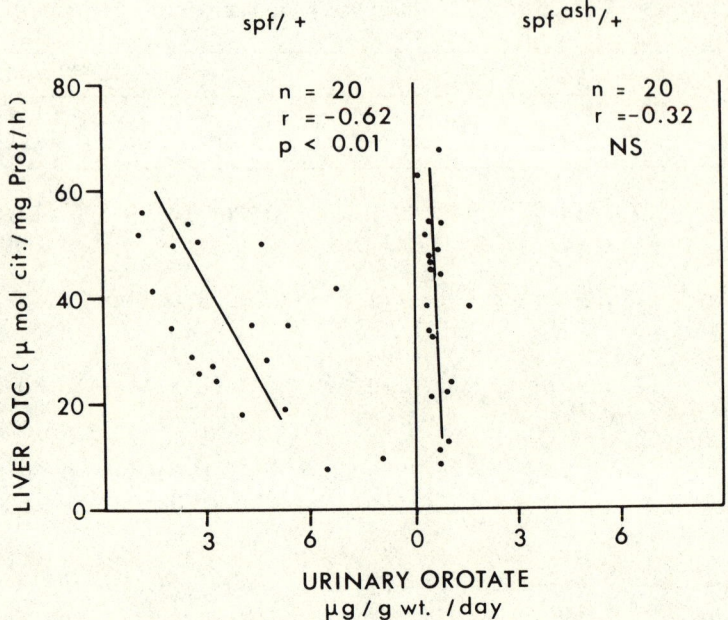

Fig. 1. Relationship of liver OTC activity to urinary orotate in spf and spf^ash heterozygotes.

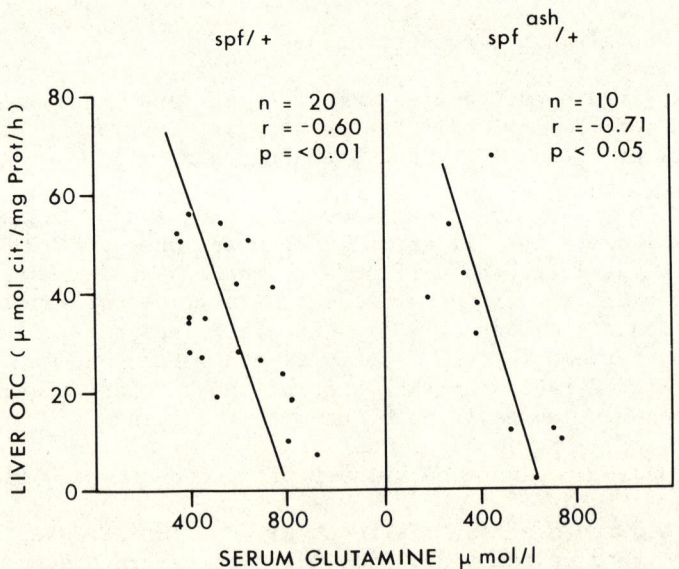

Fig. 2. Relationship of liver OTC activity to serum glutamine in spf and spf^ash heterozygotes.

Table 2. Serum NH$_3$ and glutamine concentrations in female groups of mutant mice

	Spf		Spf^{ash}	
Groups	Serum NH$_3$[a]	Serum glutamine[a]	Serum NH$_3$	Serum glutamine
Females, normal	116 ± 9[b] (9)	367 ± 32 (9)	106 ± 14 (10)	378 ± 30 (10)
Females, heterozygous	136 ± 11 NS (18)	546 ± 36** (18)	91 ± 14 NS (11)	492 ± 58 NS (11)
Females, homozygous	264 (2)	872 (2)	154 (2)	630 (2)

[a] µmol/l
[b] X̄ ± SEM (no. of animals given in parenthesis)
** $p < 0.01$. NS: non significant

decrease of urea nitrogen excretion, with an increase in urinary hippurate indicates a shift in the metabolism of NH$_3$ caused by benzoate. Except for urinary orotate, these changes seem to affect equally both normal and mutant males.

DISCUSSION

The comparison of the two spontaneous animal models of OTC deficiency i.e. spf and its allelic from spf^{ash} was undertaken to decide as to which of these strains should be bred for experimentation, to investigate the possibilities of nutritional therapy of urea cycle disorders. Excessive excretion of urinary orotate and higher levels of plasma NH$_3$ and glutamine are important features of OTC deficiency in humans (7). Alteration of these indices could be interpreted as success or failure of the particular therapeutic approach. While these variables are seen to increase more or less significantly in the affected male hemizygotes and heterozygote females of the spf strain, our studies prove that these are not expressed significantly in the spf^{ash}/+ females. Only serum glutamine levels were seen to have a significant correlation with the activity of liver OTC in spf^{ash} heterozygotes.

The lack of differences in excretion of urinary orotate in spf^{ash} females could be a source of practical hindrance in carrying out breeding plans, and distinguishing between the progeny with the same ease as can be done in the spf females (5,6). However, the differences in the expressions of the enzyme deficiency between these two models would continue to be an interesting subject of

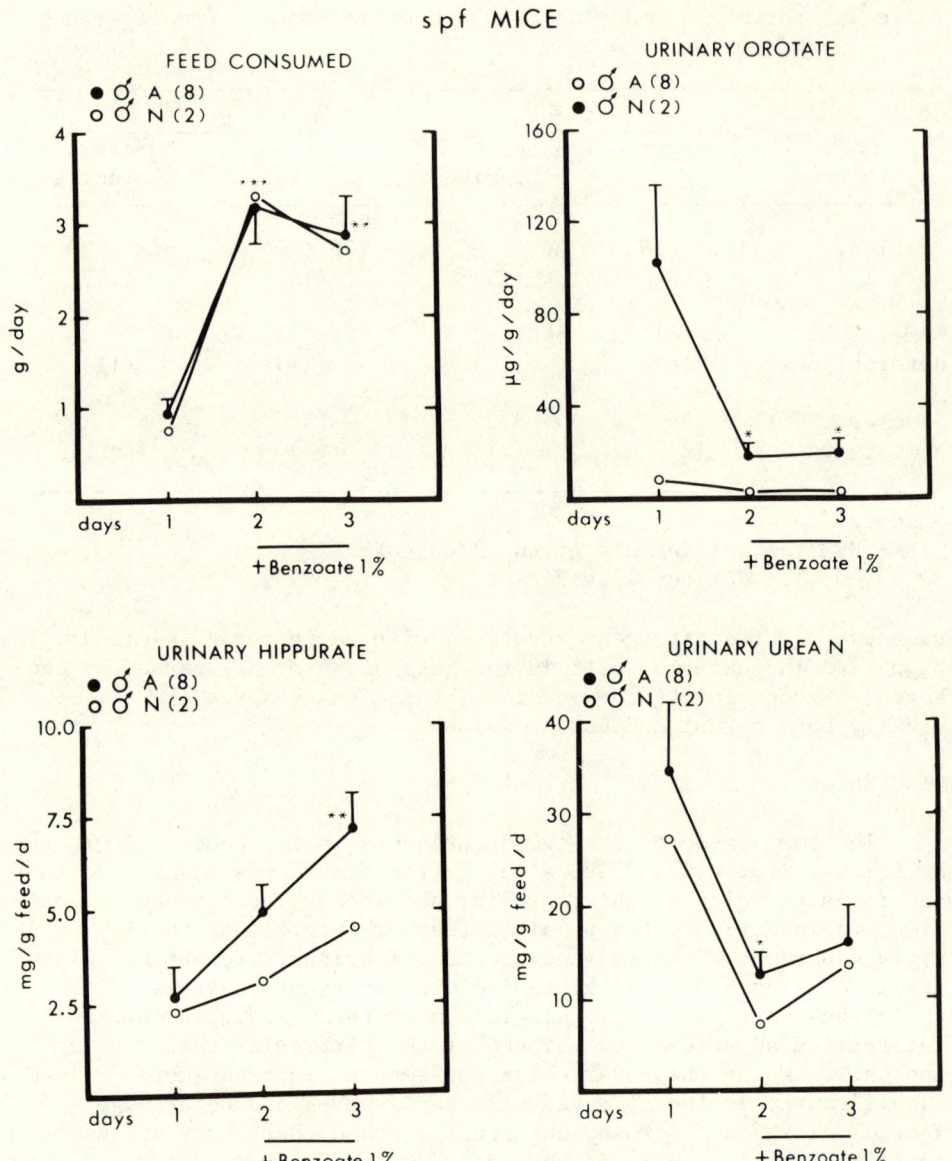

Fig. 3. Effect of sodium benzoate on dietary intake and urinary nitrogenous metabolites in spf/Y males.

research, designed to find out similarity with the expression of human forms of disease (8). It is encouraging to note that research on the comparative enzymology of the two models is already underway (19).

The preliminary trial of the administration of 1 % sodium benzoate in spf mice demonstrated that it has patterns of excretion of urinary nitrogenous metabolites similar to that seen in clinical trials on children (20). While urinary hippurate increased due to the conjugation of benzoate by glycine (21), urea nitrogen excretion decreased. The significant decrease in orotate excretion in affected mice indicates that excess NH_3 is diverted from the urea cycle and the pyrimidine pathway to increased synthesis of hippurate through glycine. Another important finding in this study was the significant increase in spontaneous feed consumption by the mutant mice. This phenomenon could be helpful in increasing protein tolerance in affected children who are in need of growth supporting levels of protein.

At present, the spf^{ash}/Y mice maintained at Purdue are separated by the morphological phenotype. A progeny test is employed to distinguish between the two types of females. Any female that produces at least six sons, none of which are mutant is assumed to be +/+ (22). Both these methods could be improved by a biochemical index. The present studies did not demonstrate the efficiency of the criterion of orotate excretion in spf^{ash} females. However, it could be interesting to study the effects of metabolic loading etc. in distinguishing heterozygotes from normal females.

In spite of differences in pathologic expression in the serum and urinary metabolites, we feel that both spf and spf^{ash} mice would be equally important as animal models. They have great potential in the investigation of etiology, specific pathology and nutritional therapy of congenital hyperammonemias.

ACKNOWLEDGEMENTS

The authors wish to acknowledge technical assistance by Ms. Diane Leblanc, Ms. Martine Caty and Mr. Luc Parent. The secretarial work by Ms. Louise d'Amours and Mrs. Sylvie Tassé is appreciated.

REFERENCES

1. M. C. Green, Mutant genes and linkages, in : "Biology of the Laboratory Mouse, 2nd Ed.," E. L. Green, ed., Mc Graw Hill, New York, p. 116 (1966).
2. R. Demars, S. L. Levan, B. L. Trend, and L. B. Russel, Abnormal ornithine carbamyl transferase in mice having the sparse-fur mutation, Proc. Natl. Acad. Sci., USA 23:1693 (1976).
3. L. L. Hulbert and D. P. Doolittle, Abnormal skin and hair : a sex linked mutation in the mouse, Genetics 28:529 (1971).

4. D. P. Doolittle, L. L. Hulbert, and C. Cordy, A new allele of the sparse-fur gene in the mouse, J. Hered. 65:194 (1974).
5. I. A. Qureshi, J. Letarte, and R. Ouellet, Ornithine transcarbamylase deficiency in mutant mice. I. Studies on the characterization of enzyme defect and suitability as animal model of human disease, Pediat. Res. 13:807 (1979).
6. I. A. Qureshi, J. Letarte, and S. R. Qureshi, Congenital hyperammonemia, in : Handbook : "Animal Models of Human Disease", J. C. Jones, D. B. Hackle, and G. Migaki, eds., Registry of Comparative Pathology, Armed Forces Institute of Pathology, Washington D.C. (1981), in press.
7. I. Adachi, A. Tanimura, and M. Asahina, A colorimetric determintation of orotic acid, J. Vitaminol. 9:217 (1963).
8. A. S. Goldstein, N. J. Hodgenraad, J. D. Johnson, K. Fukanage, E. Swierczewski, H. Cann, and P. Sunshine, Metabolic and genetic studies of a family with ornithine transcarbamylase deficiency, Pediat. Res. 8:5 (1974).
9. G. Ceriotti, Ultramicro determination of plasma urea by reaction with diacetylmonoxime antipyrine without deproteinization, Clin. Chem. 17:400 (1971).
10. K. Tomokuni and M. Ogata, Direct colorimetric determination of hippuric acid in urine, Clin. Chem. 18:349 (1972).
11. J. H. Hutchinson and D. H. Labby, New method for the microdetermination of blood ammonia by use of the cation exchange resin, J. Lab. Clin. Med. 60:170 (1962).
12. T. Welbourne, M. Weber, and N. Bank, The effect of glutamine on ammonium excretion in normal subjects and in patients with renal disease, J. Clin. Invest. 56:1852 (1972).
13. G. Ceriotti, Optimal conditions for ornithine carbamyl transferase determination, Clin. Chim. Acta 4:97 (1973).
14. I. A. Qureshi, J. Letarte, and R. Ouellet, Study of enzyme defect in a case of ornithine transcarbamylase deficiency Diabete Métabol. 4:239 (1978).
15. O. H. Lowry, N. J. Rosebrough, A. L. Farr, and R. J. Randall, Protein measurement with the Folin-Phenol reagent, J. Biol. Chem. 191:265 (1961).
16. G. W. Snedecor and W. G. Cochran, "Statistical Methods", 2nd ed. Iowa University Press, Ames, Iowa, pp. 541-575 (1967).
17. B. Levin, V. G. Oberholzer, R. L. Sinclair, Biochemical investigation of hyperammonemia, Lancet 2:170 (1969).
18. R. G. F. Gray, J. A. Black, V. H. Lyons, and R. J. Pollitt, Ornithine transcarbamylase deficiency, Pediat. Res. 10:918. (1976).
19. P. Briand, L. Cathelineau, P. Kamoun, D. Gigot, and M. Penninckz, Increase of ornithine transcarbamylase protein in sparse fur mice with ornithine transcarbamylase deficiency FEBS Lett. 130:65 (1981).
20. S. Brusilow, J. Tinker, and M. L. Batshaw, Amino-acid acylation : A mechanism of nitrogen excretion in inborn errors of urea synthesis, Science 207:659 (1980).

21. J. Kao, C. A. Jones, J. R. Fry, and J. N. Bridges, Species differences in the metabolism of benzoic acid by isolated hepatocytes and kidney tubules, Life Sci. 23:1221 (1978).
22. D. P. Doolittle, Personal communication.

SPARSE-FUR MUTATION : A MODEL FOR SOME HUMAN ORNITHINE TRANSCARBAMYLASE DEFICIENCIES

P. Briand, L. Cathelineau

Laboratoire de Biochemie Génétique, Hôpital Necker
149 rue de Sèvres, 75730 Paris Cedex 15, France

INTRODUCTION

Recently a new human enzymatic variant[1] was found in a patient with ornithine transcarbamylase (OTC ; E.C.2.1.3.3.) deficiency one of the more common genetic defects occurring in the urea cycle. This mutant enzyme had a decreased affinity with an abnormal Km value for ornithine. The maximal velocity (Vmax) varied with pH as in a normal enzyme but the sigmoid curve obtained was shifted towards alkaline pHs. Furthermore at its optimum pH (pH 9) the Vmax of the mutant enzyme was greater than the Vmax of the normal enzyme at its optimum pH (pH 7.8). Two hypotheses were proposed : either the mutant enzyme was more stable than a normal one at alkaline pH or a much higher level of OTC protein due to a regulating mechanism was characteristic of this mutant. Unfortunately we had no hepatic material left after the enzymatic analysis to investigate this problem.

We have been especially interested in mouse sparse-fur (spf) in which OTCase activity exhibits the same properties as the human mutant previously described[2,3]. Since OTC has been purified from bovine[4] rat[5,6] and human[7] livers but never from mouse, we decided first to purify normal mouse OTCase and to obtain specific antiserum for quantitative immunological assay of mouse spf OTCase. Secondly to have a better knowledge on the molecular basis of this mutant type we compared some of its enzymatic properties, with those of the normal mouse OTCase.

Table 1. Purification of normal mouse OTCase.

Fraction	Total activity (units)	Total protein (mg)	Specific activity (units/mg)	Yield (%)	Purification
Triton X-100 supernatant from 10 livers	2640	2016	1.3	100	
S.PALO Sepharose 6B fraction	2020	6.2	326	76.5	250

RESULTS AND DISCUSSION

Determination of OTC Protein in Liver Homogenates from Normal and spf Mice.

All the mice mentioned are either males (+/Y) or (spf/Y). (spf/Y) males were obtained from heterozygous females kindly supplied by I.A. Qureshi and J. Letarte, fertilized by wild males. The purification of normal mouse liver OTCase was realized by affinity chromatography with the immobilized transition state analog δ N-(phosphonacetyl)L-ornithine (δPALO)[8]. The synthesis of δPALO was performed according to Penninckx and Gigot[9]. The purification was either performed from total homogenate or from isolated mitochondria with yields between 75 to 85 %. A summary of one purification is given in Table 1. The purity of OTCase preparations was checked by acrylamide gel electrophoresis (Fig. 1). Using the purified mouse OTCase we obtained a rabbit monospecific antiserum and a quantitative immunological assay was performed to determine the amount of OTC protein in liver homogenates and mitochondria from normal and spf mice. Radial immunodiffusion clearly demonstrates (Fig. 2) that there is more OTCase cross reactive material (CRM) in spf hepatic homogenates than in wild type (spf CRM/control CRM / m = 1.2 $\pm$ 0.06 sem).

The same ratio values were found when radial immunodiffusion was performed with isolated mitochondria[10] indicating that mature intramitochondrial OTCase[11-15] is increased in spf mice. We cannot completely exclude a lower reactivity of the anti-OTC against the mutant enzyme but the enzymatic properties described later are in agreement with a punctual mutation which would not perturbe the immunoreactivity of the protein. We now try to obtain antiserum against SDS treated OTCase. This anti-serum would have the same

Fig. 1. SDS polyacrylamide gel pattern of purified normal mouse OTCase. 13 µg of protein were applied on to the gel and subjected to electrophoresis (10 V. cm 1 h).

reactivity against SDS treated normal and spf OTCase, the subunits of which have identical molecular weight.

Molecular Weight and Enzymatic Characteristics of Purified Normal Mouse OTCase and Partially Purified spf OTCase.

The molecular weight of normal mouse OTCase was 98,000 as determined by Sephadex G 200 filtration (Fig. 3) and the subunit molecular weight was 36,000 (Fig. 4). These results are in agreement with a trimeric structure which has been reported for rat[5,6], bovine[4], and human OTCase[7]. The purification of spf mouse OTCase by affinity chromatography gives a preparation with minor impurities, when checked by acrylamide gel electrophoresis. The subunit molecular weight was 36,000. The specific activity of normal and spf OTCase are respectively 300 ± 30 and 400 ± 30 µmoles/min/mg of OTCase at their respective optimum pHs (8.0 and 9.0). Because spf OTCase was not completely purified, the true value of the enzyme concentration was determined by Mancini immunodiffusion[16]. The Km for ornithine decreases in spf and normal mouse at alkaline

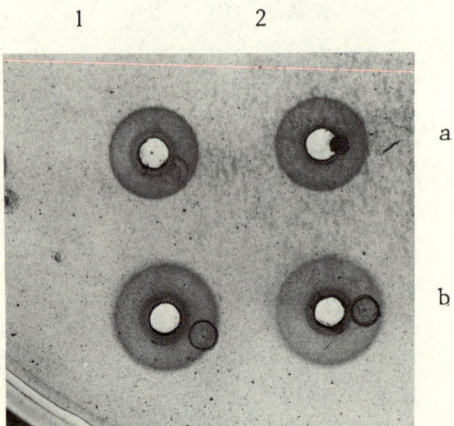

Fig. 2. Radial immunodiffusion of liver homogenate and mitochondria from spf and normal mouse ; identical amounts of protein (1 mg liver extracts (1) and 0.2 mg mitochondria (2) from normal (a) and spf (b) mouse were placed on an agarose gel containing anti-mouse OTCase).

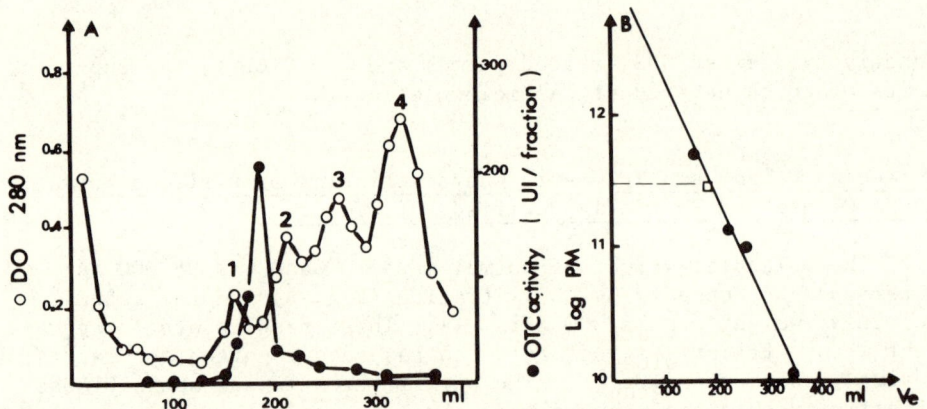

Fig. 3. Gel filtration on Sephadex G-200. Purified normal mouse OTCase was chromatographed on a 1.5 x 80 cm column of sephadex G 200 in NaCl 50 mM, Tris HCl 10 mM pH 7.5 mercaptoethanol 1 mM. Fractions were assayed for enzyme activity. Marker enzymes were chromatographed in the same buffer and fractions were assayed for protein amount.

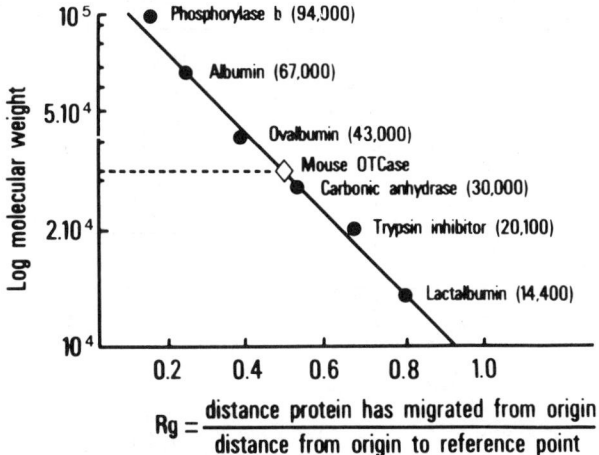

Fig. 4. SDS-polyacrylamide gel pattern of purified OTCase of purified OTCase (8 µg) and markers were subjected to electrophoresis[10].

pH, but it is near to 5 times the normal value in spf OTCase at pH 8 (Table II). The Km for carbamyl phosphate does not vary as a function of pH and seems identical for the two enzymes (0.15 mM).

Inactivation of OTCase by Butanedione.

Kalousek and Rosenberg recently reported in a brief abstract[17] that human and bovine OTCase are inactivated by butanedione and phenylglyoxal. An extensive study of the inactivation by butanedione of bovine OTCase is reported by Marshall and Cohen[18]. Butanedione is an arginine specific reagent when used in the presence of borate[19]. We studied the inactivation by butanedione of purified normal mouse OTCase and of normal and spf OTCase in homogenates. Identical results were found after 5' of incubation in butanedione 20 mM borate 50 mM. Butanedione inhibited normal and spf OTCase and this inhibition was partially reversed by carbamyl phosphate (CP) (Fig. 5). These results are in agree-

Table 2.

pH	Km for ornithine mM (+)	(spf)	Km for Zwitterion mM (+)	(spf)
7	3	3.7	0.06	0.07
7,5	1.2	-	0.11	-
8	0.4	2.3	0.06	0.39
8,5	0.28	-	0.10	-
8,9	0.1	-	0.06	-
9	0.2	-	0.13	-
9,3	0.2	-	0.16	-

(+) normal OTCase   (spf) spf OTCase   - not determined

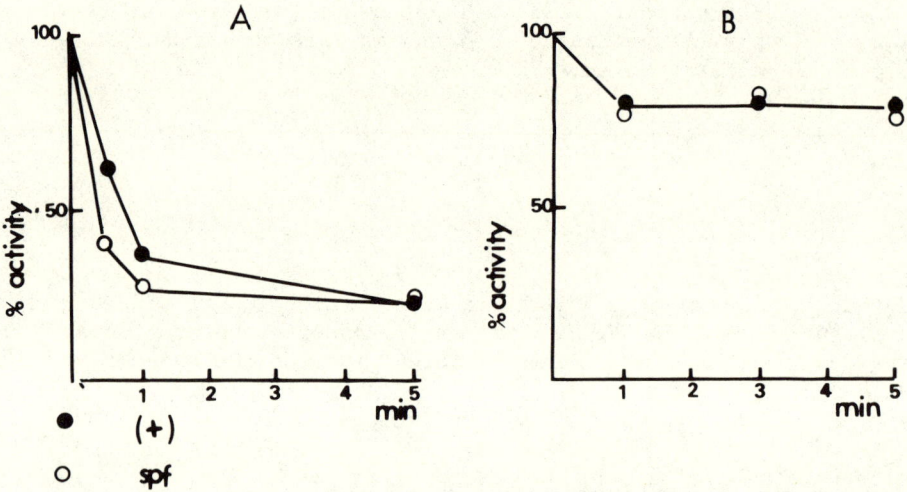

Fig. 5   Inactivation by butanedione

A) The percent of initial OTCase activity after incubation with butanedione 20 mM is plotted against the time of incubation with butanedione prior to assay of OTCase activity. Assay is realized in borate 50 mM TEA 5 mM pH 8.5   ● normal OTCase
○ spf OTCase
B) The percent of initial OTCase activity after incubation with butanedione 20 mM carbamyl phosphate 20 mM is plotted against the time of incubation
● normal OTCase   ○ spf OTCase

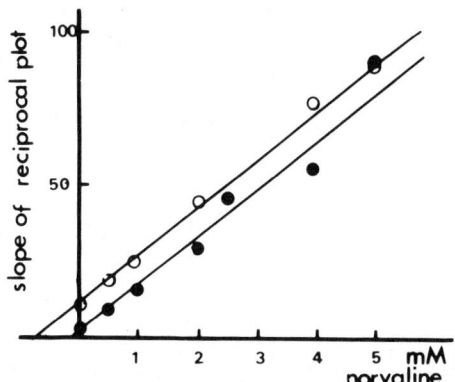

Fig. 6. Inhibition of OTCase activity by norvaline.
Inhibition by norvaline with ornithine varied and carbamylphosphate was kept constant at 5 mM. The slope of reciprocal plots is represented against norvaline concentration.
normal OTCase = ●
spf OTCase = ○

ment with the presence of an arginine residue at the CPsite of fixation which seems unmodified in spf strain.

Inhibition of OTCase Activity by Norvaline, Phosphate and δ PALO.

L Norvaline is found to be a competitive inhibitor of ornithine with a Ki of 0.1 mM for normal OTCase and 0.8 mM for spf OTCase (Fig. 6). This is in agreement with the hypothesis of a mutation affecting the ornithine fixation site and resulting in modifications of the affinity for ornithine and its analogues. On the contrary the competitive inhibition by phosphate is not modified in spf OTCase (Ki = 3.5 mM) (Fig. 7).

Inhibition by δ PALO the ligand used for the affinity chromatography, is reduced for spf OTCase as compared to normal. The variation of inhibition versus pH is reversed but inactivation is always lower for spf OTCase (Fig. 8). This explains the very low purification yield of the mutant enzyme which has to be completely purified in the classical way.

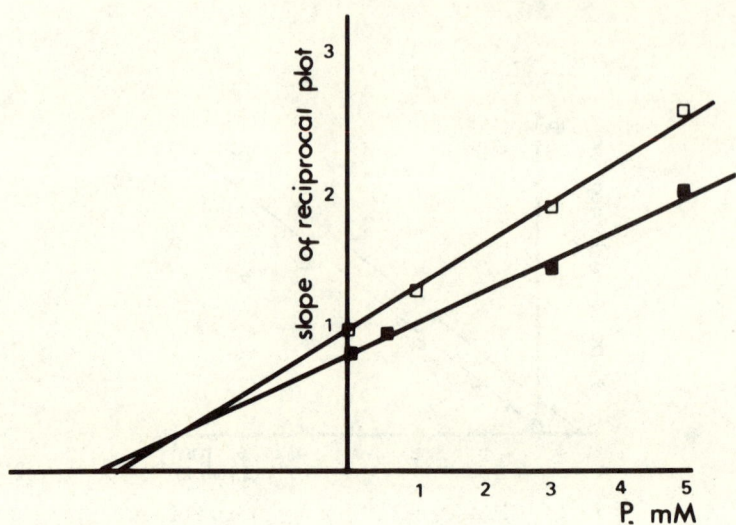

Fig. 7. Inhibition of OTCase activity by phosphate. Inhibition by phosphate with carbamyl phosphate varied and ornithine was kept constant at 5 mM. The slope of reciprocal plots is represented against phosphate concentration.
■ normal OTCase   □ spf OTCase

CONCLUSION

In conclusion, mouse OTCase is a trimer of identical subunits. Spf mutation does not affect the molecular weight and the affinity for CP of which site of fixation contains an arginine residue. On the contrary, the affinity for ornithine and norvaline is decreased. These results indicate that spf mutation is a punctual one affecting the fixation site of ornithine. Furthermore we have demonstrated that the enhancement of V max at alkaline pH is the result of two events : first an enhancement OTC protein amount and secondly an increased specific activity. The analogy firstly between enzymatic characteristics and molecular weight of mouse and human OTCase and secondly between enzymatic properties of spf OTCase of the human mutant type previously studied in homogenates allows us to think that studies on purified spf OTCase must yield further information on the human mutant type.

Similar analyses are now performed on the spf ash mouse strains 3,20 which presents reduced activity (5 %) in order to have a better knowledge of a corresponding human mutant type.

Furthermore in vitro syntheses of these two muted OTCase are now investigated.

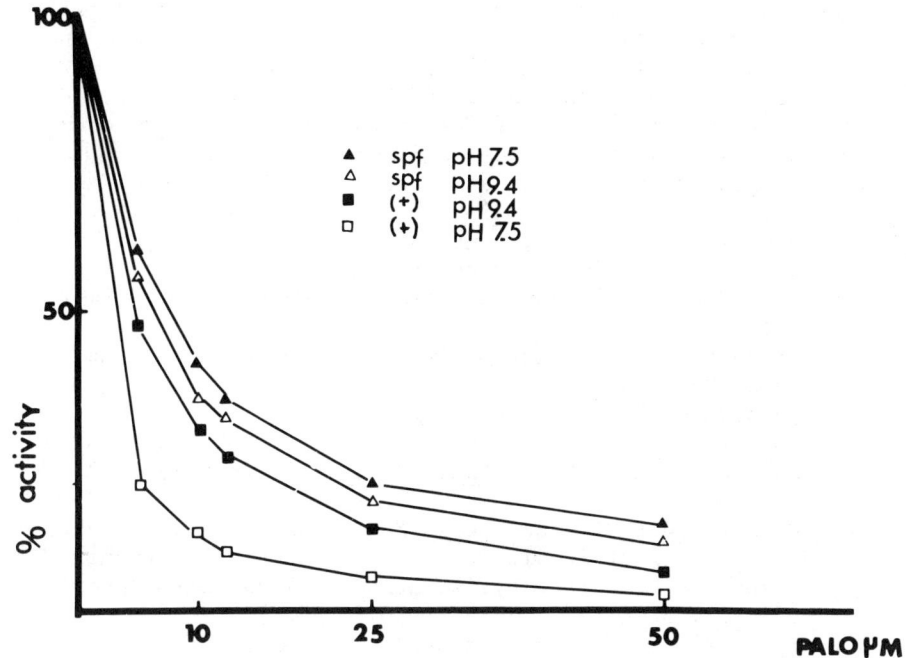

Fig. 8. Inhibition of OTCase activity by δ PALO. The percent of initial OTCase activity against δ PALO concentration is given at pH 7.5 and 9.4.

REFERENCES

1. L. Cathelineau, P. Briand, F. Petit, J. P. Nuyts, J. P. Farriaux and P. P. Kamoun, Kinetic analysis of a new human ornithine carbamyl transferase variant, Biochim. Biophys. Acta, 614: 40-45 (1980).
2. I. A. Qureshi, J. Letarte, R. Ouellet, Ornithine transcarbamylase deficiency in mutant mice. Studies on the characterization of enzyme defect and suitability as animal model of human disease, Pediat. Res. 13:806-811 (1979).
3. R. De Mars, S. L. Levan, B. L. Trend, and L. B. Russel, Abnormal ornithine carbamoyl transferase in mice having the sparse fur mutation, Proc. Natl. Acad. Sci. 73:1693-1697 (1976).
4. M. Marshall and P. P. Cohen, Ornithine transcarbamylase from streptococcus faecalis and bovine liver, J. Biol. Chem. 247:1641-1653 (1972).
5. C. J. Lusty, R. L. Jilka and E. H. Nietsh, Ornithine transcarbamylase of rat liver, J. Biol. Chem. 254:10030-10036 (1979)
6. N. J. Hoogenraad, J. M. Sutherland and G. J. Howlett, Purifica-

tion of ornithine transcarbamylase from rat liver by affinity chromatography with immobilized transition-state analog, Analytical Biochem. 101:97-102 (1980).
7. F. Kalousek, F. Baudouin and L. E. Rosenberg, Isolation and characterization of ornithine transcarbamylase from normal human liver, J. Biol. Chem. 253:3939-3944 (1978).
8. N. J. Hoogenraad, Synthesis and properties of δ-N (phosphonacetyl)-L-ornithine, Arch. Biochem. Biophy. 188:137:144 (1978).
9. M. Penninckx and D. Gigot, Synthesis and interaction with Escherichia Coli L-ornithine carbamoyl transferase of two potential transition-state analogues, Febs Lett. 88:94-96 (1978).
10. H. H. Hogeboom, Fractionation of cell components of animal tissues, Methods Enzymol. 1:16-19 (1955).
11. J. G. Conboy, F. Kalousek and L. E. Rosenberg, In vitro synthesis of a putative precursor of mitochondrial ornithine transcarbamoylase, Proc. Natl. Acad. Sci. 76:5724-5727 (1979).
12. J. G. Conboy and L. E. Rosenberg, Post translational uptake and processing of in vitro synthesized ornithine transcarbamoylase precursor by isolated rat liver mitochondria, Proc. Natl. Acad. Sci. 78:3073-3078 (1981).
13. M. Mori, S. Miura, M. Tatibana and P. P. Cohen, Characterization of a protease apparently involved in processing of pre-ornithine transcarbamylase of rat liver, Proc. Natl. Acad. Sci. 77:7044-7048 (1980).
14. T. Morita, M. Mori, M. Tatibana and P. P. Cohen, Site of synthesis and intracellular transport of the precursor of mitochondrial ornithine carbamoyltransferase, Biochem. Biophy. Res. Commun. 99:623-629 (1981).
15. M. Mori, S. Miura, M. Tatibana and P. P. Cohen, Processing of a putative precursor of rat liver ornithine transcarbamylase, a mitochondrial matrix enzyme, J. Biochem. 88: 1829-1836 (1980).
16. G. Mancini, J. P. Vaerman, A. O. Carbonera and J. F. Heremans, 1964, XI colloq. Bruges 1963. p. 370. Elsevier, Amsterdam.
17. F. Kalousek and L.E. Rosenberg, 1978, Fed. Proc. 37. 1310. Abstract 236.
18. M. Marschall and P. P. Cohen, Evidence for an exceptionally reactive arginyl residue at the binding site for carbamyl phosphate in bovine ornithine transcarbamylase, J. Biol. Chem. 255:7301-7305 (1980).
19. J. F. Riordan, Functional Arginyl residues in carboxypeptidase A. Modification with Butanedione, Biochemistry, 12:3915-3923 (1973).
20. D. P. Doolittle, L. L. Hulbert and C. Cordy, A new allele of the sparse fur gene in the mouse, J. Hered. 65:194-195 (1974).

III. BASIC BIOCHEMISTRY

1) Regulation

REGULATION OF THE N-ACETYLGLUTAMATE CONTENT OF RAT HEPATOCYTES BY
THE GLUTAMATE CONCENTRATION

H. Zollner

Institute für Biochemie der Universität Graz
Schubertstrasse 1, A-8010 Graz, Austria

INTRODUCTION

It is now generally accepted that the AcGlu concentration plays an important role in the regulation of urea synthesis (1-6). The mechanisms which control the AcGlu content are, however, not completely understood. AcGlu synthetase is specifically activated by arginine (7) but the mitochondrial arginine concentration is much higher than the Ka of arginine for isolated AcGlu synthetase (5). Aoyagi et al. (8) showed that the acetyl-CoA concentration affects the AcGlu content of isolated rat liver cells and Shigesada and coworkers (3) consider glutamate concentration to be an important factor in determining the AcGlu content.

In this paper it is shown that AcGlu content of isolated rat hepatocytes is affected by the ammonia concentration and evidence is provided for a mediating role of glutamate.

EXPERIMENTAL

Liver cells were prepared from rats fed on different diets (standard laboratory chow, boiled egg-white or glucose 20% w/v for four days, or were starved for 48h) according to the method of Berry and Friend (9) as described by Krebs (10). The cells were incubated for 30 min at 37°C in Krebs-Henseleit buffer containing 1 mM $CaCl_2$, 2 mM lactate and 2% dialysed bovine serum albumin (fraction V). Mitochondria were isolated according to Myers and Slater (11) only sucrose was replaced by mannitol, and were incubated in a medium that

---

Abbrevation. AcGlu, N-acetyl-L-glutamate

Table 2. Effect of ammonia and ornithine on the AcGlu content of hepatocytes. Initial concentrations of substrates were 10 mM NH_4Cl and 3 mM ornithine.

Diet		AcGlu (nmol/g wet wt.)			
	Additions:	None	NH_4Cl	Ornithine	NH_4Cl+Ornithine
Standard		35 ± 3	77 ± 12*	37 ± 4	89 ± 8*
Glucose 20%		12 ± 4	43 ± 3*	13 ± 3	80 ± 6*,**
Egg-white		60 ± 18	89 ± 13	62 ± 17	109 ± 12
Starved		20 ± 6	68 ± 9	45 ± 14	100 ± 10*

* 2P < 0.01 relative to control
** 2P < 0.01 relative to cells incubated in the presence of ammonia
Results from reference 6.

was essentially the same as that used by McGivan et al. (1).

AcGlu degradation was measured in a cytosolic supernatant as described by Reglero et al. (12). Metabolites were measured in the total cell suspension. Intracellular amounts were determined after centrifugation of the cells through silicone oil (13). Mitochondrial constituents were determined after disruption of the cells according to the digitonin procedure of Zuurendonk and Tager (14). AcGlu was determined by method of McGivan (1), citrulline according to Archibald (15). The other metabolites were determined by standard spectrophotometric (16) and fluorometric procedures (17).

RESULTS AND DISCUSSION

Effect of ammonia and ornithine on the AcGlu content of hepatocytes

Carbamyl phosphate synthetase and ornithine transcarbamylase, the enzymes which catalyze the first two steps of urea synthesis, are located exclusively mitochondrial (18). It is therefore to be expected that the rate of citrulline formation of isolated mitochondria is equal or higher than the rate of urea formation of hepatocytes. Using ammonia as the nitrogen source it was however found that this anticipation is not always realized. Hepatocytes isolated from rats fed on a protein free diet synthesize urea at a higher rate than liver mitochondria isolated from animals fed the same diet (Table 1).

The hypothesis proposed to explain the observed differences was that isolated mitochondria contain less AcGlu - an allosteric activator of carbamyl phosphate synthetase - than hepatocytes. A comparison of the AcGlu content of hepatocytes with that of isolated liver

Table 1. Comparison of urea synthesis of hepatocytes with citrulline formation of isolated mitochondria.
Initial substrate concentrations were 10 mM NH_4Cl and 3 mM ornithine.

Diet	Urea	Citrulline
	μmol/g wet wt.	
Standard	2.4 ± 0.2	1.16 ± 0.1
Glucose 20%	2.1 ± 0.3	0.45 ± 0.07
Egg-white	3.6 ± 0.3	3.3 ± 0.2
Starved	2.8 ± 0.1	0.78 ± 0.07

mitochondria showed that there were only minor differences, however, it was found that AcGlu increased during the incubation of hepatocytes with ammonia (Table 2). This effect was most pronounced with liver cells from glucose fed rats but was also found in cells from rats fed on laboratory chow, egg-white or starved for 48h. Up to 4 mM ammonia an almost linear relationship between AcGlu and the ammonia concentration was found. When ornithine was added to liver cells in the presence of ammonia a further increase in AcGlu was observed whereas ornithine alone had no effect except that its presence prevented AcGlu decrease in cells from starved animals.

In hepatocytes from glucose fed rats the initial rate of AcGlu formation induced by the addition of ammonia was 6-7 nmol/g wet wt. x min and a new steady state was attained 10 minutes after the addition of ammonia. With ornithine present together with ammonia a somewhat higher rate of AcGlu formation was found and it lasted 15 min to reach a steady state.

The AcGlu content was also affected by the lactate concentration. When the lactate concentration of the medium was increased from 2 to 9 mM the AcGlu content of cells from starved rats incubated in the presence of ornithine + ammonia was enhanced approximately twofold.

Mechanism of the ammonia effect

For the explanation of the effect of ammonia on the AcGlu content the following two possible mechanisms can be offered: 1. activation of AcGlu synthetase; 2. inhibition of AcGlu degradation.

The available evidence favours the suggestion that stimulation of AcGlu synthesis leads to the observed increase of AcGlu. The change

Table 3. AcGlu hydrolysis rates of cytosolic supernatant.
Initial AcGlu concentration was 20 mM.

Diet	AcGlu hydrolysis µmol/g wet wt. x h
Standard	13.6 $\pm$ 1.5
Glucose 20%	7.8 $\pm$ 1.4*
Egg-white	15.6 $\pm$ 1.8
Starved	26.8 $\pm$ 8.1*

* 2P < 0.005 relative to standard diet

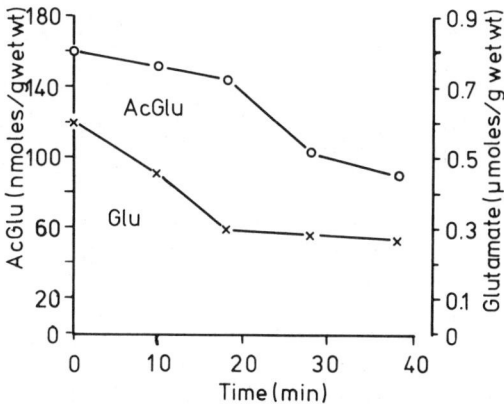

Fig. I. Decrease of AcGlu in dependence on mitochondrial glutamate. The AcGlu content of hepatocytes from starved rats was increased by incubation with 10 mM NH_4Cl for 10 min; cells were then washed free of ammonia and the decrease of AcGlu and glutamate followed.

of the catalytical activity of AcGlu degradation induced by changes in the protein content of the diet is opposite to what is expected from the changes in the AcGlu content (cf. table 2 and 3). Feeding a protein free diet decreased the AcGlu concentration of hepatocytes as well as the capacity for AcGlu degradation. The highest AcGlu content was found in hepatocytes from egg-white fed rats, the AcGlu degradation capacity, however, was intermediate, whereas the highest AcGlu degradation rate was found in starved animals when the AcGlu content was intermediate. In addition no significant effect of ammonia or ammonia + ornithine on AcGlu degradation was found.

Incubation of hepatocytes with ammonia increased the mitochondrial glutamate concentration. In cells from starved rats mitochondrial glutamate is increased from 7.5 mM to 54 mM (13) and in cells from glucose fed rats from 3.6 mM to 23 mM (6). A comparison of the changes of the mitochondrial glutamate concentration with the Km value of AcGlu synthetase, which is reported to be between 3 and 5 mM (7,19) shows that it is possible that the increase in AcGlu is due to a stimulation of AcGlu synthesis via an increase of the substrate concentration.

This conclusion is supported by the results of experiments in which the AcGlu content was compared with the mitochondrial glutamate concentration. In hepatocytes, whose AcGlu content had been increased by incubation with ammonia, the decrease of AcGlu which was observed after the cells were washed free of ammonia was preceded by a decrease in mitochondrial glutamate (Fig. I). A similar dependence of AcGlu on glutamate was found when the ammonia induced increase in AcGlu was

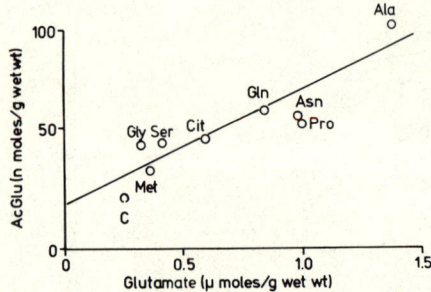

Fig. 2. Correlation of AcGlu with mitochondrial glutamate. Hepatocytes from starved rats were incubated for 30 min in the presence of different amino acids at a concentration of 10 mM each.

studied (6). In experiments where the mitochondrial glutamate concentration was increased by incubation of hepatocytes with different amino acids a significant correlation between the AcGlu and glutamate content was found (r = 0.88, 2P < 0.01) (Fig. 2).

Mechanism of the ornithine effect

Ornithine when added together with ammonia is not converted to glutamate by hepatocytes because ammonia protects ornithine against transamination by decreasing α-ketoglutarate and increasing the rate of citrulline formation (13). Accordingly no increase in mitochondrial glutamate was found when ornithine was added to cells incubated in the presence of ammonia (6). Thus the further increase of AcGlu observed upon the addition of ornithine to cells incubated in the

Table 4. Effect of ammonia and ornithine on intracellular acetyl-CoA and CoA-SH. Hepatocytes from starved rats were used. Concentrations: 10 mM NH_4Cl, 3mM ornithine.

Additions	Acetyl-CoA	CoA-SH
	nmol/g wet wt.	
Control	47 ± 3	114 ± 26
NH_4Cl	26 ± 4*	93 ± 16
NH_4Cl+ornithine	48 ± 11**	119 ± 31

* 2P < 0.02 relative to control
** 2P < 0.05 relative to cells incubated with NH_4Cl

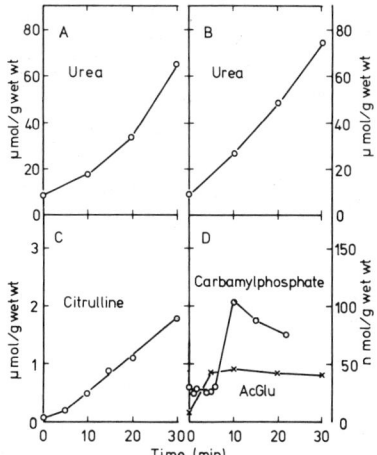

Fig. 3. Relationship between AcGlu and activity of the urea cycle in hepatocytes from glucose fed rats.
Concentrations: 10 mM NH_4Cl A,B,C,D; 3 mM ornithine A,B,C.
B: cells were preincubated for 10 min in the presence of 10 mM glutamine and urea synthesis was measured after the addition of ammonia and ornithine.

presence of ammonia cannot be explained on the basis of an increase in glutamate.

Glucose synthesis of hepatocytes with pyruvate as the gluconeogenic precursor is inhibited by ammonia and this inhibition is reversed by addition of ornithine (13). It was assumed by Meijer et al. (13) that acetyl-CoA is increased by ornithine stimulating flux through pyruvate carboxylase. As shown in table 4 ornithine affects the acetyl-CoA content of rat liver cells in the predicted way. Assuming that 50% of the cellular acetyl-CoA is mitochondrial (20), the concentration in the matrix was calculated to be between 0.2 and 0.4 mM. A comparison of these values with the kinetic constants of AcGlu synthetase shows that the mitochondrial acetyl-CoA concentration is below the reported Km value of 0.6 - 0.7 mM (7,19). The concentration of acetyl-CoA in mitochondria may even be lower than the calculated one as an appreciable amount of acetyl-CoA is bound to acetyl-CoA metabolizing enzymes (21). These considerations indicate that acetyl-CoA mediates the ornithine effect and it is concluded in agreement with the findings of Aoyagi et al. (8) that acetyl-CoA is a very effective regulator of AcGlu synthesis.

Relationship between AcGlu and urea synthesis

In the experiments depicted in Fig. 3 the kinetic of urea-,

citrulline-, and carbamyl phosphate synthesis were compared with AcGlu formation. During the period in which AcGlu increased the initial rate of urea -, citrulline-, and carbamylphosphate formation was low and increased gradually. Maximal rates were found after a new steady state concentration of AcGlu was attained.

Incubation of rat liver cells from glucose fed rats with 10 mM glutamine increased the AcGlu content to 70 nmol/g wet wt. The lag in urea synthesis of cells whose AcGlu content was raised by preincubation with glutamine is abolished (Fig. 3,B). These results show that the ammonia induced increase in AcGlu influences the rate of urea synthesis.

CONCLUSIONS

The results reported in this paper prove the existence of a short term control mechanism of urea synthesis which responds very sensible and rapid to changes in the ammonia concentration. The mediator of the ammonia effect is glutamate. The consequence of an increase in glutamate is stimulation of AcGlu synthesis and aspartate formation . In this way the activities of carbamyl phosphate synthetase and argininosuccinate synthetase, the two enzymes of the urea cycle with the lowest capacity, are modulated.

REFERENCES

1. J. D. McGivan, N. M. Bradford, and J. Mendes-Mourao, The regulation of carbamoyl phosphate synthase activity in rat liver mitochondria, Biochem. J. 154:415 (1976).
2. T. Saheki, T. Katsunuma, and M. Sase, Regulation of urea synthesis in rat liver, J. Biochem. 82:551 (1977).
3. K. Shigesada, K. Aoyagi, and M. Tatibana, Role of acetylglutamate in ureotelism, Eur. J. Biochem. 85:385 (1978).
4. T. Saheki, T. Ohkubo, and T. Katsunuma, Regulation of urea synthesis in rat liver, J. Biochem. 84:1423 (1978).
5. H. E. S. J. Hensgens, A. J. Verhoeven, and A. J. Meijer, The relationship between intramitochondrial N-acetylglutamate and activity of carbamoylphosphate synthetase (ammonia), Eur. J. Biochem. 107:197 (1980).
6. H. Zollner, Regulation of urea synthesis, Biochim. Biophys. Acta 676:170 (1981).
7. M. Tatibana, K. Shigesada, and M. Mori, Acetylglutamate synthetase, in:"The Urea Cycle", S. Grisola, R. Baguena, and F. Mayor, eds., Wiley, New York (1976).
8. K. Aoyagi, M. Mori, and M. Tatibana, Inhibition of urea synthesis by pent-4-enoate associated with decrease in N-acetyl-L-glutamate concentration in isolated rat hepatocytes, Biochim. Biophys. Acta 587:515 (1979).

9. M. N. Berry, and D. S. Friend, High-yield preparation of isolated rat liver parenchymal cells, J. Cell. Biol. 43:506 (1969).
10. H. A. Krebs, N. W. Cornell, P. Lund, and R. Hems, Isolated liver cells as experimental material, in:"Regulation of Hepatic Metabolism, Alfred Benzon Symposium VI", F. Lundquist, and N. Tygstrup, eds., Academic Press, New York (1974).
11. D. K. Myers, and E. C. Slater, The enzymic hydrolysis of adenosine triphosphate by liver mitochondria, Biochem. J. 67:558 (1957).
12. A. Reglero, J. Rivas, J. Mendelson, R. Wallace, and S. Grisola, Deacylation and transacetylation of acetyl glutamate and acetyl ornithine in rat liver, FEBS Lett. 81:13 (1977).
13. A. J. Meijer, J. A. Gimpel, G. Deleeuw, M. E. Tischler, J. M. Tager, and J. R. Williamson, Interrelationships between gluconeogenesis and ureogenesis in isolated hepatocytes, J. Biol. Chem. 253:2308 (1978).
14. P. F. Zuurendonk, and J. M. Tager, Rapid separation of particulate components and soluble cytoplasm of isolated rat-liver cells, Biochim. Biophys. Acta 333:393 (1974).
15. R. M. Archibald, Determination of citrulline and allantoin and the demonstration of citrulline in blood plasma, J. Biol. Chem. 156:121 (1944).
16. H. U. Bergmeyer, "Methoden der enzymatischen Analyse", 2nd ed., Verlag Chemie, Weinheim (1970).
17. J. R. Williamson, and B. E. Corkey, Assays of intermediates of the citric acid cycle and related compounds by fluorometric enzyme methods, Methods Enzymol. 13:434 (1969).
18. J. G. Gamble, and A. L. Lehninger, Transport of ornithine and citrulline across the mitochondrial membrane, J. Biol. Chem. 248:610 (1973).
19. F. X. Coude, L. Sweetman, and W. L. Nyhan, Inhibition by propionyl-coenzyme A of N-acetylglutamate synthetase in rat liver mitochondria, J. Clin. Invest. 64:1544 (1979).
20. E. A. Siess, D. G. Brocks, H. K. Lattke, and O. H. Wieland, Effect of glucagon on metabolite compartmentation in isolated rat liver cells during gluconeogenesis from lactate, Biochem. J. 166:225 (1977).
21. G. J. Barritt, G. L. Zander, and M. F. Utter, The regulation of pyruvate carboxylase activity in gluconeogenic tissues, in: "Gluconeogenesis - its regulation in mammalian species", R. W. Hanson, and M. A. Mehlman, eds., Wiley, New York (1976).

ACKNOWLEDGEMENT

This study was supported by Fonds zur Förderung der wissenschaftlichen Forschung, Projekt Nr. 3598.

ENZYME REGULATION OF N-ACETYLGLUTAMATE SYNTHESIS IN MOUSE AND RAT LIVER

Masamiti Tatibana, Susumu Kawamoto, Tomoko Sonoda and Masataka Mori

Department of Biochemistry, Chiba University School of Medicine, Inohana, Chiba 280, Japan

ABSTRACT

N-Acetylglutamate (AGA) synthetase of mammalian liver is known to be stimulated by low concentrations of arginine. The arginine sensitivity of the synthetase was found to show postprandial changes in the liver of DD-Y mice. AGA synthetase activity assayed in the absence of arginine changed only slightly during and after the feeding. With 1 mM arginine, the activity increased and reached a peak value 9 h after the start of feeding. The activation ratios were about 2 and 6 at 0 and 9 h, respectively. Similar changes occurred with 0%, 20%, and 60% casein diets. When the enzymes were partially purified, the respective activation ratios remained the same, suggesting no involvement of a readily dissociable low molecular weight compound. Treatment of mice with cycloheximide did not abolish the increase in the activation ratio. A homogeneous preparation of the synthetase was obtained by a 30,000-fold purification from sonicated mitoplasts of rat liver mitochondria. The molecular weight was 160,000, as estimated on sucrose density gradient centrifugation, with subunits of 57,000 on SDS-gel electrophoresis. The enzyme had a hydrophobic nature, was stabilized by Triton X-100, and contained little phospholipid. Although the molecular basis of the increase in the activation ratio remains to be established, the modification of the nature of AGA synthetase introduces another aspect into the elaborate regulation of urea synthesis.

INTRODUCTION

N-Acetyl-L-glutamate is a specific and obligatory allosteric activator of mitochondrial carbamoyl phosphate synthetase I in the liver of ureotelic animals.[1,2] On the basis of several lines of evidence[3-8] we have proposed a control mechanism for urea biosynthesis mediated by the hepatic level of acetylglutamate. This control mechanism has been supported by findings in other laboratories both in vitro using isolated mitochondria,[9,10] isolated mitoplasts[11] and isolated hepatocytes,[12] and in vivo[13,14] (see also Ref. 15 for a review).

We detected acetylglutamate synthetase, the enzyme which catalyzes the formation of acetylglutamate from glutamate and acetyl CoA, in the mitochondria of mammalian liver, and partially purified it.[4] This enzyme is activated specifically by arginine.[4,16,17] A recent report of a hyperammonemic patient with deficiency of this enzyme[18] clearly shows that the enzyme plays a critical role in ammonia detoxication in humans.

Several factors which regulate acetylglutamate synthesis in the liver include intracellular concentrations of arginine[6] and glutamate[6,19] as well as the level of the acetylglutamate synthetase activity.[9,17] Ammonia has also been postulated to be a regulatory factor[20] and a recent report[19] showed that the effect is mediated by increases in the levels of glutamate.

We report here that the sensitivity of acetylglutamate synthetase to arginine activation undergoes marked changes after ingestion of food and that these changes appear to be due to modification of the enzyme molecule itself. The observations indicate the presence of a heretofore unknown mechanism for the regulation of the synthetase activity in the liver. We also provide a preliminary account of our recent studies on the purification and characterization of acetylglutamate synthetase from rat liver mitochondria.

REGULATION OF N-ACETYLGLUTAMATE SYNTHETASE IN MOUSE LIVER

Postprandial Change in Acetylglutamate Synthetase Activity in Mouse Liver Mitochondria

Mice were trained to eat laboratory chow (protein content, 21%) during a 3 h period each day. After 4 to 5 days, the animals were killed at intervals after a meal and liver mitochondria were prepared. Acetylglutamate synthetase activity in the sonicated mitochondria was measured in the presence and absence of 1 mM arginine (Fig. 1). The activity assayed in the presence of arginine increased remarkably after the meal, reached a maximum 9 h after the start of feeding and then decreased gradually. The activity assayed in the absence of arginine was low at the start and remained at similar levels thereafter. The extent of activation, which was 1.6-fold at the start of feeding, increased to 5.5-fold in 9 h. Other sets of similar experiments, where conditions of lighting were controlled, gave results indicating that the postprandial changes in the activity are related to the ingested food and not to lighting conditions or intrinsic diurnal rhythms.

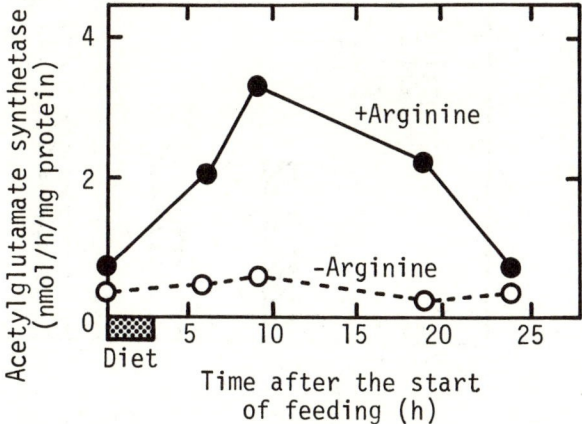

Fig. 1. Postprandial changes in acetylglutamate synthetase activity in mouse liver. Each group consisted of 3 mice. Mitochondria were isolated from the pooled 3 livers and sonicated. Acetylglutamate synthetase activity was assayed essentially as described previously[17] in a system (100 μl) containing 1.0 mM L-[^{14}C]glutamic acid (2.5 Ci/mol), 0.5 mM acetyl CoA, 1.0 mM EDTA, 50 mM Tris-HCl (pH 8.2), and the sonicate, with (●) and without (○) 1 mM L-arginine. The reaction was carried out for 30 min at 25°C and the [^{14}C]acetylglutamate formed was isolated by successive chromatography on ion exchange resins and paper.[3]

Partial degradation of arginine in the reaction mixture did occur during the enzyme assay. However, several experiments showed that the remaining arginine was sufficient in amount to bring about a full activation of the synthetase.

Effects of Protein Content in a Single Diet on Postprandial Changes in Acetylglutamate Synthetase Activity

Mice kept on a scheduled feeding were either fasted or fed diets containing different amounts of casein and 8 h after the start of feeding or fasting, acetylglutamate synthetase activity was measured both in the presence and absence of arginine. The extent of the enzyme activation by arginine remained low in the liver of the fasted animals (Fig. 2). After the feeding of various diets, the enzyme activity in the absence of arginine increased, whereas the activity in the presence of arginine increased even further, bringing about a marked increase in the extent of arginine activation. It is notable that the increase was observed even when the food contained no casein. A 60% casein diet brought about a greater increase in the activities, with and without arginine. When mice were kept on a 0% casein diet continuously for a few days, there was no similar increase in the arginine sensitivity of the synthetase after ingestion of the same diet.

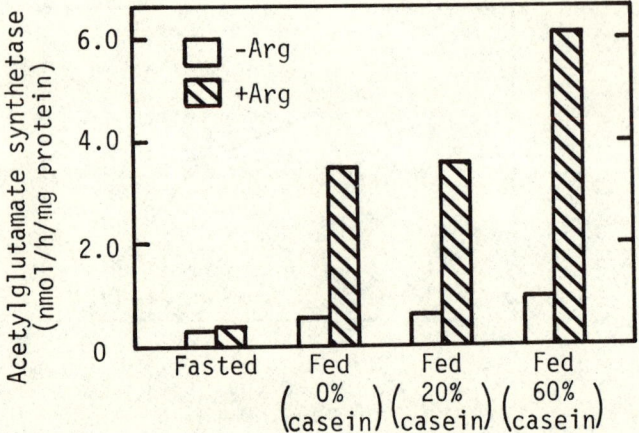

Fig. 2. Effects of protein content in a single diet on acetylglutamate synthetase activity. Mice were fed the usual laboratory chow for 3 days. On the 4th day each group of 3 mice were further fasted or fed a 0, 20 or 60% casein diet[21] for 3 h. Mice were killed at 29 h of fasting or 8 h after the start of feeding.

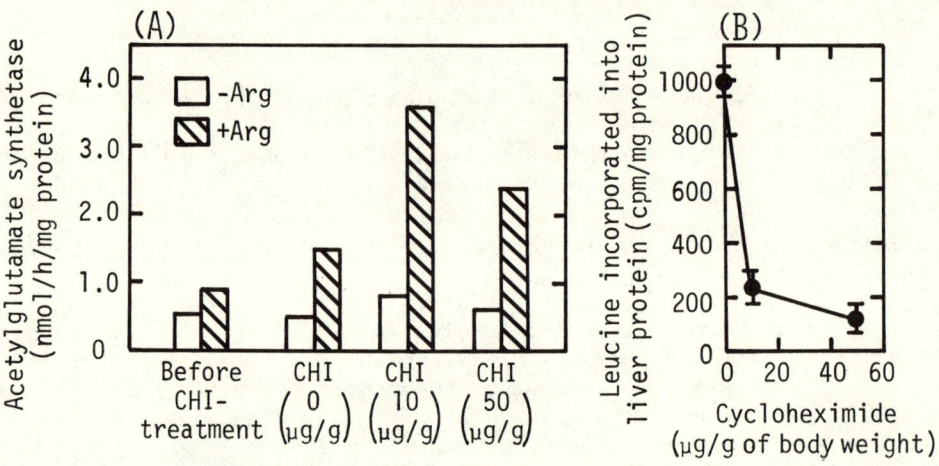

Fig. 3. Effects of cycloheximide on postprandial changes in acetylglutamate synthetase. Mice were maintained as in Fig. 2. Feeding and treatments with cycloheximide were done as described in the text. [^{14}C]Leucine (0.17 µCi per g of body weight, i.p.) was given at 0 time. Each group consisted of 3 mice. The enzyme activity (A) or incorporation of [^{14}C]leucine into the total liver protein (B) was measured at 8 h. CHI: cycloheximide.

Effects of Cycloheximide on Postprandial Changes in Acetylglutamate Synthetase Activity

Mice were given cycloheximide and [^{14}C]leucine intraperitoneally at 0 time, fed for 3 h, given the antibiotic at 3 and 6 h and then killed at 8 h. As shown in Fig. 3B, the leucine incorporation into total liver protein was inhibited strongly by the cycloheximide treatment. However, there was no inhibition of the diet-dependent increase in the sensitivity of acetylglutamate synthetase to arginine activation; actually it stimulated the increase, to a significant extent (Fig. 3A). Therefore, new synthesis of the enzyme protein or of other proteins probably is not required for the postprandial increase in the arginine sensitivity of the synthetase.

Purification of Acetylglutamate Synthetase from Livers of Fasted and Fed Mice

The enzymes from the livers of fasted and fed mice were partially purified and compared. The fed mice were treated with cycloheximide, since this antibiotic further increases the sensitivity of the enzyme to arginine (Fig. 3). Typical results of the purification are summarized in Table 1. The extent of arginine activation of the enzyme from fasted animals remained low throughout the purification, whereas that of the enzyme from fed and cycloheximide-treated animals remained high. Therefore, the change in sensitivity of the enzyme to arginine may be due to alteration in the inherent nature of the enzyme. Involvement of a readily dissociable low molecular weight compound(s) is unlikely. Upon gel filtration on a Sephacryl S-300 column, the enzyme from fasted animals was eluted in a single peak with an estimated molecular weight of 250,000, while the enzyme from fed animals was eluted with the same or a little smaller molecular weight. The enzyme from fasted animals appeared to have a slightly broader peak. The profiles suggested that the change in arginine sensitivity of the synthetase is neither accompanied by association-dissociation of the enzyme nor its profound proteolysis.

Discussion

We have presented evidence[3-8] that acetylglutamate, in addition to ornithine,[22-24] ammonia,[19-20] intramitochondrial ATP/ADP ratio[25] and Ca^{2+},[26] is a mediator which regulates urea biosynthesis in mouse and rat liver. The intramitochondrial level of acetylglutamate is regulated by the rate of its synthesis[3,7] and by the rate of the degradation.[27] Factors so far known to regulate the synthesis of acetylglutamate are the intracellular levels of arginine and glutamate[6,19] and the amount of acetylglutamate synthetase.[9] The postprandial changes in the extent of arginine activation of the synthetase, shown in this paper may represent a hitherto unknown regulatory mechanism in acetylglutamate synthesis. The increased sensitivity may contribute to prevention of a rapid decrease in acetylglutamate levels, rather than to the postprandial early increase in the level.

Table 1. Purification of acetylglutamate synthetase from livers of fasted mice (A) and of fed mice (B). (A) 32 mice were maintained on the usual laboratory chow and then were fasted for 33 h. Acetylglutamate synthetase was purified essentially as described previously.[17] (B) 15 mice were fed the chow for 4 days as in Fig. 2, given cycloheximide (50 μg per g of body weight, i.p.) at 0, 3 and 6 h after the start of the last feeding, and were killed at 8 h.

(A)

Purification step	Total activity (-arginine) (nmol/h)	Total protein (mg)	Specific activity (-arginine) (nmol/h/mg protein)	Activation by arginine (fold)
Mitochondrial extract	166	268	0.62	1.5
Ammonium sulfate fractionation (0-40%)	157	35.2	4.46	1.8
Sephacryl S-300 chromatography	45.0	6.82	6.60	1.4

(B)

Purification step	Total activity (-arginine) (nmol/h)	Total protein (mg)	Specific activity (-arginine) (nmol/h/mg protein)	Activation by arginine (fold)
Mitochondrial extract	40.3	112	0.36	5.2
Ammonium sulfate fractionation (0-40%)	117	31.4	3.72	4.8
Sephacryl S-300 chromatography	68.1	4.76	14.3	6.6

PURIFICATION AND CHARACTERIZATION OF RAT LIVER ACETYLGLUTAMATE SYNTHETASE

Purification

We recently succeeded in an over 30,000-fold purification of the synthetase from the mitochondrial extracts. The final preparation was apparently homogeneous, as examined by sodium dodecylsulfate gel electrophoresis, although the purification procedures employed can be further improved. Critical steps heretofore taken include:

(1) In the earlier phase of the purification studies, we used livers of Wistar strain rats[17] as the source of enzyme, but later used Sprague-Dawley rats as the liver mitochondria from the latter

contain a higher activity of enzyme. (2) The synthetase is
distributed in the inner membrane and matrix fraction of liver
mitochondria,[16] thus, the liver mitochondria were treated with
digitonin to remove the outer membrane[28] and the mitoplasts obtained
are sonicated. (3) The enzyme is so unstable at low protein
concentrations that some measures for stabilization are essential in
the purification and storage. We examined a variety of agents and
conditions for this purpose and found that Triton X-100 and certain
other detergents are effective in preventing the enzyme from
inactivation. Cholic acid and deoxycholic acid inactivated the
enzyme. (4) Another effective means of stabilization is to prevent
the enzyme solution from direct contact with a glass surface. Thus
the glassware used for the enzyme purification, storage and assay was
coated with silicone. Plastics including polyethylene are also
usable. (5) Based on the observed nature of the enzyme, hydrophobic
chromatography with the use of ω-aminooctyl-Sepharose and phenyl-
Sepharose proved to be effective.

The final homogeneous preparation catalyzed the formation of
acetylglutamate from 0.5 mM acetyl CoA and 1.0 mM L-glutamate at a
rate of 18.7 μmol/min/mg at 25°C, in the presence of 1 mM L-arginine.
The ratio of the activities with and without 1 mM arginine was 3.7.

Molecular Weight

The molecular weight was estimated to be 160,000 by sucrose
density gradient centrifugation. Upon polyacrylamide gel electrophoresis in sodium dodecyl sulfate, the enzyme migrated as a single
protein band with a molecular weight of 57,000, indicating that the
synthetase is composed of three identical subunits or three subunits
of identical size.

Substrate Specificity

The synthetase was previously shown to have a relatively high
substrate specificity toward L-glutamate and acetyl CoA.[16,17]
Minor activity was found only for propionyl CoA as an acyl donor
and glycine as an acyl acceptor. In view of the importance of side
reactions in pathologic states[29] more rigorous substrate specificity
studies with purified preparations are underway.

Activation by L-Arginine and Other Agents

An important property of the synthetase is that it is activated
specifically by L-arginine. No other low molecular weight compounds
including other amino acids, guanidino compounds and urea cycle
intermediates exhibited significant activation.[16,17] However,
protamine sulfate did stimulate synthetase activity. The effect is
nearly additive to that of arginine. Furthermore, Triton X-100,
which is a stabilizer for the labile synthetase and has been used
in its purification, did stimulate the enzyme activity. A maximal

activation (about 4-fold) was effected at a concentration of 0.1% (w/v). The activating effect was not additive to that of arginine. The agent increased the Vmax without changing $\underline{K_m}$ values for the substrates, as in the case of arginine.[16] This common kinetics suggests that a kind of "unmasking" of the active site is involved in the mechanism of activation by the two agents, although their different chemical natures do not support the idea of totally common mechanisms.

Discontinuities in Arrhenius Plot of the Synthetase Reaction

Stabilization of the synthetase by Triton X-100, its inactivation on contact with a glass surface and binding to phenyl-Sepharose suggest peculiar molecular properties of this enzyme, including a hydrophobic nature. Consistent with these properties is a unique temperature dependence of the catalytic activity (Fig. 4). The maximal activity was obtained at 30°C; the activity was lower at 37°C. When the enzyme reaction was carried out initially at 37°C and the tube was transferred into a bath at 30°C, the reaction rate was increased. Thus, the lower rate at 37°C was not due to irreversible enzyme inactivation.

Discontinuities in Arrhenius plot of an enzyme at temperatures above about 10°C is often ascribable to phase transition of environmental lipids (for a Review, Ref. 30). The phospholipid content of a highly purified preparation, as assayed by organic phosphorus analysis, was insignificant. This suggests that the phase transition of this enzyme at 30°C as well as at 15°C depends on inherent components, not environmental lipids, of the enzyme protein.

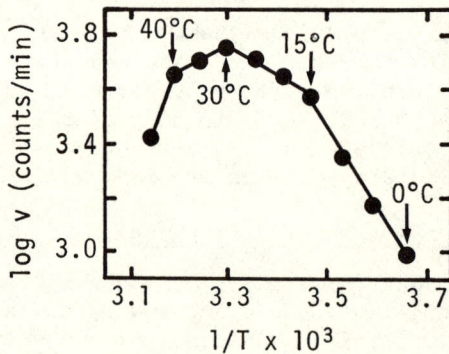

Fig. 4. Arrhenius plot of rat liver acetylglutamate synthetase reaction. The enzyme reaction was for 15 min in the presence of 1 mM L-arginine.

Acknowledgements. We thank Dr. C. Uchiyama of our laboratory for helpful discussion, Dr. H. Ishida for participation in the initial phase of this work, M. Ohara for critical reading of the manuscript and K. Shinya for secretarial services. This work was supported in part by the Scientific Research Grant (577136) from the Ministry of Education, Science and Culture of Japan.

REFERENCES

1. S. Grisolia and P. P. Cohen, J. Biol. Chem. 204:753 (1953).
2. L. M. Hall, R. L. Metzenberg, and P. P. Cohen, J. Biol. Chem. 230:1013 (1958).
3. K. Shigesada and M. Tatibana, J. Biol. Chem. 246:5588 (1971).
4. K. Shigesada and M. Tatibana, Biochem. Biophys. Res. Commun. 44:1117 (1971).
5. M. Tatibana and K. Shigesada, Adv. Enzyme Regul. 10:249 (1972).
6. M. Tatibana and K. Shigesada, in:"The Urea Cycle," S. Grisolia, R. Báguena, and F. Mayor, eds., John Wiley and Sons, New York, p.301 (1976).
7. K. Shigesada, K. Aoyagi, and M. Tatibana, Eur. J. Biochem. 85:385 (1978).
8. K. Aoyagi, M. Mori, and M. Tatibana, Biochim. Biophys. Acta 587:515 (1979).
9. J. D. McGivan, N. M. Bradford, and J. Mendes-Mourão, Biochem. J. 154:415 (1976).
10. A. J. Meijer and G. M. van Woerkom, FEBS Lett. 86:117 (1978).
11. C. W. Cheung and L. Raijman, J. Biol. Chem. 255:5051 (1980).
12. H. E. S. J. Hensgens, A. J. Verhoeven, and A. J. Meijer, Eur. J. Biochem. 107:197 (1980).
13. T. Saheki, T. Katsunuma, and M. Sase, J. Biochem. 82:551 (1977).
14. P. M. Stewart and M. Walser, J. Biol. Chem. 255:5270 (1980).
15. A. J. Meijer, Trends Biochem. Sci. 4:83 (1979).
16. M. Tatibana, K. Shigesada, and M. Mori, in:"The Urea Cycle," S. Grisolia, R. Báguena, and F. Mayor, eds., John Wiley and Sons, New York, p.95 (1976).
17. K. Shigesada and M. Tatibana, Eur. J. Biochem. 84:285 (1978).
18. C. Bachmann, S. Krähenbühl, J. P. Colombo, G. Schubiger, K. H. Jaggi, and O. Tönz, N. Engl. J. Med. 304:543 (1981).
19. H. Zollner, Biochim. Biophys. Acta 676:170 (1981).
20. T. Saheki, T. Ohkubo, and T. Katsunuma, J. Biochem. 84:1423 (1978).
21. A. E. Harper, J. Nutr. 68:405 (1959).
22. H. A. Krebs and K. Henseleit, Z. Physiol. Chem. 210:33 (1932).
23. T. Saheki and N. Katsunuma, J. Biochem. 77:659 (1975).
24. N. S. Cohen, C. W. Cheung, and L. Raijman, J. Biol. Chem. 255:10248 (1980).
25. R. J. A. Wanders, G. M. van Woerkom, R. F. Nooteboom, A. J. Meijer, and J. M. Tager, Eur. J. Biochem. 113:295 (1981).

26. A. J. Meijer, G. M. van Woerkom, R. Steinman, and J. R. Williamson, J. Biol. Chem. 256:3443 (1981).
27. T. Morita, M. Mori, and M. Tatibana, J. Biochem., in press
28. C. Schnaitman and J. M. Greenawalt, J. Cell Biol. 38:158 (1968).
29. F. X. Coude, L. Sweetman, and W. L. Nyhan, J. Clin. Invest. 64:1544 (1979).
30. J. K. Raison, Bioenergetics 4:285 (1973).

ACUTE GLUCAGON TREATMENT IN RATS FED VARIOUS PROTEIN DIETS

EFFECT ON N-ACETYL GLUTAMATE CONCENTRATION

L. Cathelineau, D. Rabier, F.X. Coudé

Laboratoire de Biochimie Génétique, Hôpital Necker-
Enfants Malades, F-75730 Paris Cedex 15, France

INTRODUCTION

Carbamoyl phosphate synthetase (ammonia) (CPS) which is localized in mitochondria has an absolute requirement for N-acetyl glutamate (NAGA) as activator. Since the important work of Shigesada and Tatibana[1] the role of its intramitochondrial concentration upon the regulation of ureogenesis has been stressed by numerous workers : the levels of NAGA show a positive correlation with the ability of the liver or of the isolated hepatocytes to synthetize citrulline[1,2,3,4]. NAGA is closely localized into mitochondria, because it is destroyed in the cytosol by a powerful deacylase activity[5]. Then, NAGA content in mitochondria must be appreciated either by measurements in the whole liver or by assay on isolated mitochondria. By the use of one of these two methods, it has been reported that low NAGA concentrations are obtained with short[6] or prolonged[3] hypo or aprotidic diets, with acute changes from 70 % to 5 % proteins in the diet[2], with acute accumulation of toxic metabolites in the liver, propionate[7,8] pent-4-enoate[9] and valproate[10]. Low NAGA concentrations are also observed during the last days of the gestation in rat females[11]. In contrary, high NAGA levels are obtained with short[6] and prolonged[3] hyperprotidic diets, with acute changes from 5 % to 70 % or protein in the food[2], with the administration in vivo of arginine[6] and also observed during the delivery of rat females[11]. Prolonged hyperprotidic diets increase the rate of N-acetyl glutamate synthetase activity[12] while short term changes of protein amount in nutriments have no effect[6] on it.

In conclusion, NAGA levels are under the dependence of acute or prolonged events in protein metabolism, exogenous supply or endogenous catabolism and anabolism. It was, then, justified to postulate that this important factor of urea cycle regulation would be under the dependence of acute and prolonged hormonal control. In early works[13,14,15], it has been reported that glucagon stimulates the capacity of mitochondria to synthetize citrulline when it is administered acutely to the whole animal or when it is added to isolated hepatocytes. It has been concluded in two papers[13,14] that the effect of glucagon on citrullinogenesis was due to the increase of ATP concentration due to an increased rate of mitochondrial respiration[16]. In those works, NAGA concentrations had not been measured. More recently, the effect of acute glucagon administration on the NAGA levels has been investigated in three different laboratories : Hengsens et al[17] and Cathelineau et al[18] found a significant increase of NAGA under the stimulation by glucagon, while, Titheradge and Haynes[19] were unable to obtain a modification of NAGA concentration.

In order to establish further arguments upon the role of NAGA concentration changes during the hormonal treatment we studied the stimulation by glucagon on rats submitted during several days to various diets in regard to their protein content. We postulated that the effect of glucagon would be more clear if the basal level of NAGA in control animals was low (as is obtained with an aprotidic diet[3]).

PREPARATION OF THE ANIMALS

Four groups of Wistar rats weighing 200 g were submitted to four kinds of diets during three consecutive days : a standard diet (19 % protein) a carbohydrate diet (0 % protein) a carbohydrate + lipids diet (0 % protein) and a hyperprotidic one (70 % protein). Then a group of animals was injected intraperitoneally with glucagon (NOVO) 20 µg for 100 g body weight and another group (control animals) received ClNa 9 % in the same manner. Fifteen minutes later, the animals were sacrificed, the livers removed and the mitochondria prepared according to the method of Hogeboom[20].

RESULTS

1. Variations in the capacity of mitochondria to synthetize citrulline

Citrulline synthesis by intact mitochondria was measured by a method earlier described[21], in three kinds of medium, the first with succinate 10 mM, the second with ATP, the third with ATP 2 mM + N-acetylglutamate 5mM. Citrulline synthesis in intact mitochondria is supported by the addition of HCO_3^-, NH_4^+ and ornithine.

ATP, which is necessary for the activity of carbamyl phosphate synthetase is supplied into the mitochondria in two manners : a) when mitochondria are incubated with an oxidative substrate, as succinate, ATP is synthetized into the mitochondria during the incubation. In those conditions, the rate of citrulline formation depends from the rate of ATP synthesis which in turn depends from the mitochondrial respiration ; b) when mitochondria are incubated without an oxidative substrate and ATP added to the medium +2-4 dinitrophenol + oligomycine it must enter into the mitochondria[22]. In those conditions the rate of citrulline formation does not depend from the mitochondrial respiration ; ATP concentrations are identical in all mitochondria.

It is observed (fig. 1) that :

a) In rats fed with a standard diet, citrulline synthesis is increased significantly in the two media, with succinate or with ATP, by the glucagon treatment of animals. But in the "ATP medium" the difference between control rats and glucagon treated rats is low (while significant).

b) In rats submitted during 3 days to a carbohydrate diet, citrulline synthesis is very low in control rats. Nevertheless in the "ATP medium" citrulline synthesis is higher than in "succinate medium". Citrulline synthesis is dramatically increased by glucagon treatment in the two media. It is enhanced 6 fold over in presence of succinate and 4 fold over in presence of ATP as compared to control rats. This dramatic increase of citrullinogenesis in mitochondria of treated rats when they are incubated with ATP establishes clearly that glucagon acts on another mechanism than oxidative phosphorylation.

Furthermore, the addition of NAGA into the "ATP medium" with mitochondria isolated from untreated rats mimics quite exactly the glucagon treatment. This addition has no or poor effect on mitochondria isolated from treated rats. Then, it must be assumed, that in this group of animals, the ability of mitochondria to synthetize citrulline is low in control rats because of low NAGA concentration which is restored partly by the glucagon treatment.

c) In rats with carbohydrates + lipids, citrulline synthesis of control rats is lower than in rats with a carbohydrate diet. Glucagon poorly stimulates the capacity of citrulline synthesis. Addition of NAGA in the ATP medium enhances dramatically citrullinogenesis. Then it is concluded that in this group of animals, NAGA concentration is lower than in the rat fed with a carbohydrate diet, and that glucagon treatment acts poorly on this concentration.

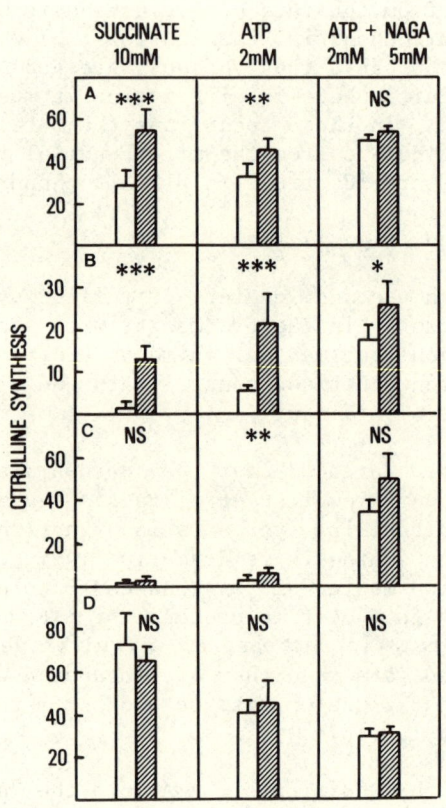

Fig. 1. Citrulline synthesis was measured according to the method of Charles et al[24]. Oligomycine 10 μg/ml and 2.4 dinitrophenol 0,04 mM were always added in the medium containing ATP or ATP + NAGA. ▢ controls and ▨ + glucagon. A/standard diet, B/carbohydrate diet, C/carbohydrates + lipids, D/hyperproditic diet. The results are expressed as nanomoles/m/mg protein. Each result is the mean ± SD from 5 or more experiments. ✱✱✱ P < 0,001 ✱✱ P < 0,01 ✱ P < 0,02

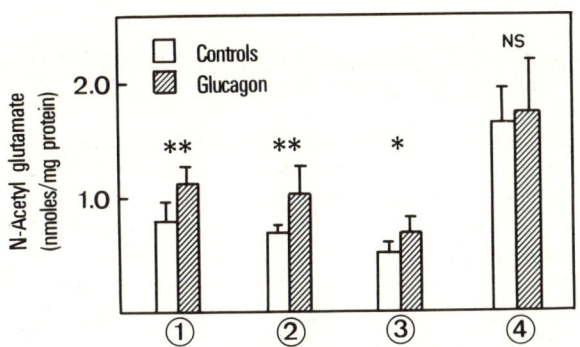

Fig. 2. N-acetyl glutamate concentrations in mitochondria from rats submitted to various diets : each result is the mean + SD from 5 experiments or more. **P<0,001 *P<0,01. ① standard diet ② carbohydrate diet ③ carbohydrates + lipids diet ④ 70 % protein.

d) In control rats submitted to a hyperprotidic diet, citrullinogeneses is quite 2 fold over the rate observed in control rats receiving a standard diet, in presence of succinate, lower in presence of ATP, and further lower in presence of NAGA. Glucagon treatment has no effect.

2) N-acetyl-glutamate concentration

NAGA was measured into mitochondria by the method of Meijer and Van Voerkom[4]. This method gives results higher than those of others[3,19]. Nevertheless the variations of NAGA concentrations are exactly those which may explain the variations of the capacity to synthetize citrulline (Fig. 2).

a) NAGA concentration is 0.80 + 0.16 in control rats fed with a standard diet. It is increased by 35 % in glucagon treated rats.
b) NAGA concentration is 0.66 + 0.10 in rats fed with a carbohydrate diet and increases by 50 % under glucagon treatment.
c) It is 0.54 + 0.09 with carbohydrates + lipids and it increases only by 27 % in treated rats.

d) In rats with hyperprotidic diets control rats NAGA concentrations are 1.68 ± 0.30 and are not modified by the treatment.

3. Relationship between N-acetyl glutamate concentrations and the capacity of mitochondria to synthetize citrulline

Citrullinogenesis shows a positive relation with the level of N-acetyl glutamate (fig. 3). Mixing the results of control and treated rats, this relation is found with rats submitted to a standard diet and rats fed with a carbohydrate diet. But the ability to synthetize citrulline remains lower in the last group of rats than in the rats fed with a standard food whatever the concentration of N-acetyl glutamate (probably, because of lower activity of CPS : see later).

4. In conclusion

a) When glucagon treatment has an effect on the NAGA levels, it is able to stimulate citrullinogenesis but when it has no effect or a poor effect on NAGA concentrations citrulline synthesis, in turn, is not stimulated.

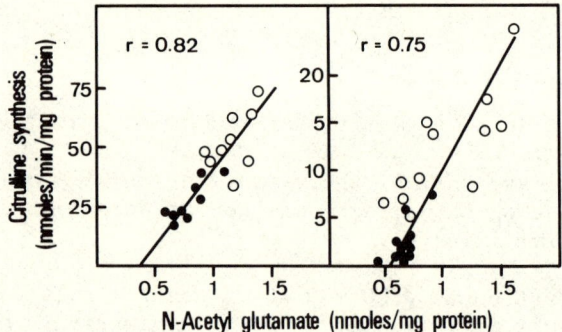

Fig. 3. Relations between the capacity of mitochondria to synthetize citrulline and the intramitochondiral concentration of N-acetyl glutamate ● : control rats ○ : glucagon treated rats. Left : standard diet. Right : carbohydrate diet.

b) Then NAGA seems to have an important role in the acute effect of glucagon on citrulline synthesis : nevertheless, the increase of ATP concentration by glucagon stimulation has been demonstrated 14,16,17,19 and cannot be underestimated. In vivo the respective role of these two events, increase of ATP and increase of N-acetyl glutamate probably depends from the basal metabolic conditions of the animal.

POSSIBLE MECHANISM FOR THE GLUCAGON EFFECT ON N-ACETYL GLUTAMATE CONCENTRATION

1. Variation of Enzymatic activities under the acute stimulation by glucagon

Ornithine transcarbamylase (OTC) activity is not very sensitive to the variations of protein content in the diet. Carbamylphosphate synthetase I (CPS) is slightly increased by hyperprotidic food and decreased by hypoprotidic food. But the N-acetyl glutamate synthetase activities varies widely with the changes of protein content in the diet (Table 1). Glucagon has no effect on these three activities.

2. Variations of the concentration of the substrates and effector of N-acetyl glutamate synthetase

a) Variations in glutamate and arginine concentrations. It has been reported that, in rats submitted during two hours to a hyperprotidic diet, glutamate concentrations are increased in the liver [6]. Glucagon treatment has not the same effect. In contrary, glutamate concentrations are significantly lowered in rats fed with the standard diet under the hormonal treatment (fig. 4) if we assume that 1 mg of mitochondrial protein represents 1 µl of mitochondrial volume, the concentration of glutamate is 1.4 mM in control rats and falls near 0.8 mM in glucagon treated rats. The same glutamate decrease under acute glucagon treatment has been reported by Siess et al[23], with the use of isolated hepatocytes. In rats submitted to carbohydrates and carbohydrates + lipids diets the decrease of glutamate concentration is not significant.

A slight decrease of the arginine concentration is found in rats fed with a standard and carbohydrate diets under hormonal stimulation (fig. 4). But in control rats the concentration is found to be near 40 M, much more higher than the Ka of arginine for N-acetyl glutamate synthetase which is 10 µM. Then, we assume that the slight increase observed by glucagon treatment is not able to have a role on the enhancement of NAGA concentration.

Table 1. Intramitochondrial enzymatic activities of urea cycle.

Enzymes	Animals	70% protein	standard diet	carbo hydrate diet	carbohydrates + lipids diet
CPS	controls	78+20	61+14	32+9	36+15
	+glucagon	69+15	51+15	38+13	42+7
OTC	controls	2500+540	2040+725	2130+611	1742+330
	+glucagon	2600+712	2450+1069	2107+426	2147+388
NAGA synthetase	controls	45.9+23.1	23.8+7.4	11.2+4.3	7.6+3.4
	+glucagon	46.0+16.1	23.1+3.6	10.5+3.7	8.1+3.7

Activities were measured on disrupted mitochondrias. CPS was measured according to the method of Brown and Cohen[25]. OTC by the method of Snodgrass[26] and N-acetyl glutamate synthetase by the method of Coude et al[27]. Results are the mean ± SD from five or more experiments. CPS and OTC activities are expressed as nanomoles/m/mg protein and N-acetyl glutamate synthetase as nanomoles/hour/mg protein.

b) <u>Variations in Acetyl CoA concentrations</u>. In an early work, Siess et al[23] had obtained an enhancement of acetyl CoA levels into mitochondria from isolated liver hepatocytes incubated with glucagon. The same increase is obtained in this work by the in vivo administration of the hormone. This increase is limited to the rats fed with standard and carbohydrate diets while it is not observed in rats with carbohydrates + lipids. The enhancement is 31 % for the first animal group and 34 % for the second. Control levels are 0.35 mM in rats with standard diet and 0.32 mM in rats with carbohydrate diets. They are increased respectively to 0.46 mM and 0.47 mM by glucagon treatment. The km of acetyl CoA for N-acetyl glutamate is reported to be 0,7 mM[12]. Then, it must be assumed that a variation of acetyl CoA concentration taking place under this molarity must be responsible for an important effect on N-acetyl glutamate synthetase velocity. It is of interest to remark that when acetyl CoA concentration is not modified by glucagon treatment as it is in the rats with carbohydrates + lipids, the concentration of N-acetyl glutamate is also poorly affected.

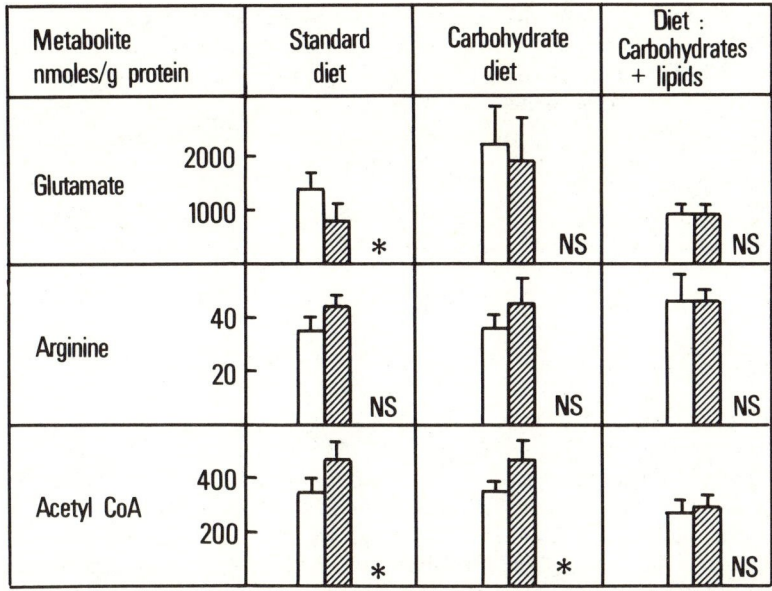

Fig. 4. Variations of intramitochondrial concentrations of the substrates or effector of N-acetyl glutamate synthetase activity. Glutamate and arginine were measured into sulfosalicylic extracts of just prepared mitochondria, by amino-acid chromatography. Acetyl CoA was assayed on HClO$_4$ extracts of just prepared mitochondria by the method of Williamson[28]. Each result is the mean ± SD of 5 experiments or more * P <0,01.

CONCLUSION

Several mechanisms have been envisaged by Hengsens et al[17] to explain the increase of N-acetyl glutamate concentration by glucagon. An increase of glutamate, of arginine, of acetyl CoA concentrations into the mitochondria or an inhibition of N-acetyl glutamate efflux out of the mitochondria. While this last possibility has not been explored in this work, we think that the increase of acetyl CoA must be the best hypothesis to explain the mechanism of glucagon action on NAGA concentration.

Glucagon treatment is known to stimulate also the mitochondrial step of gluconeogenesis, the pyruvate carboxylation[13,16]. It was, then justified to assume that an increase of ATP concentration under glucagon treatment was the cause of both the synchronous increases of pyruvate carboxylase and carbamyl phosphate synthetase velocities. With the results, we must postulate that the increase of acetyl CoA which is a potent activator of pyruvate carboxylase and the substrate of N-acetyl glutamate synthetase, must have also a key role in the glucagon regulation of gluconeogenesis and ureogenesis.

ACKNOWLEDGMENTS

This work was supported by a grant from I.N.S.E.R.M. n° 803004.

REFERENCES

1. K. Shigesada and M. Tatibana, Role of Acetyl glutamate in ureotelism, J. Biol. Chem. 246:5588 (1971).
2. T. Saheki, T. Katsunuma and M. Sase, Regulation of urea synthesis in rat liver, J. Biochem. 85:551 (1977).
3. J. D. Mc Givan, N. H. Bradford and J. Mendes-Mourao, The regulation of carbamoyl phosphate synthetase activity in rat liver mitochondria, Biochem. J. 154:415 (1976).
4. A. J. Meijer and G. M. Van Voerkom, Control of the rate of citrulline synthesis by short term changes in N-acetyl glutamate levels in isolated rat liver mitochondria, Febs Lett. 86:117 (1978).
5. A. Reglero, J. Rivas, J. Mendelson, R. Wallace and S. Grisolia, Deacylation and transacetylation of acetyl glutamate and acetylornithine in rat liver, Febs Lett. 81:13 (1977).
6. M. Tatibana and K. Shigesada, Regulation of urea biosynthesis by acetyl glutamate arginine system in :"The urea cycle". S. Grisolia, R. Baguena, and F. Mayor, Eds. John Wiley and sons, New York.
7. D. Rabier, L. Cathelineau, P. Briand and P. Kamoun, Propionate and succinate effects on acetylglutamate biosynthesis by rat liver mitochondria, Biochem. Biophys. Res. Comm. 91:456 (1979).
8. P. Stewart and M. Walser, Failure of the normal ureagenic response to amino-acids in organic acids loaded rats, J. Clin. Invest. 66:484 (1980).
9. K. Aoyagi, M. Mori and M. Tatibana, Inhibition of urea synthesis by Pent-4-enoate associated with decrease in N-acetyl-L-glutamate concentration in isolated hepatocytes. Biochim. Biophys. Acta 587:515 (1979).
10. F. X. Coudé, D. Rabier, L. Cathelineau, G. Grimber, P. Parvy and P. Kamoun, A mechanism for valproate induced hyperammonemia. Ped. Res. 15:974 (1981).
11. L. Cathelineau, D. Rabier, F. Petit and P. Kamoun, Physiological and hormonal variations of acetyl glutamate and citrullinogenesis in rat liver mitochondria, Enzyme (in press).

12. K. Shigesada and M. Tatibana, N-acetyl glutamate synthetase rat liver mitochondria, Eur. J. Biochem. 84:285 (1978).
13. R. K. Yamazaki and G.S. Graetz, Glucagon stimulation of citrulline formation in isolated hepatic mitochondria, Arch. Biochem. Biophys. 178:19 (1977).
14. J. Bryla, E. J. Harris and J. A. Plumb, The stimulatory effect of glucagon and dibutyryl cyclic AMP on ureogenesis and gluconeogenesis in relation to the mitochondrial ATP content. Febs Lett. 80:443 (1977).
15. K. G. Triebwasser and R. A. Freedland, The effect of glucagon on ureogenesis from ammonia in isolated mitochondria, Biochem. Biophys. Res. Comm. 76:1159 (1977).
16. R. K. Yamazaki, Glucagon stimulation of mitochondrial respiration, J. Biol. Chem. 250:7924 (1975).
17. H. E. Hengsens, A. J. Verhoeven and A. J. Meijer, The relationship between N-acetylglutamate and activity of carbamoyl phosphate synthetase (ammonia). The effect of glucagon, Eur. J. Biochem. 107:197 (1980).
18. L. Cathelineau, D. Rabier, F. Petit and P. Kamoun, Rôle de l'acetyl glutamate dans la stimulation de la citrullinogénèse par le glucagon, C.R. Acad. Sc. Paris, 291:625 (1980).
19. M. A. Titheradge and R. C. Haynes, The hormonal stimulation of ureogenesis in isolated hepatocytes through increases in mitochondrial ATP production, Arch. Biochem. Biophys. 201:44 (1980).
20. G. H. Hogeboom, Fractionation of cells components in animal tissues in "Methods in Enzymology", S. Colowick and N. Kaplain eds. Academic Press 1:16 (1955).
21. L. Cathelineau, F. Petit, F. X. Coudé and P. Kamoun, Effect of propionate and pyruvate on citrulline synthesis and ATP content in rat liver mitochondria, Biochem. Biophys. Res. Comm. 90:327 (1979).
22. W. J. Graafmans, R. Charles and J. M. Tager, Mitochondrial citrulline synthesis with exogenous ATP, Biochim. Biophys. Acta 153:916 (1968).
23. E. A. Siess, D.G. Broks, H. K. Lattke and O. H. Wieland, Effect of glucagon on metabolite compartmentation in isolated rat liver cells during gluconeogenesis from lactate, Biochem. J. 166:225 (1977).
24. R. Charles, J. Tager and E. Slater, Citrulline synthesis in rat liver mitochondria, Biochem. Biophys. Acta 131:29 (1967).
25. G. Brown and P. P. Cohen, Comparative biochemistry of urea synthesis, J. Biol. Chem. 234:1769 (1959).
26. P. J. Snodgrass, The effect of pH on the kinetics of human liver ornithine carbamyl transferase, Biochemistry 7:3047 (1968).
27. F.X. Coudé, L. Sweetman and W. L. Nyhan, Inhibition by propionyl coenzyme A of N-acetyl glutamate synthetase in rat liver mitochondria, J. Clin. Invest. 64:1544 (1979).

28. J. R. Williamson and B. E. Corkey, Assays of intermediates of the citric acid cycle and related compounds by fluorometic enzyme method, in :"Methods in Enzymology", S. Colowick and N. Kaplan Eds.

THE RELATION BETWEEN THE DEVELOPMENTAL TIMING OF BIRTH AND

DEVELOPMENTAL INCREASES IN UREA CYCLE ENZYMES

W.H. Lamers, P.G. Mooren, W. Oosterhuis, H. Lunstroo,
A. De Graaf and R. Charles

Department of Anatomy and Embryology, University of
Amsterdam, Mauritskade 61, 1092 AD Amsterdam
The Netherlands

SUMMARY

Experimental studies of rat development have shown that the first appearance and possibly part of the preweaning increase of many liver-specific enzymes is due to developmental maturation of the liver itself, but that the major increases in enzyme activity in the perinatal and preweaning periods are due to changing hormone levels. To establish the general nature of this type of developmental regulation we have studied enzyme activity changes during development of the spiny mouse, a closely related murine rodent which has apart from a different developmental timing of the moment of birth, a similar developmental timescale as the rat. Comparison of these altricial (rat) and precocial (spiny mouse) modes of development shows that the first appearance of enzyme activity in the liver (but also in the pancreas and small intestine) is developmental stage-specific, but that the first major increase in enzyme activities awaits the (differently timed) perinatal period. The preweaning increase in enzyme activity appears to be at least partly a developmental stage-specific event. These comparative studies therefore confirm the regulatory model of enzyme accumulation which was derived from experimental studies in the rat. Developmental profiles of human liver enzymes resemble those of the spiny mouse more than they do those of the rat.

INTRODUCTION

The developmental profile of enzymes describes the acquisition of enzyme activity as a function of the developmental age of the

organism involved. In rat liver, such enzymic profiles typically
show changes in enzyme activity in the perinatal and in the preweaning period (1,2). Similar profiles have been found in other
organs which are vital for immediate postnatal survival and which
are, as the liver, derived from the embryonic foregut, i.e. the
pancreas (3), the small intestine (4) and the lung (5). The
question was then raised whether the first appearance and to what
extent the subsequent increases in enzyme activity were to be
attributed to exogenous factors like changing hormone levels and
the occurrence of birth, or to endogenous factors in the relevant
organs, i.e. to the developmental stage.

In previous studies (2,6-9) we have shown that developmental
changes of enzyme activity in rat liver are related with developmental changes in the circulating levels of corticosterone and
glucagon (Fig. 1). However, hypophysectomy of the fetus did not
prevent the perinatal appearance of enzyme activity (7,8).
Therefore, it was proposed that the first appearance of liver-
specific enzyme activity in the developing rat was due to developmental maturation of the liver itself and that subsequent increases
in enzyme activity were mainly due to changing hormone levels (2).
However, since adrenalectomy of sucklings attenuated but did not
prevent the preweaning increase in enzyme activities (8), this
phase of the developmental profile could, at least in part, be

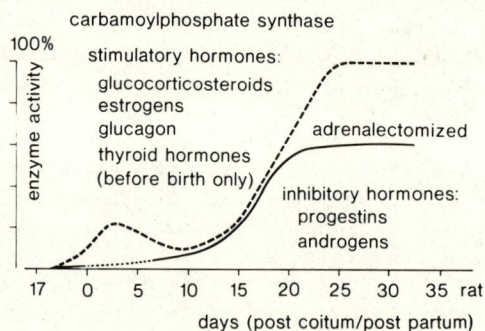

Fig. 1. Developmental profile of CPS activity in rat liver. The
enzyme is first detectable at 4 days ante partum and
shows typical perinatal and preweaning increases in
activity. By comparing the effects of hormone administration on enzyme activity with the developmental
profiles of these hormones (not shown) it can be de-
duced that only corticosterone and glucagon have major
effects on enzyme activity in this period.

developmental stage-specific, too. The process of birth seemed to affect the developmental profile of enzymes mainly by associated changes of hormone levels (2) and did not seem to affect hormone-induced enzyme accumulation (7). The above model of perinatal regulation of enzyme levels was derived from experimental studies in the rat which involved highly unphysiological circumstances for the fetuses and neonates involved. Therefore, we have looked for an animal model that is comparable to the rat model in its developmental timescale, but has a different developmental timing of birth. The spiny mouse, Acomys cahirinus seems to meet the requirements for such a model. The spiny mouse is a semi-desert murine rodent that has a relatively long gestation period (approx. 39 days), associated with a small (1-4) litter size (10). The neonates show an advanced stage of development at birth, having a fur coat, open eyes and ears, and are capable of locomotion and thermoregulation. Self-feeding begins after the first postnatal week (10).

The late intra-uterine developmental stages of the spiny mouse therefore seem to be comparable to the early extra-uterine developmental stages of the rat. Comparison of the developmental profiles of rat and spiny mouse liver enzymes may therefore answer the questions a) whether or not the first appearance of enzyme activity is developmental stage-specific and b) to what extent the developmental timing of birth and weaning determine the developmental profiles of enzyme activities. In this study we show that in both species, the first appearance of enzyme activity in the liver is developmental stage-specific, but that the first major increase in enzyme activity awaits the perinatal period. The preweaning increase in enzyme activity may, in part, be developmental stage-specific. Furthermore, it will be shown that this type of developmental regulation of enzyme activity is not confined to the liver, but is also present in other organs derived from the caudal end of the embryonic foregut, i.e. lungs, small intestine and pancreas.

MATERIAL AND METHODS

Animals

The rats used in this study were obtained from the TNO animal farm is Zeist (The Netherlands). Fetal age was determined by dated matings. Birth usually occurred at the beginning of day 22.

The spiny mice used in this study were part of a laboratory colony begun in 1979 from 3 breeding couples which were a kind gift of Prof. Dr. A.E. Renold, University of Geneva. The animals were fed pelleted rat chow (Hope Farms diet R.M.H.-B) and sunflower seeds. As the exact day of conception was usually not known, fetal age was read from a graph relating body weight to gestational age (10, cf. 11).

Assays

Carbamoyl-phosphate synthetase (CPS), ornithine transcarbamoylase (OTC), arginase (ARG), glucose-6-phosphatase (G6P) and tyrosine aminotransferase (TAT) activities and CPS and DNA content in liver were estimated as described before (6, 8, 12). Phosphatidate phosphatase and cholinephosphate cytidyl transferase activities in lung tissue were estimated as described (13, 14). Sucrase and lactase activities in small intestine were estimated as described (15), except that the glucose oxidase-peroxidase solution was heated for 1 hour at 56° C to remove residual sucrase activity and that the development of colour was stopped after 10 min. by heating

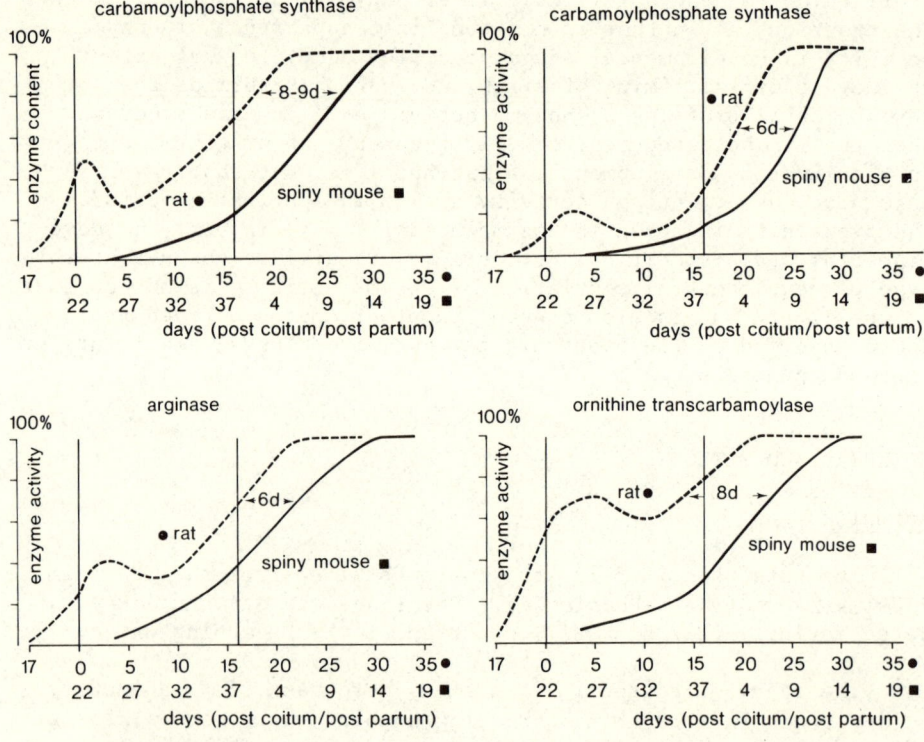

Fig. 2. Developmental profiles of rat and spiny mouse urea cycle enzymes.

in a boiling waterbath. Amylase activity in the pancreas was estimated as described (16).

Graphs

Activities (contents) are depicted as best-fitting curves of averages of 4-8 estimates every day in the perinatal period, every other day in the second and third postnatal week and twice weekly thereafter. The estimates of CPS, OTC, ARG, TAT and G6P activities and DNA content in rats are derived from earlier studies (6, 8). The activities of sucrase and lactase in developing rats are derived from (4), while the activities of amylase in rats are derived from (3). Activities were normalized to a fraction of 100 %, this being the highest average activity measured in the period studied. Figs. 2-7 : rat (------) ; spiny mouse (———).

RESULTS

The rat urea cycle enzymes CPS, OTC and ARG are first detectable at 17-18 days post coitum and show characteristic increases in enzyme activity in the perinatal period ("late fetal" and "neonatal" phases) and in the third postnatal week ("late suckling" phase (1)), when maximal activities are reached (Fig. 2). The spiny mouse urea cycle enzymes are first detectable at 25-26 days post coitum and gradually increase in activity thereafter to reach maximal levels in the second postnatal week (Fig. 2). In the case of CPS, enzyme activity and enzyme content follow similar courses (Fig. 2). Comparison of the developmental profiles shows that in the spiny mouse, in contrast to the rat, no distinct perinatal phase can be discerned. Spiny mice start eating solid food in their second postnatal week. Changes in enzyme activity in this period can therefore be compared to those in the preweaning period of the rat. Thus, it seems that the distinct perinatal and preweaning phases of enzyme accumulation in the rat have fused in the spiny mouse. Furthermore, it can be seen that both the first appearance of, and the preweaning increase in enzyme activities in the spiny mouse lag approx. 1 week behind those of the rat.

Another example of this 6-8 days lag is provided by the developmental profile of the DNA content of liver (Fig. 3). DNA content of rat liver decreases in the prenatal period due to the maturation and disappearance of the erythroid cell population and due to an increase in the size of the hepatocytes (17). Upon weaning a further increase in hepatocyte cell size is seen. In fig. 3 it can be seen that the developmental profile of liver-DNA content in the

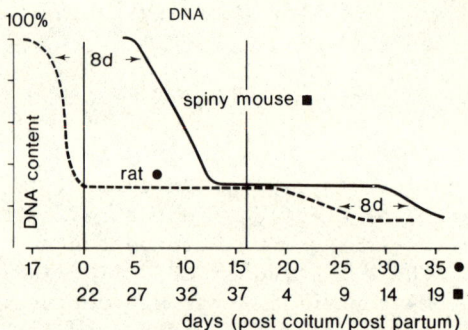

Fig. 3. Developmental profile of rat and spiny mouse DNA content of the liver.

spiny mouse follows a course similar to that of the rat albeit with a 8-day lag.

To further support the hypothesis of fusion of the distinct perinatal and preweaning phases of enzyme profiles of rat liver during spiny mouse development, 2 enzymes which show dramatic increases in activity in the neonatal period of the rat, viz. TAT and G6P were studied. In Fig. 4, it can be seen that while both enzymes reach developmental maxima in activity shortly after birth in both species, the developmental profiles of the spiny mouse enzymes are monophasic instead of distinctly biphasic as in the rat. The perinatal increase in enzyme activities in the spiny

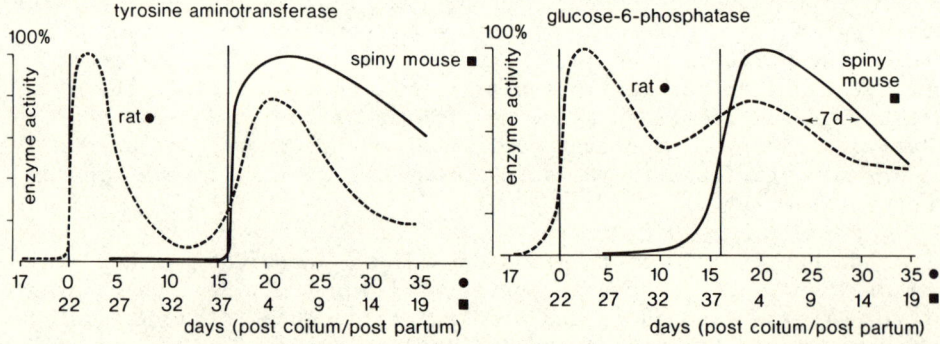

Fig. 4. Developmental profiles of rat and spiny mouse enzymes which reach maximal developmental levels in the perinatal period.

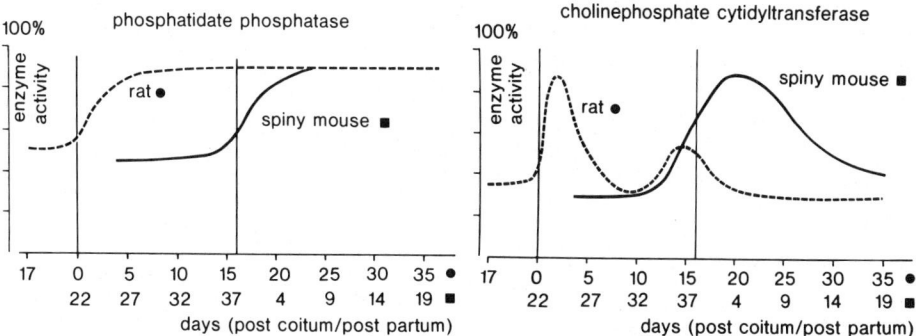

Fig. 5. Developmental profiles of rat and spiny mouse lung enzymes. Note the 16-day lag period in the perinatal phases of the profiles of the spiny mouse compared to the rat and the monophasic (right panel) developmental profile in the spiny mouse compared to the biphasic profile in the rat.

mouse therefore lags 2 1/2 weeks behind the perinatal increase in the rat. The preweaning phase of the developmental profile cannot be discerned from the perinatal increase in the spiny mouse. However, the weaning phase of the developmental profile of both enzymes, characterized by a decline in enzyme activity, again lags 7-10 days behind in the spiny mouse compared to the rat.

To find out whether these characteristic differences in developmental profiles of rat and spiny mouse liver also exist in other organs of endodermal origin, developmental profiles of a few organotypic enzymes of lung (Fig. 5), small intestine (Fig. 6) and pancreas (Fig. 7) were established. From Figs. 5-7 it can be concluded that, when applicable, the first appearance of enzyme activity in the spiny mouse lags approx. one week behind that of the rat, that the first major increase in enzyme activity awaits the perinatal period and that the distinct perinatal and preweaning phases of enzyme activity change in the rat seem to have fused into a single, more extended phase in the spiny mouse.

DISCUSSION

Rats and spiny mice can be considered as, respectively, altricial and precocial representatives of the same murinoid subfamily of myomorphic rodents (20). Although altrical and precocial patterns of development are wide-spread, their simultaneous existence within one taxonomic group is extremely rare. The only other well-

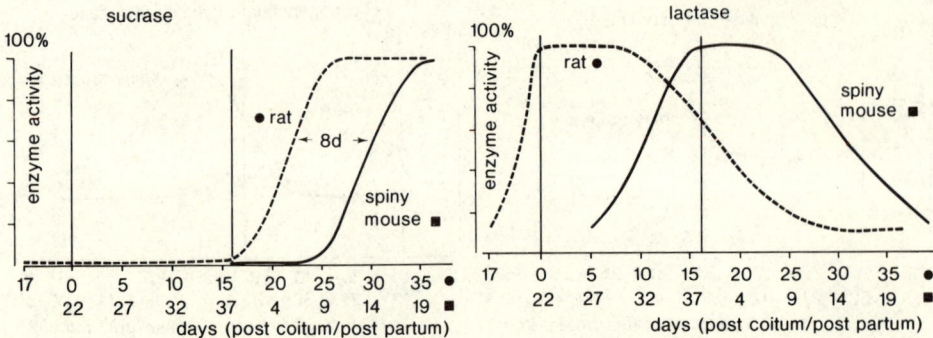

Fig. 6. Developmental profiles of rat and spiny mouse small-intestinal enzymes. Lactase, which has maximal activity at birth, and sucrase, which appears only in the preweaning period, show the characteristic lag time in the spiny mouse compared to the rat of 2 and 1 weeks, respectively.

known example is found in the leporinoid subfamily of lagomorphic rodents : the rabbit (Oryctolagus) and the hare (Lepus), respectively. Such greatly different patterns of development in such closely related species offer good opportunities for a comparative approach of the analysis of determinants in developmental processes.

In contrast to the rat (18), embryonic development of the spiny mouse has not been extensively studied. The time course of development of the rat and the spiny mouse seems to concur

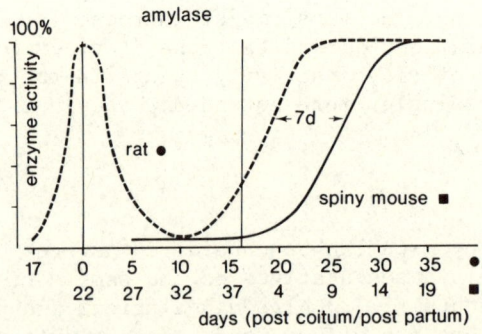

Fig. 7. Developmental profiles of rat and spiny mouse pancreatic amylase. Note the one-week lag period in the spiny mouse compared to the rat.

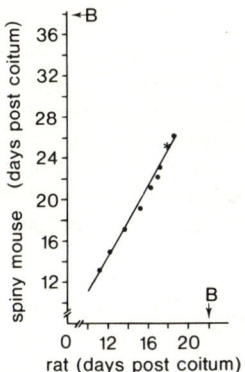

Fig. 8. Time course of embryonic development in the rat and in the spiny mouse. Horizontally, rat postcoital days and vertically spiny mouse postcoital days are depicted. At 11 days post coitum, the developmental stages of both species are comparable. Thereafter, the spiny mouse develops at 60 % of the rate of the rat, so that a rat at 18 days post coitum (i.e. when organotypic enzyme activity is first detectable) has the same developmental stage as a spiny mouse of 25 days post coitum (*). Data are derived from (10).

during the first 10 postcoital days (cf. 18,19). However, in Fig. 8, it can be seen that thereafter, i.e. between the beginning of organogenesis (day 11 post coitum) and the first appearance of organotypic enzymes (day 18 and 25 post coitum in the rat and the spiny mouse, respectively), development in the spiny mouse proceeds at approx. 60 % of that of the rat, thus explaining the one-week lag period. So far one set of studies dealing with insulin accumulation in the fetal pancreas of rat and spiny mouse seems to support such a lag period in the developmental stage-specific appearance of organotypic proteins in the spiny mouse (20,21). If the 3-day difference in the time course of organogenesis as well as the moment of birth (cf. 10,21) in this study is taken into account, the lag period is again 6-8 days.

If this one-week retardation in development is taken into account, a concurrent time course of development from the first appearance of organotypic enzyme activity to the preweaning increase appears to exist in the rat and in the spiny mouse. This hypothesis is supported by a concurrent time course of development of the skin, sense organs, locomotor system and brain in this period (10) ; see also (22). However, those changes in organotypic enzyme activity which are related to the perinatal period, do not follow a concurrent time course in both species, but await the process of birth which occurs 16 days (approx. 9 developmental days) later in the spiny mouse than in the rat. Thus, the first appearance of enzyme activity in the fetus as well as (at least part (cf. 8)) of the preweaning increase in enzyme activity in the suckling seem to be developmental stage-specific, while the perinatal increases in enzyme activity seem to depend on the timing of birth with the associated changes in hormone levels. Currently, we are measuring the serum levels of adrenal, thyroidal and pancreatic hormones in the spiny mouse.

The fact that the developmental increases in enzyme activity in the perinatal period start at a different developmental time point, but follow similar time courses, appears to imply that the process of birth extends over a greater time period than that of the expulsion from the uterus alone. In fact, a successful survival of birth may require both anticipatory (prenatal) and adaptive (postnatal) changes in organ function. Furthermore, in view of the fact that those organs which show most typical perinatal changes in enzyme activity, as lung, liver, pancreas and small intestine, are vital for postnatal survival, it implies that rats are born at the earliest feasible developmental stage and that in spiny mice those vital organs are ready to start functioning some time before the process of birth starts. This view is supported by the fact that successful survival of premature birth of the rat never precedes anticipated birth by more than one day. Premature birth in the spiny mouse is not a rare event though, and can be easily recognized by low birth weight combined with closed eyes (see also (22)).

One must consider, of course, the applicability of these models to the human situation. Human neonates have mainly altricial characteristics at birth, despite a prolonged fetal phase of development (second and third trimester) (cf. 23). It is therefore tempting to speculate that human development will show characteristics of both the altricial and precocial modes of murine development. Urea cycle and other organotypic enzymes are first detectable in the liver at the end of the first trimester of pregnancy (24), i.e. in a developmental stage which is morphologically comparable to that of a rat at 16-17 days post coitum. Most enzyme activities gradually increase to approx. 50 % of their adult value during the next two trimesters. Postnatal developmental profiles are, understandibly, rather incomplete, but suggest a gradual increase during the first postnatal months (25). Some enzymes like TAT, have very low activities in the prenatal period only (24). The time course of the profiles is therefore more comparable to those of the spiny mouse than to those of the rat. Human ontogenesis may therefore be classified as a third model with characteristics of both altricial and precocial modes of development.

These models provide good examples of heterochrony in mammals, i.e. they point to changes in the onset or timing of development so that the appearance or rate of development of a feature in one species is accelerated or retarded relatively to the appearance or rate of development of the same feature in another, related species during phylogeny (26). One can wonder in which species the rate of development changes, as the precocial developmental profiles of enzymes are similar to those of the metamorphic period of Anurans, while the altricial developmental profiles bear a resemblance to that of a neotene Urodele (27). One is therefore tempted to speculate that birth in the rat has to be considered as a comparatively precocious event involving the abolishment of the fetal period proper. This precocious birth necessitates a precocious maturation of organs indispensible for postnatal survival and thus leads to a clearly discernable perinatal phase in developmental profiles of organotypic enzymes.

REFERENCES

1. O. Greengard, Biochemical actions of hormones, Vol. 1., Academic Press, New York (1970).
2. W. H. Lamers and P. G. Mooren, Mech. Ageing Dev. 15:93 (1981).
3. T. Takeuchi, M. Ogawa, C. Furihata, T. Kawashi and T. Sugimura, Biochim. Biophys. Acta 497:657 (1977).
4. S.J. Henning and N. Kretchmer, Enzyme 15:3 (1973).
5. W. Oosterhuis et al., to be published.
6. W. H. Lamers and P. G. Mooren, Biol. Neonate 37:113 (1980).
7. W. H. Lamers and P. G. Mooren, Biol. Neonate 37:264 (1980).
8. W. H. Lamers and P. G. Mooren, Mech. Ageing Dev. 15:77 (1981).
9. W. H. Lamers and P. G. Mooren, Biol. Neonate, in press.
10. F. Dieterlen, Z. Saugetiere 28:193 (1962).
11. A. W. A. Gonzalez, Anat. Rec. 52:117 (1932).
12. R. Charles, A. de Graaf, A. F. Moorman, Biochim. Biophys. Acta 629:36 (1980).
13. W. M. Maniscalco, C. M. Wilson, I. Gross, L. Gobran, S.A. Rooney, J. B. Warshaw, Biochim. Biophys. Acta 530:333 (1978).
14. S. A. Rooney, T. S. Wai-Lee, L. Gobran, E. K. Motoyama, Biochim. Biophys.Acta 431:447 (1976).
15. T. T. Ngo and H. M. Lenhoff, Anal. Biochem. 105:389 (1980).
16. M. Caska, K. Birath and B. Brown, Clin. Chim. Acta 26:437 (1969).
17. O. Greengard, M. Federman, W.E. Knox, J. Cell Biol. 52:261 (1972).
18. F. Keibel, "Normen Tafeln zur Entwicklungsgeschichte der Wirbeltiere", Heft 15, Gustav Fisher Verlag, Jena (1937).
19. W. Ruch, Rev. Suisse Zool. 74:566 (1967).
20. L.B. Rall, R. L. Pictet, R. H. Williams, W. J. Rutter, Proc. Nat. Acad. Sci. USA 70:3478 (1973).
21. M. De Gasparo, Gen. Comp. Endocrinol. 41:499 (1980).
22. E. Flückiger and P. Operschall, Rev. Suisse Zool. 69:297 (1962).
23. L. Bolk, "Das Problem der Menschwerdung", Gustav Fisher Verlag. Jena (1926).

24. O. Greengard, Pediat. Res. 11:669 (1977).
25. N. C. R. Räihä, "The urea cycle", Wiley, New York (1975).
26. S. J. Gould, "Ontogeny and Phylogeny", Harvard University Press, Cambridge, Ma, USA (1977).
27. W. H. Lamers et al., J. Exp. Zool., accepted for publication.

STUDIES ON THE ENZYMES OF UREA CYCLE INTERMEDIATES IN NORMAL AND

INFARCTED MYOCARDIAL TISSUE OF RAT

B. Sadasivudu, M. Swamy and G.N. Rao

Department of Biochemistry, Osmania Medical College
Hyderabad A. P., India

SUMMARY

   The high content of glutamine and the high activity of aspartic acid aminotransferase signifies that myocardium has an appreciable amount of nitrogenous metabolism. It is well established that in extra hepatic tissues the ammonia is mainly disposed of in the form of glutamine. It has been demonstrated by Russian workers that the urea content in coronary sinus blood was high in patients suffering from varying degrees of myocardial infarction. A study of three enzymes belonging to urea cycle, namely, argininosuccinate synthetase, argininosuccinase and arginase has been made in the myocardium of normal rats and in rats subjected to myocardial infarction. It was observed that the myocardial tissue did not exhibit any activity of argininosuccinate synthetase while appreciable activity of argininosuccinase and arginase was observed. The arginase activity was significantly higher in the left side of the heart than in the right side. The activity of this enzyme was found to be 1 1/2 times more in the infarcted tissue. Though the argininosuccinase activity was almost the same in both sides of the heart it was, however, found to be higher in the infarcted tissue. A similar difference in the activity of ornithine aminotransferase was found between the left and right sides of the myocardium and the infarcted tissue appears to have a significant metabolism of arginine. Its enhancement in infarcted tissue together with increased activity of ornithine aminotransferase may help in the production of glutamate and in the removal of ammonia.

EFFECTS OF ARGININE-FREE MEALS ON UREAGENESIS IN CATS

P.M. Stewart, M. Batshaw, D. Valle and M. Walser

Johns Hopkins University School of Medicine, Baltimore
Maryland, U.S.A.

SUMMARY

Cats given a single arginine-free meal have been reported to develop severe hyperammonemia, attributed to impaired function of ornithine aminotransferase (OAT). We found that cats which developed hyperammonemia following an arginine-free meal had low hepatic ornithine levels. However, the average sum of hepatic ornithine plus arginine plus citrulline rose, indicating that some ornithine synthesis via OAT took place, and hyperammonemia failed to occur in cats with higher hepatic ornithine levels. OAT activity and kinetic constants were comparable to values reported in the rat. Furthermore, dietary supplementation with ornithine caused only occasional and transient hyperornithinemia. Thus, OAT can function in the cat. The K_a of N-acetylglutamate (AGA) synthetase for arginine was five times higher in cats than in rats, but AGA content and citrullinogenesis by intact mitochondria were the same following arginine-free or arginine-containing meals. Other kinetic parameters of AGA synthetase and carbamylphosphate synthetase were similar to values in the rat. We conclude that low levels of hepatic ornithine are probably responsible for making some cats susceptible to hyperammonemia following this stimulus.

DYNAMISM OF RAT LIVER ORNITHINE METABOLISMS IN RELATION TO DIETARY

HIGH-PROTEIN STIMULATION OF THE UREA CYCLE

Takeo Matsuzawa, Naofumi Sugimoto, and Isao Ishiguro

Department of Biochemistry
School of Medicine
Fujita-Gakuen University, Toyoake, Aichi 470-11, Japan

INTRODUCTION

Reports on hereditary deficiencies of the urea cycle enzyme, e.g. hyperammonemia with homocitrullinuria[1] and ornithine carbamoyl transferase(OCT) deficiency[2], led us to the view that ornithine transport into the mitochondrial matrix represents an important factor controlling citrulline synthesis and plays a role in the regulation of carbamoyl phosphate synthesis[3]. We have long been interested in whether ornithine is derived from arginine or proline, i.e., whether ornithine, produced from arginine in the cytosol, is transported into the mitochondria, or whether ornithine is derived from proline via Δ^1-pyrroline-5-carboxylate(P5C) in the mitochondrial matrix. Once the urea cycle is activated, ornithine is continuously transported into the mitochondrial matrix from cytosol and released from mitochondria into the cytosol("ornithine flux"); this transport system is composed of a citrulline-exchange transporter and an H^+ antiporter[4,5]. We now present some *in vitro* mitochondrial studies on ornithine transport and citrulline formation from proline, and an analysis of *in vivo* arginine and ornithine metabolisms. This investigation was undertaken to elucidate the characteristics of "ornithine flux" and its relation to dietary stimulation of urea cycle.

MATERIALS AND METHODS

1. Assay of Ornithine Transport Activity

In a final volume of 1.9 ml, the assay mixture contained 1.0 ml of isolation medium, 10 mM HEPES buffer, 5 mM L-norvaline, 1mM L-ornithine, 0.1 µCi of L-[U-^{14}C]ornithine(200 mCi/mmol), and 10-15 mg

protein of the inner membrane inverted vesicles. It was incubated at
0°C or 37°C for 10 min, and ornithine was allowed to permeate into
vesicles. Then 10 μmol(0.1 ml) of counter-exchange substrate(succinate or citrulline) was added, and the mixture was further incubated
at 0°C or 37°C for 30 min. The reaction was stopped by chilling in
ice, and the mixture was centrifuged at 56000 X $\underline{g}$ for 30 min at 4°C.
The packed vesicles were washed several times with isolation medium,
suspended in 1.5 ml of H_2O, poured into a 10 ml liquid scintillator
(PPO, 2.75 g; POPOP, 0.05 g; Triton X-100, 166.5 ml; toluene, 333.5
ml), and the radioactivity was counted.

2. Reaction System for Citrulline Formation from Proline

In a final volume of 2 ml, the mixture contained 0.3 M mannitol,
20 mM Tris-HCl buffer(pH 7.5), 5 mM potassium phosphate buffer(pH
7.5), 10 mM KCl or NH_4Cl, 5 mM $NaHCO_3$, 2.5 mM $MgCl_2$, 5 mM N-acetyl
glutamate, 0.2 mM EDTA, 10 mM L-proline and about 5 mg of mitochondria as protein. After 7 min, 37°C temperature equilibration, the
reaction was started by adding L-proline; it was continued for 15 min
under constant shaking and stopped by adding 2 ml of 10% $HClO_4$. The
neutralized supernatant was subjected to analysis of the metabolites.

3. Experiments on In Vivo Arginine and Ornithine Metabolism

Details regarding the animals, feeding conditions, and administration and analyses of [^{14}C]arginine and [^3H]proline have been described elsewhere[6]. The livers were freeze-clamped[7], homogenized in
5% $HClO_4$ and centrifuged at low speed. The neutralized supernatant
was subjected to analysis of the metabolites.

4. Metabolite Assays

Citrulline was determined by the method of Archibald[8] after
prior urea hydrolysis by urease; ornithine and P5C by our method[9]
and urea with diacetylmonooxime thiosemicarbazid[10]; glutamate and
aspartate by the enzymatic method[11,12]; and acetoacetate and 3-hydroxybutyrate with 3-hydroxybutyrate dehydrogenase[13,14].

RESULTS AND DISCUSSION

1. Ornithine Transport Systems

Bradford and McGivan[5], who used citrulline-loaded mitochondria,
were the first to present evidence for an ornithine-citrulline antiporter in rat liver mitochondria. Our present study yielded evidence
for the existence of two types of ornithine transport systems, an
ornithine-H^+ and an ornithine-citrulline exchange system; it was uncovered by salicylate in ornithine-loaded mitochondrial inner membrane vesicles. The uptake of labeled ornithine by the vesicles was
quite rapid and proportional to the amount of membrane protein added;

Table 1. Release of Pre-loaded Ornithine from Inverted
Vesicles with Counter-exchange Substrates

Additions	% Ornithine remaining in the vesicles	
	0°C	37°C
Rat liver:		
none, (n=4)		
none	100	100
+ succinate	68.6 $\pm$ 1.0	65.4 $\pm$ 5.2
+ citrulline	76.5 $\pm$ 4.0	62.9 $\pm$ 5.3
5 mM salicylate, (n=3)		
none	100	100
+ succinate	72.1 $\pm$ 7.0	82.2 $\pm$ 1.0*
+ citrulline	77.5 $\pm$ 2.7	66.8 $\pm$ 4.3

*$|t| > t_{0.05}$ significant. Inverted inner membrane vesicles, prepared by the method of Chan et al.[18], were incubated for 10 min with labeled ornithine at the indicated temperatures. For the exchange reaction, counter-exchange substrate (succinate or citrulline) was added, followed by further 30-min incubation.

in rat liver, ornithine permeation(nmol) per mg of protein was 4.24 $\pm$ 0.05 (n=11). We posit that the rapid permeation of ornithine into the vesicles is mediated by an ornithine-H^+ exchange carrier, because it was clearly stimulated when 5 mM salicylate was present during 37°C incubation(Table 1). In the presence of succinate, salicylate appeared to enhance the efflux of protons from the vesicles; 5 mM salicylate inhibited the exchange of preloaded ornithine with succinate. On the other hand, salicylate had no effect on the release of preloaded ornithine when the counter-exchange substrate was citrulline(Table 1). These observations lead us to suggest that there are two types of ornithine transport systems in the inner membrane of rat liver mitochondria; an ornithine-H^+ and an ornithine-citrulline exchange system.

2. Citrulline Synthesis from Proline in Isolated Mitochondria

We have data indicating that NH_4^+ ion stimulates both proline oxidation[15] and P5C accumulation in isolated rat liver mitochondria, and that the addition of ADP enhances aspartate formation which is accompanied by the rapid disappearance of P5C accumulation. Citrulline formation was markedly faster than glutamate synthesis; it reached a plateau within 30 to 45 min(Fig. 1A). Coupled mitochondria exhibited maximum citrulline formation activity in the presence of

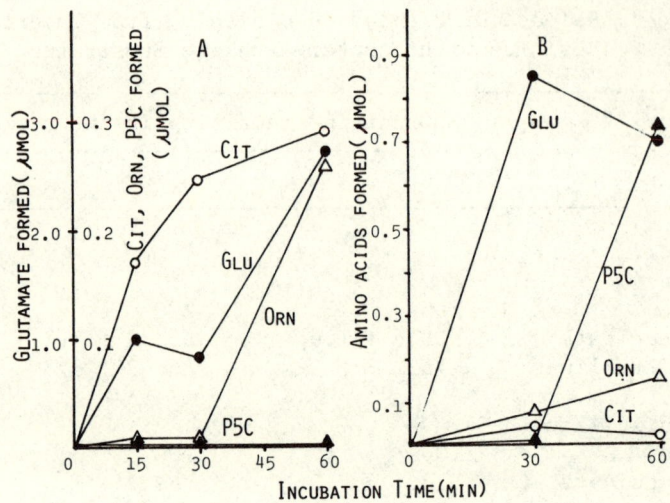

Fig. 1. In vitro citrulline synthesis from proline, using isolated rat liver mitochondria. Mitochondria were prepared by routine procedures and the overlying lysosomes were removed. For details regarding the reaction system, see Materials and Methods. The ordinates indicate differences in amino acid formation (formation with NH_4Cl minus formation with KCl). A, aged Shizuoka Wistar rats; B, young JCR Wistar rats.

NH_4Cl and ADP; this activity was completely inhibited by canaline, a specific inhibitor of ornithine oxoacid aminotransferase. The amount of citrulline formed was at least one tenth less than synthesized glutamate (Fig. 1A). Smith et al.[16] have demonstrated ornithine formation from proline, using isolated rat liver mitochondria, however, regulatory and metabolic characterization remained to be done. We found that this pathway is stimulated by NH_4^+ ion, but is critically limited by mitochondrial carbamoyl phosphate synthetase activity (Fig. 1B). The question of whether this pathway functions in vivo still needs to be answered.

3. Diet-induced Changes in the Level of Liver Metabolites

We first determined the hepatic ornithine level and the subcellular distribution, using a specific enzymatic method[9] and the freeze-clamping technique[7] (Table 2). The hepatic ornithine level increased proportional to the dietary protein content[6]. Cytosol yielded about 80%, mitochondria 10% ornithine. The calculated approximate ornithine concentrations in cytosol and the mitochondrial matrix are shown in Table 2. At pH 7.5 of OCT, Km for ornithine was 1-2 mM; for carbamoyl phosphate it was 10-20 μM[19]. On the other hand, Km values at pH 7.5 of ornithine oxoacid aminotransferase were

Table 2. Ornithine Level in Rat Liver and Dietary Protein Content

Dietary protein content	Ornithine[a] (nmol/g liver)	Calculated ornithine level[b]	
		In cytosol	In mitochondria
		(mM)	
5% casein	284.1 ± 81.5	0.42	0.51
25% casein	397.0 ± 54.0	0.59	0.71
70% casein	765.1 ± 193.3	1.13	1.37

[a] Each value represents the mean ± SD of 4 rats.
[b] In cytosol: the measured mean ornithine level was multiplied by 0.8 and divided by 0.54; in mitochondrial matrix: the measured mean ornithine level was multiplied by 0.1 and divided by 0.056.

1.4 mM for ornithine and 0.26 mM for 2-oxoglutarate(unpublished results). Thus for the OCT reaction, an intramitochondrial ornithine level exceeding 1 mM should be maintained; because the intramitochondrial level depends on the cytosol level, the hepatic ornithine level must be raised to above 0.56 μmol/g liver. At 4 and 28 hr, the early transitional phase after the change-over from a 25% to a 70% casein diet, we observed marked accumulation of the intermediates(citrulline and ornithine); thereafter, they decreased, reaching new steady-state levels by 52 hr(Fig. 2). The new steady-state levels of ornithine and glutamate were higher than in the controls(25% casein diet); the new steady-state citrulline level was as low as in the control. The levels of these two groups of amino acids at 4 and 28 hr, exhibited a see-saw pattern; the change in the citrulline level followed on those of ornithine and glutamate. During the transitional phase (4-28 hr after the dietary change), 2-oxoglutarate decreased; urea, ammonia, and 3-hydroxybutyrate increased; P5C was unchanged[6]. The see-saw changes in the level of intermediates support the hypothesis that the ornithine transport system for the urea cycle is an ornithine-citrulline exchange system. Our observations on the diet-induced changes in the metabolite levels lead us to posit that upon dietary stimulation of the urea cycle, the mitochondrial part is activated first, and that this is followed by stimulation of the cytosol half-cycle.

4. Diet-induced Changes in Arginine and Ornithine Metabolism

Based on the assumption that hepatic metabolite levels are constant during a given 3-min period, we attempted to elucidate the kinetic behavior of ornithine metabolism, using pulse-labeling data obtained with L-[U-^{14}C]arginine and L-[2,3,4,5-^{3}H]proline. An es-

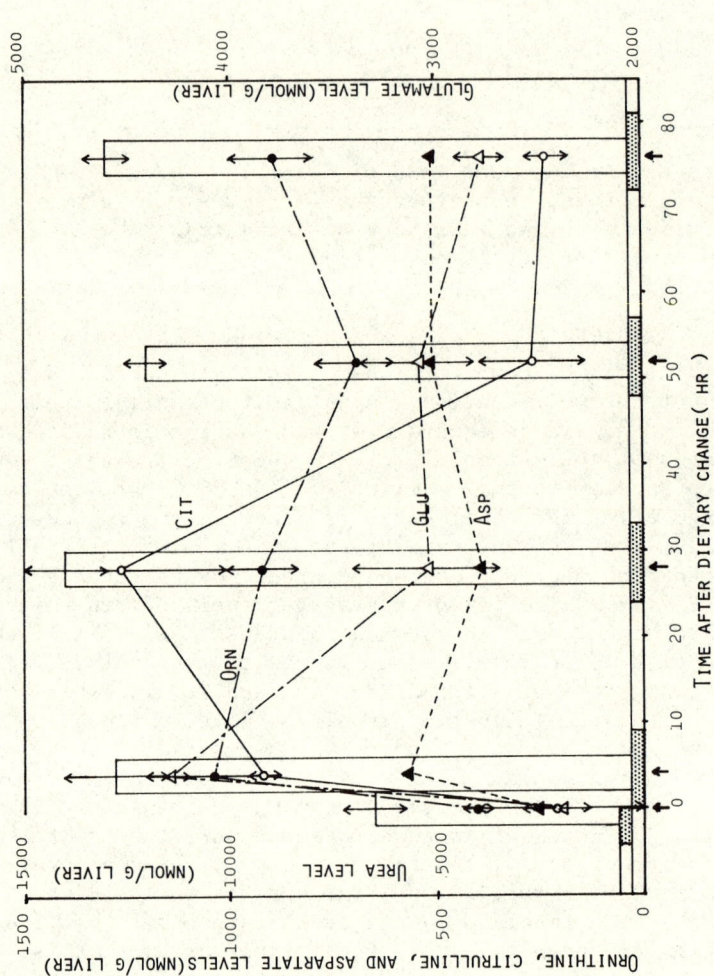

Fig. 2. Diet-induced changes in rat hepatic metabolite levels after the change-over from a 25% to a 70% casein diet. The arrows at the bottom indicate the time of killing. Feeding(dotted horizontal bars at the bottom) was from 9:00 to 18:00 in the dark. All livers were freeze-clamped 4 hr after feeding. Open vertical bars indicate the hepatic urea levels. Values were given as Mean ± S.D.

Table 3. Equations used for Calculating the Rates of Ornithine Catabolism, Urea Output and Rotation of "Ornithine Flux"

The rate of ornithine catabolism:
$$-\frac{d[\text{Orn}]}{dt} = k_{orn}[\text{Orn}] \quad [1]$$

The rate of urea output from liver:
$$\frac{d[\text{Urea}]}{dt} = k_{urea}[\text{Urea}] \quad [2]$$

The approximate rotary rate of "ornithine flux":
$$\text{RPM} = \frac{d[\text{Urea}]}{dt} \cdot \frac{1}{[\text{Orn}]} \quad [3]$$

The first-order rate constant of the ornithine catabolism:
$$k_{orn} = 2.303 \log \frac{(\text{Radioactivity of original ornithine fraction})}{(\text{Radioactivity of ornithine fraction})} \cdot \frac{1}{\text{labeled period(min)}} \quad [4]$$

The first-order rate constant of the urea disappearance in liver:
$$k_{urea} = 2.303 \log \frac{(\text{Radioactivity of original ornithine fraction}/5)}{(\text{Radioactivity of urea fraction})} \cdot \frac{1}{\text{labeled period(min)}} \quad [5]$$

The original ornithine fraction included ornithine fraction plus ornithine-derived amino acid fractions.
From Matsuzawa and Ishiguro[6].

Table 4. Changes in the Rate of Ornithine Catabolism, Urea Output, and Rotation of "Ornithine Flux"

Conditions	Rate of Orn catabolism	Rate of urea output	Rotary rate of "Orn flux" (RPM)
	(nmol/min/g liver)		
Control			
Clamped portal vein	72 ± 7	54 ± 30	0.2 ± 0.1
Open portal vein	67 ± 8	3328 ± 146	10.4 ± 1.3
Hours after change-over from 25 to 70% casein diet			
28	103 ± 16[a]	5299 ± 398[a]	9.5 ± 2.8[b]
52	79 ± 15	4469 ± 1121[a]	11.6 ± 3.4[b]
76	107 ± 15[a]	5811 ± 810[a]	11.5 ± 2.0[b]

[a]Significantly increased ($p < 0.05$); [b]not significant.
From Matsuzawa and Ishiguro[6].

tablished model for such an approach can be seen in the protein turnover study of Segal and Kim[17]. We used the equations shown in Table 3 to determine the rate of ornithine catabolism, the rate of hepatic urea output and the approximate rotary rate of the "ornithine flux". Our results are listed in Table 4. The calculated in vivo rate of ornithine catabolism was 70-110 nmol/min/g liver; it increased about 1.5-fold during the transitional phase. The calculated in vivo rate of urea output nearly ceased in the control rats with clamped portal vein; postprandially, it was about 3 μmol/min/g liver in the control rats with open portal vein. The rate of urea output increased 1.6-fold during ingestion of the 70% casein diet; it was 50 times faster than the rate of ornithine catabolism. The approximate rotary rate of the ornithine carbon skeleton in the urea cycle, i.e., "ornithine flux", was 10-12 rpm and showed no significant changes throughout the transitional phase(Table 4). This constancy of the rotary rate suggests that stimulation of the urea cycle is accomplished not by an accelerated spin of the cycle, but by an increase in the metabolite flux. The radioactivity in the ornithine and arginine fractions derived from [^{3}H]proline was negligible. This finding indicates that in rats, citrulline synthesis from proline is non-functional, at least immediately after the ingestion of a high-protein meal.

SUMMARY

We present evidence for the existence of two types of ornithine transport systems in rat liver mitochondrial inner membrane, an ornithine-citrulline and an ornithine-H^+ exchange system. Citrulline was quickly formed from proline in isolated mitochondria, but in postprandial in vivo experiments, no ornithine was derived. Approximate rates of some in vivo ornithine metabolisms in rats were calculated by pulse-labeling data, on the assumption that hepatic metabolite levels are constant during a given 3-min period. The rate of ornithine catabolism was 70-110 nmol/min/g liver; that of urea output was 3-6 µmol/min/g liver; the rotary rate of the "ornithine flux" was 10-12 rpm. The change from a 25% to a 70% casein diet resulted in 1.5-fold augmentation in the rate of ornithine catabolism and a 1.6-fold increase in the rate of urea output; however, the rate of the "ornithine flux" remained nearly constant. These findings suggest that stimulation of the urea cycle is accompanied not by acceleration of cycle rotation, but rather by an increase in the metabolite flux.

ACKNOWLEDGMENT

We thank Mr. Mitsuharu Takeuchi, Toshisuke Imaeda for their technical assistance in this work, and Ms. Ursula Petralia for reviewing this manuscript. This investigation was supported by the Fujita-Gakuen University Research Fund.

REFERENCES

1. V. Fell, R. J. Pollitt, G. A. Sampson, and T. Wright, Ornithinemia, hyperammonemia, and homocitrullinuria, Am. J. Dis. Child 127:752 (1974).
2. E. Takeda, and Y. Kuroda, Ammonia detoxication in ornithine carbamyl transferase deficient infant, Acta Paed. Jap. 82: 902 (1978)
3. M. Tatibana, and K. Shigesada, Regulation of urea biosynthesis by the acetylglutamate-arginine system, in:"The urea cycle," S. Grisolia, R. Bagena, and F. Mayor, eds., Wiley, New York (1976).
4. J. D. McGivan, N. M. Bradford, and A. D. Beavis, Factors influencing the activity of ornithine aminotransferase in isolated rat liver mitochondria, Biochem. J.162: 147 (1977).
5. N. M. Bradford, and J. D. McGivan, Evidence for the existence of an ornithine/citrulline antiporter in rat liver mitochondria, FEBS Lett. 113: 295 (1980).
6. T. Matsuzawa, and I. Ishiguro, Ornithine metabolism in relation to stimulation of urea cycle, induced by high protein diet, Arch. Biochem. Biophys. 208: 101 (1981).
7. R. P. Faupel, H. J. Seitz, W. Tarnowski, V. Thiemann, and C. H. Weiss, The problem of tissue sampling from experimental animals with respect to freezing technique, anoxia, stress and narcosis, Arch. Biochem. Biophys. 148: 509 (1972).
8. R. M. Archibald, Determination of citrulline and allantoin and demonstration of citrulline in blood plasm, J. Biol. Chem. 156: 121 (1944).

9. T. Matsuzawa, M. Ito, and I. Ishiguro, Enzymatic assay of L-ornithine and L-P5C in tissues, and ornithine-load test in human subjects, Anal. Biochem. 106: 1 (1980).
10. J. J. Coulombe, and L. Favreau, A new simple semimicro method for colorimetric determination of urea, Clin. Chem. 9: 102 (1963).
11. E. Bernt, and H. U. Bergmyer, L-Glutamate, in:"Methods of enzymatic analysis," H. U. Bergmyer, ed., Academic Press, New York (1965).
12. C. Pfleiderer, L-Aspartic acid and L-asparagine, in:"Method of enzymatic analysis," H. U. Bergmyer, ed., Academic Press, New York (1965).
13. J. Mellanby, and D. H. Williamson, Acetoacetate, in:"Method of enzymatic analysis," H. U. Bergmyer, ed., Academic Press, New York (1965).
14. D. H. Williamson, and J. Mellanby, D-(-)-β-Hydroxybutyrate, in: "Method of enzymatic analysis," H. U. Bergmyer, ed., Academic Press, New York (1965).
15. T. Matsuzawa, and I. Ishiguro, Transaminase-glutamate dehydrogenase-linked reaction, Abstract of 10th IUB Congress 386 (1976).
16. A. D. Smith, M. Benziman, and H. J. Strecker, The formation of ornithine from proline in animal tissues, Biochem. J. 104: 557 (1967).
17. H. J. Segal, and Y. S. Kim, Glucocorticoid stimulation of the biosynthesis of glutamic-alanine transaminase, Proc. Nat. Acad. Sci. USA. 50: 912 (1963).
18. T. L. Chan, J. W. Greenwalt, and P. L. Pedersen, Biochemical and ultrastructural properties of a mitochondrial inner membrane fraction deficient in outer membrane and matrix activities, J. Cell Biol. 45: 291 (1970).
19. M. Marshall, and P. P. Cohen, Ornithine transcarbamylase from streptococcus faecalis and bovine liver, J. Biol. Chem. 247: 1641 (1972).

REGULATION OF UREA SYNTHESIS : CHANGES IN THE CONCENTRATION OF ORNITHINE IN THE LIVER CORRESPONDING TO CHANGES IN UREA SYNTHESIS

T. Saheki, M. Hosoya*, S. Fujinami* and T. Katsunuma

Department of Biochemistry, School of Medicine, Kagoshima University Kagoshima, 890 and *Department of Biochemistry School of Medicine, Tokai University Isehara, 259-11 Japan

INTRODUCTION

The biosynthesis of urea is regulated mainly by two factors, the amounts of urea cycle enzymes and the concentrations of acetylglutamate and ornithine. Schimke[1] pointed out that the contents of all the urea cycle enzymes in the liver were directly proportional to the daily consumption of protein, then the activities of urea cycle enzymes are an important regulatory factor of the urea cycle. On the other hand, other investigators[2-4] reported that the concentration of acetylglutamate, an allosteric activator of carbamylphosphate synthesis, and of ornithine, the rate limiting intermediate, changed under various dietary conditions and suggested that these amino acids play a role in the regulation of urea synthesis. We reported that ornithine and acetylglutamate play a more important role in the regulation of urea synthesis especially shortly after the dietary change. In the liver of rats subjected to acute dietary transitions from high to low protein or vice versa, the concentrations of ornithine and acetylglutamate changed greater and prior to the activity changes of urea cycle enzymes. The rate of urea synthesis from ammonium salt as a substrate were greatly changed in the perfused liver and correlated with the changes in the concentration of ornithine in the liver after the dietary changes[5]. Arginine derived from dietary protein is thought to be the main source of ornithine and also the cause of changes in acetylglutamate[3-5]. However, other factors must be involved in the regulation of the concentration of ornithine, since the concentration of ornithine as well as acetylglutamate increased after the intraperitoneal injection of the ammonium salt without altering either arginine or protein input[6].

The difference in the concentration of ornithine in the liver

was observed between conventional and germfree mice. As compared
with germfree animals, so-called conventional mice synthesize at
least 20 to 30% larger amounts of urea. This is because of the
enterohepatic circulation of urea nitrogen, as described and studied
by many investigators[7-9]. The concentration of ornithine in the
liver of conventional mice was about two times higher than that in
the liver of germfree mice under the same dietary condition, but no
significant differences were observed in the activities of urea
cycle enzymes and the concentration of acetylglutamate between the
two kinds of mice[10].

In this paper, we report another example of the changes in the
concentration of ornithine in the liver under relatively physiological conditions and an attempt to clarify the mechanism underlaying
the phenomenon.

MATERIALS AND METHODS

Animals and Diets

In the experiment shown in Table 1 and 2: Male Wistar strain
rats weighing 90-100g were divided into groups A and B, and fed on
the diets described below. The basal diet given group A was composed
of 24.5% casein, 45.5% corn starch, 10% sucrose, 6% oil, 5% cellulose, 1% α-starch, 1% vitamin mixture and 7% salt mixture. The experimental diet given group B rats consisted of the basal diet,
but with 25,5% corn starch instead of 45,5%, and 20% non-essential
amino acid mixture. Non-essential amino acid mixture contained
10% alanine, 10% asparagine, 10% aspartic acid, 20% glutamic acid,
20% glutamine, 10% glycine and 20% sodium glutamate. Rats were
killed between 9 a.m. and 10 a.m. after 2 weeks feeding. A 24-hour
urine was collected from each rat for the determination of urea
excreted on experimental day 13.

In the experiment shown in Table 3: Rats were fed on the basal
diet for 1 week and the diet was withdrawn from 6 p.m. to 6 p.m. of
the next day (24 hours), then the basal diet was given for group A
rats and the diet containing 20 % non-essential amino acid mixture,
for group B rats. After 2 hours, rats were killed for the determination of enzymes and amino acids.

In the experiment shown in Fig. 1 and 2: Rats (190-220g body
weight) maintained on a basal diet were subjected to intraperitoneal
injection of [3-^{3}H] ornithine, 15 µCi/100 g body weight about 9 a.m.
After 10 min. they were again subjected to intraperitoneal injection
of ammonium chloride, 0.2mmol/100g body weight, or sodium chloride,
then killed at the time indicated.

Determination

The amino acids and acetylglutamate in the liver were extracted
and purified according to the method of Shigesada et al.[11]. The
method of Prescott and Jones[12] was used for the determination of

Table 1. Food intake and gain of body weight of rats fed on a basal diet (group A) and a basal diet plus 20% non-essential amino acid mixture (group B)

	Group A	Group B
Number of animals	5	5
Diet containing	24.5% casein	24.5% casein + 20% non-essential amino acid
Body weight (g) Experimental day 14	209 ± 8	181 ± 14
Food intake (g/day/head) Experimental day 1-4	13.1 ± 0.8	9.2 ± 0.9
9-13	16.3 ± 2.5	16.5 ± 2.8
Gain of body weight (g/day) Experimental day 1-4	7.1 ± 0.8	4.5 ± 1.0
9-14	7.8 ± 0.4	7.0 ± 0.5

Values are expressed as m ± S.D.

citrulline, ornithine and arginine, as described earlier[10]. The radioactivity incorporated into ornithine was determined in the ornithine fraction eluted from Dowex 50 column. The radioactivity unabsorbed to both Dowex 50 and I was also determined.

The activities of argininosuccinate synthetase and ornithine transaminase were determined by the methods of Rochovansky and Ratner[13] and Katunuma et al.[14] respectively.

RESULTS AND DISCUSSION

Effect of the Addition of the Non-essential Amino Acid Mixture to a Diet on the Concentration of Ornithine in the Liver.

Effect of the addition of the non-essential amino acid mixture (alanine, glycine, aspartate, glutamate, asparagine and glutamine) to a basal diet to 20% on the biosynthesis of urea was examined. During the experimental days 1 to 4, food intake and gain of body weight of group A rats were larger than those of group B rats, but there were no differences in food intake and gain of body weight between the two groups during the experimental day 9 to 14, as shown in Table 1. Examination of 24-hour urines during the experimental day 13 to 14 showed that group B rats excreted 4 times larger amounts of urea than group A rats. Under this dietary condition, where there was no difference in arginine intake between groups A and B rats, the concentrations of urea cycle intermediate amino acids, ornithine,

Table 2. Effect of the addition of non-essential amino acid-mixture to a basal diet to a final concentration of 20 % on urea synthesis (2 week-feeding)

	Group A	Group B	
Number of animals	5		
Urea in urine			
nmol/day/head	6.57 ± 0.03	27.0 ± 6.2	*
In liver			
Urea μmol/g liver	6.3 ± 0.9	14.7 ± 0.9	**
Ammonia μmol/liver	0.79 ± 0.21	1.17 ± 0.17	**
Ornithine nmol/g liver	184 ± 52	478 ± 75	*
Citrulline nmol/g liver	85 ± 3	169 ± 23	**
Arginine nmol/g liver	14 ± 6	44 ± 17	*
Acetylglutamate	17 ± 5	43 ± 6	**
Argininosuccinate synthetase			
μmol/min/g liver	2.0 ± 0.66	5.4 ± 0.97	**
μmol/min/mg protein	0.018 ± 0.0055	0.059 ± 0.016	**
Ornithine transaminase			
μmol/min/g liver	0.92 ± 0.20	2.45 ± 0.42	**
μmol/min/mg protein	0.0061 ± 0.0013	0.016 ± 0.0024	**

Values are expressed as m ± S.D.
* $p < 0.01$, ** $p < 0.001$

citrulline and arginine, and the concentration of acetylglutamate in the liver of group B rats were about two to three times higher than those in the liver of group A rats. The activities of argininosuccinate synthetase, thought to be a rate-limiting enzyme of urea cycle, and ornithine transaminase, an ornithine degrading enzyme, were also higher in the liver of group B rats than in the liver of group A rats. These results indicate that the addition of non-essential amino acids to the diet stimulated urea synthesis, which was accompanied by the increase in the concentrations of urea cycle intermediate amino acids and acetylglutamate, and induced the urea cycle enzyme after 2 weeks of the feeding. The induction of argininosuccinate synthetase and ornithine transaminase may be caused by the stimulation of glucagon secretion by alanine, glycine and other amino acids. Recently, Snodgrass and Lin showed the induction of urea cycle enzymes by the addition of a single amino acid to a diet and suggested that the ratio of glucagon to insulin in portal vein blood might be a critical element[15].

In order to test the effect of the addition of non-essential amino acids on urea synthesis in a short term, rats were starved for 24 hours, then fed on two kinds of the diet described above and killed after 2 hours. As shown in Table 3 the concentration of

Table 3. Effect of the addition of non-essential amino acid mixture to a basal diet on urea synthesis of rats starved for 24 hours and then fed for 2 hours

		Group A	Group B
Number of animals		4	4
In liver			
Urea	μmol/g liver	5.3 ± 1.7	10.9 ± 1.9 **
Ammonia	μmol/g liver	0.70 ± 0.17	0.92 ± 0.36
Ornithine	nmol/g liver	325 ± 153	924 ± 140 ***
Citrulline	nmol/g liver	72 ± 35	346 ± 254 *
Arginine	nmol/g liver	33 ± 5	164 ± 70 **
Acetylglutamate	nmol/g liver	36 ± 23	112 ± 34 **
Argininosuccinate synthetase			
	μmol/min/g liver	2.1 ± 0.21	2.1 ± 0.30
	μmol/min/mg protein	0.018 ± 0.002	0.023 ± 0.005
Ornithine Transaminase			
	μmol/min/g liver	1.5 ± 0.34	1.6 ± 0.40
	μmol/min/mg protein	0.0078 ± 0.0015	0.0080 ± 0.0015

* < 0.1   ** < 0.01   *** < 0.001

urea, urea cycle intermediate amino acids and acetylglutamate in the liver of group B rats were two to six times higher than those in the liver of group A rats, while the activities of argininosuccinate synthetase and ornithine transaminase were not different between the two groups. This result again indicates the more important role of ornithine and acetylglutamate in the regulation of urea synthesis on a short term basis.

The two experiments described above showed that the concentrations of ornithine and acetylglutamate increased concomitantly with the increase in urea synthesis by a relatively physiological dietary change inducing an increase in nitrogen intake but with no change in arginine intake.

<u>Effect of the Intraperitoneally Injection of an Ammonium Salt on the Concentration of Ornithine and the Metabolism of [^{3}H] Ornithine in the liver.</u>

The next step in this study was to make clear the mechanisms of increase in ornithine in the liver concomitant with the stimulation

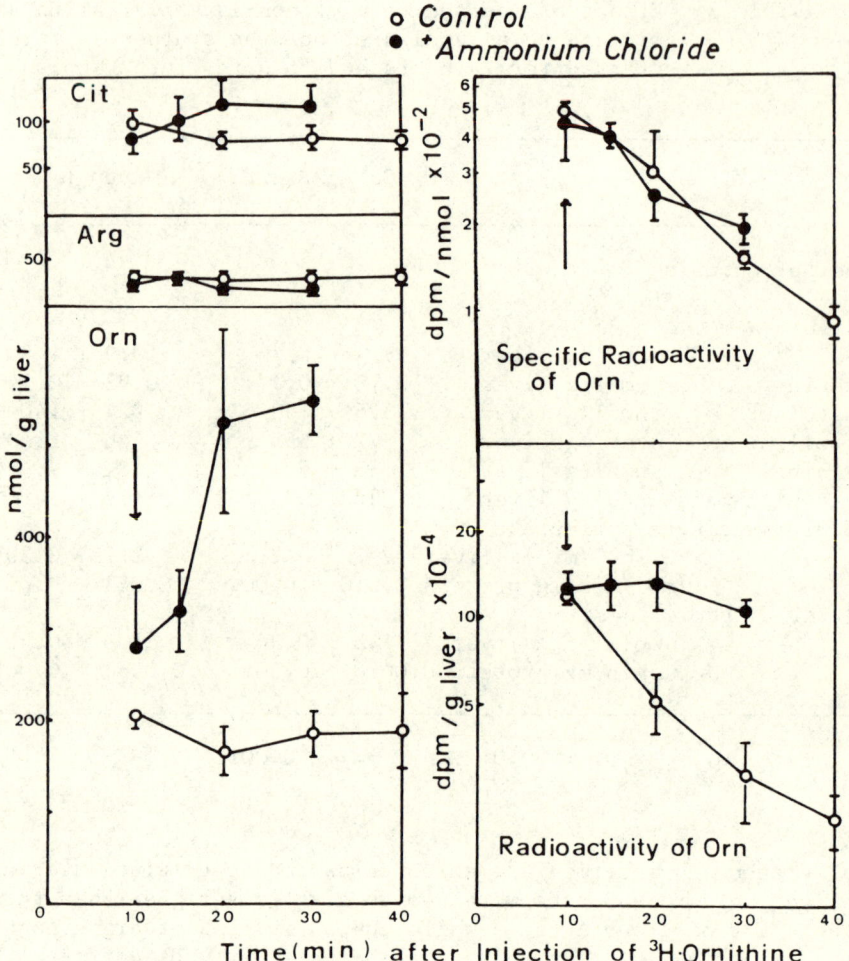

Fig. 1. Effect of the intraperitoneal injection of an ammonium salt on the concentration of ornithine and the metabolism of [^{3}H] ornithine in the liver. [3-^{3}H] ornithine, 15 μCi/100g body weight, was injected intraperitoneally into rats at 0 time and ammonium chloride (●) or sodium chloride (o), 0.2 mmol/100g body weight, was injected at 10 min (arrow shown in Fig).
Determinations were done as described in "Materials and Methods".

of urea synthesis. As reported earlier, intraperitoneal injection of an ammonium salt caused an increase in ornithine and acetylglutamate as well as ammonia and urea in the liver. We examined the fate of [^{3}H] ornithine injected intraperitoneally to rats subsequently

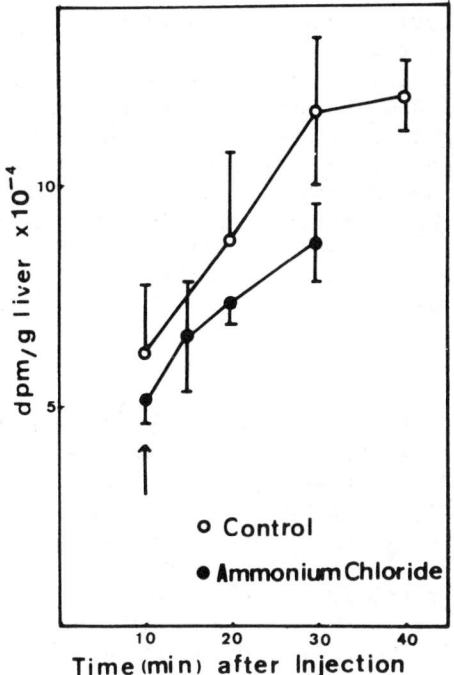

Fig. 2. Effect of the intraperitoneal injection of an ammonium salt on the radioactivity in the fraction unadsorbed to Dowex 50 and 1. Experimental procedures were described in the legend to Fig. 1 and "Materials and Methods".

treated with or without an ammonium salt. As shown in Fig. 1, the injection of ammonium chloride, 0.2 mmol/100g body weight, resulted in an increase in ornithine in the liver at 10 min after the injection and the high level continued for at least 10 min.
^{3}H ornithine injected before 10 min. of the second injection and incorporated into ornithine in the liver of the control rats decreased linearly on semilogarithmic plots (dpm/g liver versus time), while that of ammonium chloride-treated rats did not decrease up to 30 min. at all. On the contrary, the specific radioactivity of ornithine in the liver was not different between the control and the ammonium chloride-treated rats. This suggests that the increase in ornithine in the liver corresponding to the injection of an ammonium salt may be caused by inhibition of degradation of ornithine or stimulation of intake of ornithine into the liver, but not by activation of de novo synthesis of ornithine. We detected radioactivity in the fraction unadsorbed to both anion and cation ion exchange resins. The radioactivity in this fraction was distilled more than 95% with a rotary evaporator. We consider it as [^{3}H] water derived from [3-^{3}H] ornithine via glutamic semialdehyde,

glutamate and α-ketoglutarate. As shown in Fig. 2, the radioactivity in this fraction increased linearly with time up to 30 min. The rate of increase in radioactivity was not significantly different between the control and the ammonium chloride-treated rats. This suggests that inhibition of the degradation of ornithine may not be a major cause of the increase in ornithine.

These results are consistent with the previous results showing that ornithine in the perfused liver did not increase by the addition of an ammonium salt into the perfusate although acetylglutamate did increase, and suggesting that some other organ might participate in the control mechanism.[6]

SUMMARY

In this paper, we report changes in the concentration of ornithine in the liver corresponding to the changes in urea synthesis under relatively physiological dietary conditions with constant arginine intake, and an attempt to clarify the mechanism of the increase in ornithine in the liver. The concentrations of acetylglutamate and urea cycle intermediate amino acids, ornithine, citrulline and arginine, in the liver of rats fed on a diet containing additively 20% non-essential amino acid mixture (alanine, aspartate, asparagine, glutamate, glutamine and glycine) (group B) for 2 weeks were much higher than those in the liver of rats fed on a basal diet (group A). This phenomenon was accompanied by the excretion of larger amount of urea in urine and higher activity of a urea cycle enzyme, argininosuccinate synthetase, of group B rats. Similarly, after 2 hour-feeding of the diets, the concentrations of acetylglutamate and urea cycle intermediate amino acids in the liver of group B rats were higher than those of group A rats, but there was no difference in the activity of argininosuccinate synthetase between the two groups.

From these and other previous results[5, 6, 10], we conclude that the concentrations of ornithine in the liver as well as those of acetylglutamate change correspondingly to the fluctuation of urea synthesis and play an important role in the regulation of urea synthesis, especially in a short term. It is considered from the experiment with [^{3}H] ornithine that the stimulation of intake of ornithine into the liver, but neither inhibition of the degradation nor stimulation of the synthesis, may be a major cause of the increase in ornithine in the liver.

ACKNOWLEDGEMENT

We thank Ms. M. Ogawa for technical assistance and Dr. E. A. Khairallah for critical reading of the manuscript.

REFERENCES

1. R.T. Schimke, Adaptive characteristics of urea cycle enzymes in

the rat, J. Biol. Chem. 237 : 459 (1962).
2. N. Katunuma, M. Okada and Y. Nishii, Regulation of the urea cycle and TCA cycle by ammonia. In: "Advances in Enzyme Regulation", G. Weber, ed., Pergamon Press, Oxford, 4 : 317 (1966).
3. M. Tatibana and K. Shigesada, Regulation of urea biosynthesis by the acetylglutamate-arginine system. In: "The Urea Cycle", S. Grisolia, R. Baguene and F. Mayor, eds., John-Wiley and Sons, New York, 301 (1976).
4. J.W. Kramer and R. A. Freedland, Possible rate-limiting factors on urea synthesis by the perfused rat liver, Proc. Soc. Exp. Biol. Med., 141 : 833 (1972).
5. T. Saheki, T. Katsunuma and M. Sase, Regulation of urea synthesis in rat liver, charges of ornithine and acetylglutamate concentration in the livers of rats subjected to dietary transitions, J. Biochem., 82 : 551 (1977).
6. T. Saheki, T. Ohkubo and T. Katsunuma, Regulation of urea synthesis in rat liver, Increase in the concentrations of ornithine and acetylglutamate in rat liver in response to urea synthesis stimulated by the injection of an ammonium salt, J. Biochem. 84 : 1423 (1978).
7. M. Walser and L. J. Bodenlos, Urea metabolism in man, J. Clin. Invest. 38 : 1617 (1959).
8. E. A. Jones, R. A. Smallwood, A. Craigle and V. M. Rosenoer, The enterohepatic circulation of urea nitrogen, Clin. Sci. 37 : 825 (1969).
9. J.A. Gibson, N.J. Park, G. E. Sladen and A.M. Gawson, The role of the colon in urea metabolism in man, Clin. Sci. Mol. Med. 50 : 51 (1976).
10. T. Saheki, A. Ueda, M. Hosoya, T. Katsunuma, N. Ohnishi and A. Oazwa, Comparison of the urea cycle in conventional and germ-free mice. J. Biochem. 88 : 1536 (1980).
11. K. Shigesada, K. Aoyagi and M. Tatibana, Role of acetylglutamate in ureotelism. Variation in acetylglutamate level and its possible significance in control of urea synthesis in mammalian liver, Eur. J. Biochem. 85 : 385 (1978).
12. L.M. Prescott and N.E. Jones, Modified methods for the determination of carbamylaspartate, Anal. Biochem. 32 : 408 (1969).
13. O. Rochovansky and S. Ratner, Biosynthesis of urea XII. Further studies on argininosuccinate synthetase : Substrate affinity and mechanism of action, J. Biol. Chem. 242 : 3829 (1967).
14. N. Katunuma, Y. Matsuda and I. Tomino, Studies on ornithineketoacid transaminase, J. Biochem. 56 : 499 (1964).
15. P.J. Snodgrass and R.C. Lin, Induction of urea cycle enzymes of rat liver by amino acids, J. Nutr. 111 : 586 (1981).

BASIC BIOCHEMISTRY

2) <u>Enzymes</u>

SYNTHESIS AND INTRACELLULAR TRANSPORT OF MITOCHONDRIAL CARBAMYL PHOSPHATE SYNTHETASE I AND ORNITHINE TRANSCARBAMYLASE

Masataka Mori, Satoshi Miura, Tetsuo Morita and Masamiti Tatibana

Department of Biochemistry, Chiba University School of Medicine, Inohana, Chiba 280, Japan

INTRODUCTION

Carbamyl phosphate synthetase I (CPS) and ornithine transcarbamylase (OTC), the first two enzymes of urea synthesis, are localized in the liver mitochondrial matrix of ureotelic animals.[1] The two enzymes are coded by nuclear genes, synthesized on cytoplasmic 80 S ribosomes, and subsequently transported across the two mitochondrial membranes to the matrix space. Studies in our[2,3] and other laboratories[4,5] have shown that both enzymes are synthesized in larger precursor forms (pCPS and pOTC) in cell-free protein-synthesizing systems. These precursors form large aggregates and have conformations different from those of the mature enzymes.[6] We further showed that pOTC was transported into isolated rat liver mitochondria in association with the processing of pOTC to the mature form of the enzyme.[3,7,8] It has been shown that both pCPS[9] and pOTC[10] are synthesized on membrane-free polysomes, released into the cytosol and then transported rapidly into mitochondria. The present paper describes detailed kinetic studies of the synthesis, processing and intracellular transport of pCPS and pOTC in isolated rat hepatocytes. The paper also deals with the chemical nature of pOTC and its transport into isolated mitochondria in vitro.

Abbreviations: CPS, carbamyl phosphate synthetase I; pCPS, precursor of carbamyl phosphate synthetase I; OTC, ornithine transcarbamylase; pOTC, precursor of ornithine transcarbamylase; SDS, sodium dodecyl sulfate.

RESULTS

Synthesis, Intracellular Transport and Processing of pOTC and pCPS in Isolated Hepatocytes

Isolated rat hepatocytes were prepared by a modification of the method of Berry and Friend.[11] The cells were incubated with [^{35}S]methionine at 37°C for various times, and then fractionated into the cytosol fraction and particulate components (crude mitochondrial fraction) by the digitonin procedure.[12] The subcellular fractions were subjected to immunoprecipitation with anti-OTC, and the immunoprecipitates were analyzed by SDS gel electrophoresis and fluorography (Figs. 1 and 2). The labeled pOTC was found exclusively in the cytosol fraction, and the radioactivity in pOTC increased linearly for the first 10 min, then more slowly until a plateau was reached in 10-20 min of the pulse. On the other hand, the labeled OTC was found exclusively in the particulate fraction. The radioactivity in OTC appeared after a lag time of a few min and then increased steeply and almost linearly with time up to 40 min. The pulse-labeled pOTC disappeared almost completely from the cytosol in 10 min of the subsequent chase.

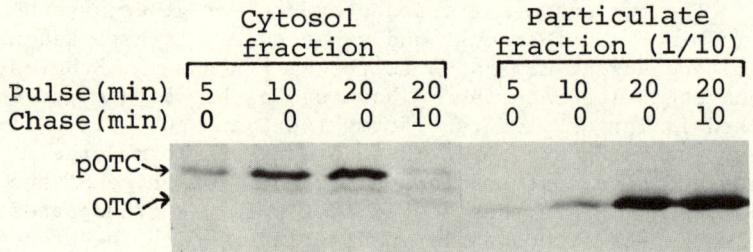

Fig. 1. Fluorographic detection of synthesis of pOTC and OTC in isolated rat hepatocytes. The hepatocytes (3.4 x 10^7 cells/ml) were pulsed with [^{35}S]methionine (190 µCi/ml) at 37°C in a volume of 0.1 ml for 5, 10, and 20 min or were pulsed for 20 min and then chased for 10 min in the presence of 20 mM unlabeled methionine. Particulate components and the cytosol fraction of the cells were separated by the digitonin procedure,[12] and the particulate components were dissolved in a buffer containing 0.1% SDS and 0.1% Triton X-100. Whole samples of the cytosol fractions and one-tenth of the particulate extracts were subjected to immunoprecipitation using anti-OTC and fixed Staphylococcus aureus cells.[3,13] The immunoprecipitates were subjected to SDS gel electrophoresis and fluorography. A portion of the fluorogram is shown.

SYNTHESIS AND INTRACELLULAR TRANSPORT

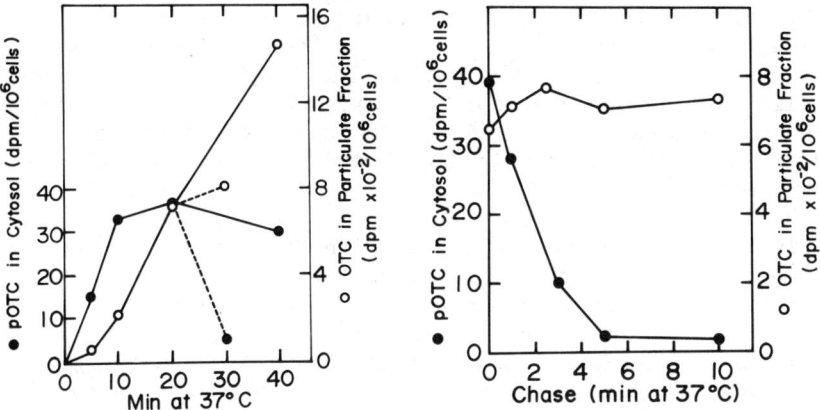

Fig. 2. (left) Synthesis of pOTC and OTC in isolated rat hepatocytes. Dotted lines show the data of the chase experiments.

Fig. 3. (right) Kinetics of pOTC disappearance from the cytosol in pulse-chase experiments. Hepatocytes were pulsed with [^{35}S]-methionine for 20 min as described in Fig. 1, chased for indicated times, and analyzed as described in Fig. 1.

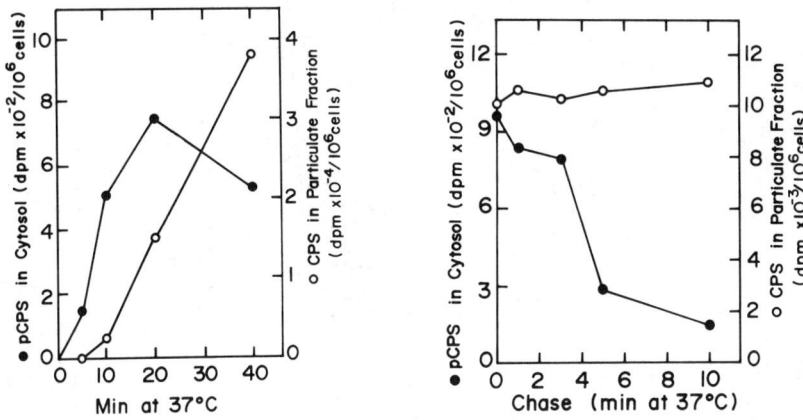

Fig. 4. (left) Synthesis of pCPS and CPS in isolated hepatocytes. Hepatocytes were incubated with [^{35}S]methionine for indicated times as described in Fig. 1, and fractionated into the cytosol and particulate fractions. pCPS and CPS were isolated by immunoprecipitation and SDS gel electrophoresis.[2,13]

Fig. 5. (right) Kinetics of pCPS disappearance from the cytosol in pulse-chase experiments. Hepatocytes were pulsed with [^{35}S]methionine for 20 min, chased for indicated times, and analyzed as described in Fig. 4.

Kinetic aspects of pOTC disappearance in pulse-chase experiments are shown in Fig. 3. The pulse-labeled pOTC disappeared from the cytosol with an apparent half life of 2 min. The actual half life is presumably somewhat shorter, because completion of the labeled nascent peptide of pOTC would continue during the chase period. The radioactivity in OTC showed a substantial increase during the chase.

Similar kinetics were obtained with CPS (Fig. 4). The labeled pCPS appeared in the cytosol fraction, and the radioactivity reached a maximum after about 20 min of pulse and then decreased somewhat. On the other hand, the radioactivity in mature CPS was first detected after 10 min in the particulate fraction and then increased linearly up to 40 min. Kinetic aspects of pCPS disappearance in pulse-chase experiments are shown in Fig. 5. The pulse-labeled pCPS disappeared from the cytosol slowly for the first 3 min and then more rapidly. The apparent delay in pCPS disappearance is presumably due to elongation and completion of the radiolabeled nascent peptide of pCPS during the chase. In fact, no delay was observed when cycloheximide was included in the chase medium. The half life of pCPS estimated from the experiments with cycloheximide or from the values after 3 min of the chase in Fig. 5, was 1-2 min.

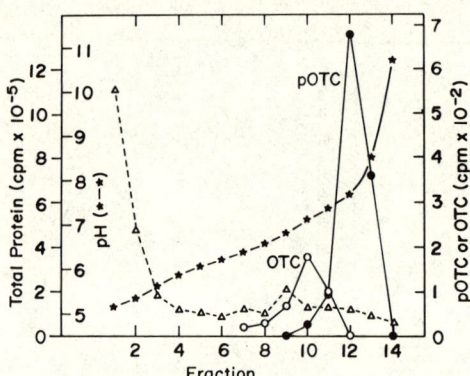

Fig. 6. Isoelectric focusing of pOTC and OTC. A discontinuous gradient (22 ml) of 5 to 60% (w/v) glycerol containing 2% ampholine (LKB Producer AB, pH 6-8) was prepared. The solution containing the labeled pOTC and OTC was introduced at the center of the gradient, and electrophoresis was carried out for 43 h at 4°C at 600 V. The labeled OTC had been prepared by incubating cell-free synthesized pOTC with isolated rat liver mitochondria. Fractions of 1.6 ml were collected and subjected to immunoprecipitation. The immunoprecipitates were analyzed by SDS gel electrophoresis and fluorography. Radioactivities in total protein ($\triangle$), pOTC ($\bullet$) and OTC ($\bigcirc$) were measured.

A High Isoelectric Point of pOTC

The peptide extensions of the mitochondrial enzyme precursors are thought to play an important role in their transport into mitochondria. However, little is known about the chemical nature of the extrapeptides. pOTC synthesized in vitro and mature OTC were subjected to isoelectric focusing (Fig. 6). OTC was focused as a single protein at pH 7.2, which is in accord with the pI value reported for the purified rat liver enzyme.[14] On the other hand, pOTC was focused at about pH 7.9. These results demonstrate that pOTC has a pI higher than that of the mature enzyme. The pI value of the 37,000 dalton product (P37), a possible intermediate in the processing,[7] was 7.6, such being somewhat lower than that of pOTC.

Hydrophobicity of pOTC and mature OTC was examined by hydrophobic chromatography with octyl-Sepharose and phenyl-Sepharose. The elution profile of pOTC from both gels was almost identical to that of OTC (data not shown). Thus, pOTC has a hydrophobicity similar to that of the mature enzyme.

Basic Proteins Inhibit pOTC Processing by Isolated Mitochondria

We showed previously that pOTC synthesized in vitro could be taken up and processed to the mature enzyme by isolated rat liver mitochondria.[3,8] To gain insight into the possible role of the basic nature of pOTC in its transport into mitochondria, effects of basic proteins on pOTC transport-processing were examined with this in vitro reconstitution system (Fig. 7). Salmon protamine (1 mg/ml) and calf thymus histone (2 mg/ml) strongly inhibited pOTC processing by isolated mitochondria. Among three types of histones tested, histone H4 was the most effective and histones H1 and H2B were less effective. Among other basic proteins and peptides tested, chicken egg lysozyme (2 mg/ml) and polyarginine (0.5 mg/ml) were moderately inhibitory, whereas horse heart cytochrome c and polylysine (2 mg/ml each) were ineffective. On the other hand, non-basic proteins including bovine serum albumin, chicken ovalbumin, rabbit muscle creatine phosphokinase, rabbit muscle aldolase, bovine liver catalase, yeast alcohol dehydrogenase, soybean trypsin inhibitor, and polyglutamate at a concentration of 2 mg/ml were not inhibitory or only slightly inhibitory (some data not shown). These results indicate that certain basic proteins inhibit pOTC processing fairly specifically.

Membrane-Perturbing Reagents Inhibit pOTC Processing by Isolated Mitochondria

Effects of membrane-interacting reagents on pOTC processing in the in vitro reconstitution system are shown in Fig. 8. pOTC processing by isolated mitochondria was strongly inhibited by all the reagents tested except procaine, including polymixin B

(0.1 mg/mg protein), flufenamic acid (0.1 mM), chlorpromazine (0.1 mM), dibucaine (1 mM), and reserpine (0.1 mM). These reagents are known to interact with phospholipids. Therefore, it is likely that perturbation of membrane phospholipid structure impairs mitochondrial uptake-processing of pOTC.

DISCUSSION

The results presented in this paper as well as those reported previously are summarized in Fig. 9. Unpublished results are also

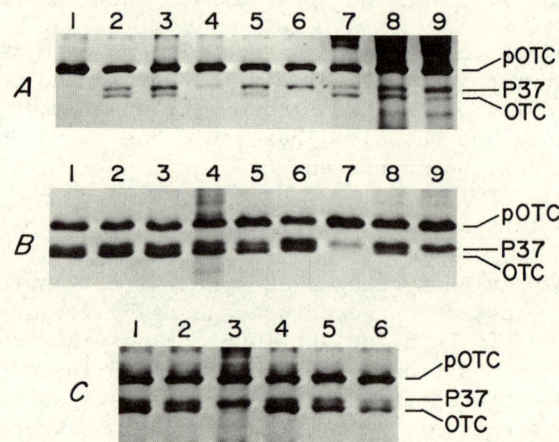

Fig. 7. Inhibition by basic proteins of pOTC processing by isolated mitochondria. The cell-free translation mixture (50 µl) containing [^{35}S]pOTC was incubated with rat liver mitochondria (50 µg of protein) at 25°C for 60 min in a total volume of 54 µl with the additions as follows.
A. Lane 1, minus mitochondria; lane 2, no addition; lanes 3 and 4, 0.2 and 1 mg/ml of salmon protamine; lanes 5 and 6, 0.5 and 2 mg/ml of calf thymus histone; lanes 7 and 8, 0.5 and 2 mg/ml of polylysine; lane 9, 0.5 mg/ml of polyarginine.
B. Lane 1, no addition; lanes 2 and 3, 0.5 and 2 mg/ml of bovine serum albumin; lanes 4 and 5, 0.5 and 2 mg/ml of chicken ovalbumin; lanes 6 and 7, 0.5 and 2 mg/ml of histone H1; lanes 8 and 9, 0.5 and 2 mg/ml of histone H2B.
C. Lane 1, no addition; lanes 2 and 3, 0.1 and 0.5 mg/ml of histone H4; lane 4, 2 mg/ml of horse heart cytochrome c; lane 5, 2 mg/ml of rabbit muscle creatine phosphokinase; lane 6, 2 mg/ml of chicken egg lysozyme. The reaction mixtures were subjected to immunoprecipitation, and the immunoprecipitates were analyzed by SDS gel electrophoresis and fluorography. Portions of the fluorograms are shown. P37, 37,000 dalton product.

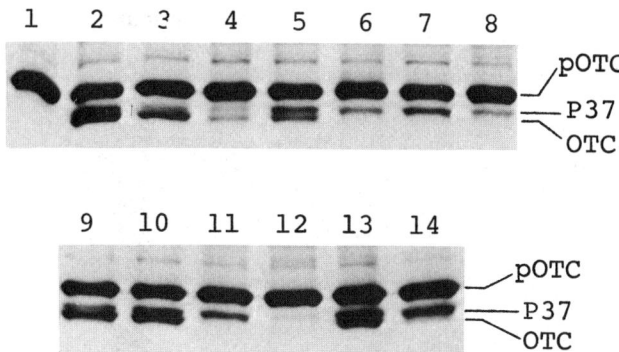

Fig. 8. Effects of membrane-interacting reagents on pOTC processing by isolated mitochondria. Incubation was performed as described in Fig. 7 with the additions as follows. Lane 1, minus mitochondria; lane 2, no addition; lanes 3 and 4, 25 and 100 μg/mg mitochondrial protein of polymixin B; lanes 5 and 6, 0.1 and 1 mM flufenamic acid; lanes 7 and 8, 0.1 and 1 mM chlorpromazine; lanes 9 and 10, 1 and 5 mM procaine; lanes 11 and 12, 1 and 5 mM dibucaine; lanes 13 and 14, 10 and 100 μM reserpine. The reaction mixtures were analyzed as described in Fig. 7.

included. This paper presents a detailed kinetic study on the synthesis and intracellular transport of CPS and OTC. The results show most clearly that these two urea cycle enzymes are synthesized in the cytosol as larger precursors, transported from the cytosol into the mitochondria and processed to the mature enzymes. The extramitochondrial pools of the precursors are small and their half lives are estimated to be 1-2 min. Raymond and Shore[16] recently reported a similar half life of pCPS in rat liver explants. The proteolytic processing of the precursors is thought to occur concomitantly with or immediately following the transport, because no precursor was detected in the particulate fraction.

The present paper shows that pOTC has a hydrophobicity similar to that of the mature enzyme, but is more basic than the mature enzyme. Furthermore, it was found that some basic proteins inhibit the uptake-processing of pOTC by isolated mitochondria. These results suggest that the basic nature of the precursor, and probably that of the peptide extension, is required for interaction of the precursor with the mitochondrial membrane, and that the basic proteins compete with the precursor for the interaction with the organelle. The results are in accord with the recent report of Anderson,[17] who showed that probable precursors for several

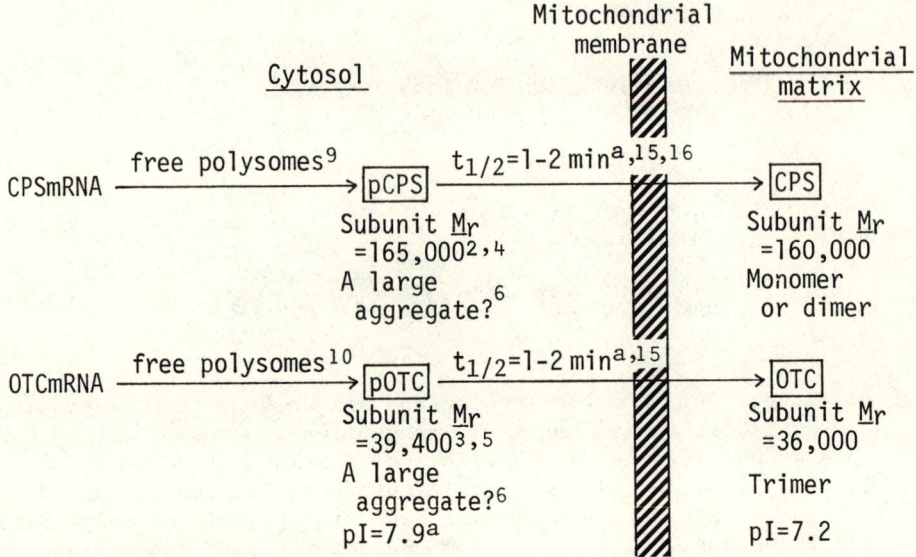

Fig. 9. Intracellular transport and processing of pCPS and pOTC.
[a]This paper. [b]Unpublished results.

mitochondrial proteins of human cells are quite basic. It is conceivable that the interaction between positively charged peptide extension and negatively charged proteins or phospholipids of the mitochondrial membrane facilitates the specific binding of the precursors to the organelle.

Finally, it is shown here that a number of membrane-interacting reagents inhibit pOTC processing by isolated mitochondria. It is likely that these reagents inhibit the transport of pOTC and thus prevent the precursor to become accessible to the protease responsible to the processing. However, direct inhibition of the processing protease by these reagents cannot be excluded. We have also shown that a high integrity of mitochondria including the outer membrane or intermembrane fluid or both in addition to the mitoplasts, is required for the pOTC transport-processing (unpublished results). It is known that in some pathological conditions such as Reye's syndrome[18] and vitamin E deficiency,[19] hepatic levels of CPS and OTC are decreased significantly. In light of the above findings, it may be reasonable to speculate that the decrease in the enzyme levels is due partly to impaired transport-processing of the enzyme precursors.

Acknowledgements. We thank Dr. F. Ikeda of this laboratory for preparing isolated hepatocytes, Drs. M. Nomoto and H. Hayashi (Gunma University) for providing us with histones, and Dr. P. P. Cohen (University of Wisconsin-Madison) for his participation in part of this work. We also thank M. Takiguchi and H. Kusumoto for technical assistance.

REFERENCES

1. J. G. Gamble and A. L. Lehninger, Transport of ornithine and citrulline across the mitochondrial membrane, J. Biol. Chem. 248:610 (1973).
2. M. Mori, S. Miura, M. Tatibana, and P. P. Cohen, Cell-free synthesis and processing of a putative precursor for mitochondrial carbamyl phosphate synthetase I of rat liver, Proc. Natl. Acad. Sci. USA 76:5071 (1979).
3. M. Mori, S. Miura, M. Tatibana, and P. P. Cohen, Processing of a putative precursor of rat liver ornithine transcarbamylase, a mitochondrial matrix enzyme, J. Biochem. 88:1829 (1980).
4. G. C. Shore, P. Carignan, and Y. Raymond, In vitro synthesis of a putative precursor to the mitochondrial enzyme, carbamyl phosphate synthetase, J. Biol. Chem. 254:3141 (1979).
5. J. G. Conboy, F. Kalousek, and L. E. Rosenberg, In vitro synthesis of a putative precursor of mitochondrial ornithine transcarbamylase, Proc. Natl. Acad. Sci. USA 76:5724 (1979).
6. S. Miura, M. Mori, Y. Amaya, M. Tatibana, and P. P. Cohen, Aggregation states of precursors for mitochondrial carbamoyl-phosphate synthetase I and ornithine carbamoyltransferase, Biochem. Int. 2:305 (1981).
7. M. Mori, S. Miura, M. Tatibana, and P. P. Cohen, Characterization of a protease apparently involved in processing of pre-ornithine carbamoyltransferase of rat liver, Proc. Natl. Acad. Sci. USA 77:7044 (1980).
8. M. Mori, T. Morita, S. Miura, and M. Tatibana, Uptake and processing of the precursor for rat liver ornithine transcarbamylase by isolated mitochondria: Inhibition by uncouplers, J. Biol. Chem. 256:8263 (1981).
9. Y. Raymond and G. C. Shore, The precursor for carbamyl phosphate synthetase is transported to mitochondria via a cytosolic route, J. Biol. Chem. 254:9335 (1979).

10. T. Morita, M. Mori, M. Tatibana, and P. P. Cohen, Site of synthesis and intracellular transport of the precursor of mitochondrial ornithine carbamoyltransferase, Biochem. Biophys. Res. Commun. 99:623 (1981).
11. M. N. Berry and D. S. Friend, High-yield preparation of isolated rat liver parenchymal cells: A biochemical and fine structural study, J. Cell Biol. 43:506 (1969).
12. P. F. Zuurendonk and J. M. Tager, Rapid separation of particulate components and soluble cytoplasm of isolated rat liver cells, Biochim. Biophys. Acta 333:393 (1974).
13. M. Mori, S. Miura, M. Tatibana, and P. P. Cohen, Cell-free translation of carbamyl phosphate synthetase I and ornithine transcarbamylase messenger RNAs of rat liver: Effect of dietary protein and fasting on translatable mRNA levels, J. Biol. Chem. 256:4127 (1981).
14. C. J. Lusty, R. L. Jilka, and E. H. Nietsch, Ornithine transcarbamylase of rat liver: Kinetic, physical, and chemical properties, J. Biol. Chem. 254:10030 (1979).
15. M. Mori, T. Morita, Y. Amaya, M. Tatibana, and P. P. Cohen, Synthesis, intracellular transport, and processing of the precursors for mitochondrial ornithine transcarbamylase and carbamyl phosphate synthetase I in isolated hepatocytes, Proc. Natl. Acad. Sci. USA, in press.
16. Y. Raymond and G. C. Shore, Processing of the precursor for the mitochondrial enzyme, carbamyl phosphate synthetase: Inhibition by p-aminobenzamidine leads to very rapid degradation (cleaning) of the precursor, J. Biol. Chem. 256:2087 (1981).
17. L. Anderson, Identification of mitochondrial proteins and some of their precursors in two-dimensional electrophoretic maps of human cells, Proc. Natl. Acad. Sci. USA 78:2407 (1981).
18. T. Brown, G. Hug, L. Lansky, K. Bove, A. Scheve, M. Ryan, H. Brown, W. K. Schubert, J. C. Partin, and J. Lloyd-Still, Transiently reduced activity of carbamyl phosphate synthetase and ornithine transcarbamylase in liver of children with Reye's syndrome, N. Engl. J. Med. 294:861 (1976).
19. K. Shimbayashi and S. Shoya, Effect of vitamin E on the urea cycle enzymes, ornithine-keto acid transaminase, and isocitric dehydrogenase in rat liver, Agr. Biol. Chem. 35:983 (1971).

ISOLATION OF ARGININOSUCCINASE FROM BOVINE BRAIN : CATALYTIC, PHYSICAL AND CHEMICAL PROPERTIES COMPARED TO LIVER AND KIDNEY ENZYMES

Kimiko Murakami-Murofushi[*] and Sarah Ratner

Department of Biochemistry, The Public Health Research Institute of The City of New York, Inc., New York N.Y. 10016 ([*]Present address : Dept. of Biochemistry Teikyo University, School of Medicine, Tokyo 173)

INTRODUCTION

Argininosuccinase together with argininosuccinate synthetase occurs in the brain of all ureotelic species as well as in other tissues[1]. Considering the levels of these enzymes in brain and the amount of citrulline supplied via the circulation, they may play a role in the formation of arginine from citrulline in normal brain metabolism.

In the human genetic disorder of nitrogen metabolism associated with a reduced activity of liver argininosuccinase, the blood level and the urinary excretion of argininosuccinate are increased, while the cerebrospinal fluid level exceeds that in blood[2,3]. Some reports would suggest that the mutational impairment involves both liver and brain enzymes[1,4].

Cultured cells such as fibroblasts and leukocytes have been employed for enzymatic assays to detect suspected mutations. As the extensive derangements in nitrogen metabolism of affected individuals indicates that the liver may be the major site of this deficiency, it is of interest for genetic, as well as metabolic reasons to determine whether the enzymes catalyzing the same reaction in different organs and tissues of the same species are identical proteins.

To pursue our interest in the question whether argininosuccinases from various tissues of the same species are identical, we have brought argininosuccinase from bovine brain to a state of physical homogeneity for comparison with the liver and kidney

Table 1. Purification of Argininosuccinase from Bovine Brain[a]

Fractionation step	Total protein (mg)	Specific activity (U/mg)	Total activity (U)	Yield (%)
1. Neutralized extract				
2. First $(NH)_2SO_4$ fractionation	155,320	0.75	116,380	100
3. Second $(NH)_2SO_4$ fractionation	85,800	1.26	108,020	93
4. Mild heat at pH 5.2	27,720	3.68	102,080	88
5. Fractionation at pH 6.6 and concentration	16,450	5.72	94,200	81
6. Fractionation at pH 5.5	8,350	9.52	79,500	69
7. DEAE-Cellulose and concentration	342	150	51,200	44
8. First gel filtration	51.5	555	28,600	24
Concentration 1	46.7	548	25,636	22
2	36.2	407	14,700	13
3	26.6	613	16,000	14
9. Second gel filtration	9.85	1,030	10,100	8.6
Concentration	8.54	1,250[b]	10,700	9.2

[a] Compiled from Murakami-Murofushi & Ratner[10].
[b] After the removal of oxidized DTT and arginine by dialysis, the total protein was 6.0 mg, the specific activity was 1,400 based on E_{280} (1%)=13.0 and the yield was 7.2 %. 79.2 kg were processed.

enzymes isolated from the same species.

Purification of Bovine Brain Argininosuccinase

The purification of argininosuccinase from fresh steer brains is summarized in Table 1. Homogeneous enzyme purified at least 2,000-fold was obtained in an overall yield of 7% based on Step 2. In order to obtain 6 mg of homogeneous enzyme by the procedures given, about 80 kg of brain was processed.

To insure full retention of catalytic activity of pure enzyme, it was necessary to add the protecting agents (5 mM DTT, 4 mM arginine and 1 mM EDTA) during long storage at -20°. The assay and experimental studies of homogeneous enzyme were conducted only after exhaustive dialysis to remove product inhibition due to arginine and to remove extraneous absorption at 280 nm due to oxidized

Table 2. Comparison of Physical Properties of Purified Bovine Brain, Liver and Kidney Argininosuccinases[a]

Property		Enzyme source		
		Brain	Liver	Kidney
Specific activity (U/mg)		1,300-1,400	1,300-1,400	1,300-1,400
$s_{20,\omega}$	Tetramer	-	9.3	9.3
	Dimer	-	5.6	-
	Monomer	-	1.9	-
Molecular weight				
	Tetramer	200,000±10,000	202,000	202,000
	Dimer	100,000	100,000	100,000
	Monomer	50,000	52,000	50,400
E_{280} (1%)		13.0	13.0	12.5

[a] Modified from Ratner et al.[6,10].

DTT formed during long storage. Under these conditions, each run had a specific activity of 1,400 using the absolute protein value E_{280} (1%)=13.0. Despite the low tissue level (liver:kidney:brain= 270:43:3), the specific activity of the purified brain enzyme was found to be the same as the values reported for the liver and kidney enzymes (5-8, and table 2).

Molecular Weight of Brain Argininosuccinase

By gel filtration, the brain enzyme emerged in a symmetrical peak as shown in Fig. 1 judged by enzymatic activity and absorption at 280 nm. The recoveries of both protein and activity from Bio-gel were in excellent yield although a small load was placed on the column.

Homogeneity was further established by 7.5% polyacrylamide disc gel electrophoresis in Tris-acetate buffer, pH 7.3. Samples of the brain, liver and kidney enzymes migrated as single bands having the same electrophoretic mobility and no contamination was observed (Fig. 2).

The molecular weight of the catalytically active enzymes were determined by gel filtration, and as shown in Fig. 3, the elution position of brain argininosuccinase corresponds to a molecular weight of 200,000 ± 10,000. Liver argininosuccinase gave the same elution position as the enzyme from brain. A value of 202,000 ± 5,000 has been found for the liver enzyme as determined by sedimentation equilibrium[5].

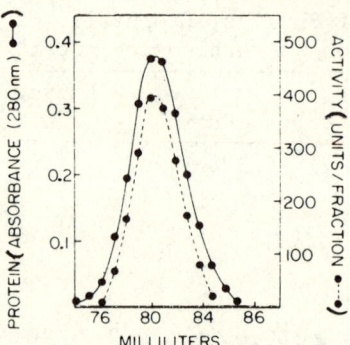

Fig. 1. Elution profile of active bovine brain argininosuccinase on a Bio-gel A-0.5m column (1.5x83.0cm). From Murakami-Murofushi & Ratner[10].

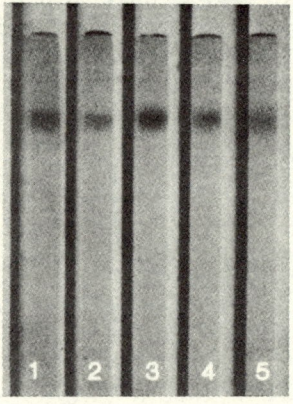

Fig. 2. Disc gel electrophoresis of active argininosuccinases. Gels 1 and 2:liver enzyme 2.5 and 1.25 µg; gels 3 and 4:2.77 and 1.38 µg of brain enzyme; gel 5:2.37 µg of kidney enzyme. Electrophoresis was conducted at 25° for 5.5 h in 0.2 M Tris-acetate buffer, pH 7.3, at 140 V and 15 mA per gel. From ref. 10.

Subunit Composition

A value of the minimum subunit molecular weight was determined by disc gel electrophoresis after dissociation with SDS in the presence of 2-mercaptoethanol (Fig. 4). The value of 50,000 agrees with

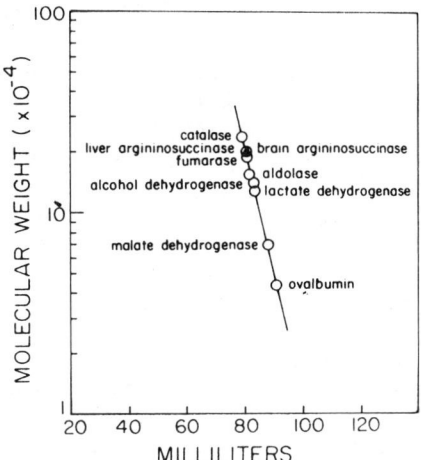

Fig. 3. Molecular weight of active argininosuccinase from bovine brain determined by gel filtration. From Murakami-Murofushi & Ratner[10].

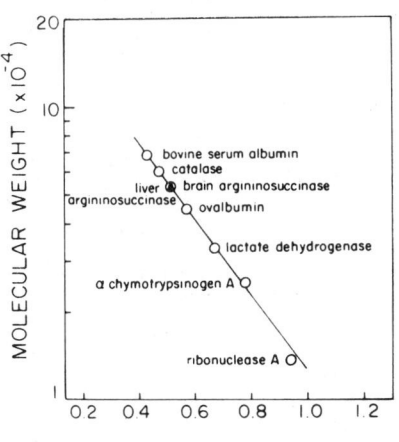

Fig. 4. Minimum subunit molecular weight of bovine brain argininosuccinase determined by electrophoresis on 7.5% polyacrylamide gels in 0.2M Tris-acetate, pH 8.5 and 0.1% SDS. From ref. 10.

values for the liver and kidney enzymes obtained by this method and by equilibrium sedimentation in 6M guanidine-HCl[7]. Thus the catalytically active brain enzyme closely resembles the enzymes from liver and kidney in molecular weight and subunit composition.

If the tetrameric brain enzyme is composed of four identical polypeptide chains, a close similarity in primary structure is expected. Figure 5 shows a comparison of the tryptic cleavage maps of the brain and liver enzymes. The 48-49 ninhydrin-positive peptides given are in excellent agreement with one quarter the total number of arginine plus lysine residues, 192 per 202,000 g[7]. The number of peptide fragments observed and the selective reagent specificity were in all cases close to or equal to the number of peptide fragments and specificity predicted on the basis of four polypeptide chains closely similar in primary structure (Table 3). The maps of cyanogen bromide cleavage products are shown in Fig. 6, indicating good accordance with prediction based on methionine residues (Table 3). As shown in Fig. 5 and Fig. 6, the peptide patterns are closely superimposable suggesting that the four monomers of the brain enzyme are similar both to each other in amino acid sequence and to those of the liver enzyme.

Cold Lability and Oligomeric Structure

Argininosuccinase with a quaternary structure loses activity on standing at low temperature in cationic buffers. Figure 7 shows that the brain enzyme lost activity as a function of time at 0° in Tris buffer at pH 8.5. The ratio of inactivation in the cold was the same

Table 3. Chromatographic Analysis of Peptide Cleavage Maps[a]

Reagent and residue specificity	Peptides found		Residue/mole
	Brain	Liver	
Tryptic peptides			
Ninhydrin	48	48-49	192[b]
Pauly reagent (histidine & tyrosine)	20	20	83
α-Nitrosyl-β-naphthol (tyrosine)	8	8	36
Platinic iodide (methionine & carboxymethyl cysteine)	17	18	67
Cyanogen bromide peptides			
Ninhydrin	15	14	51[c]

[a]From Murakami-Murofushi & Ratner[10].
[b]Number of the sum of arginyl and lysyl residues per mole (202,000 g)
[c]Number of methionyl residues per mole (202,000 g).

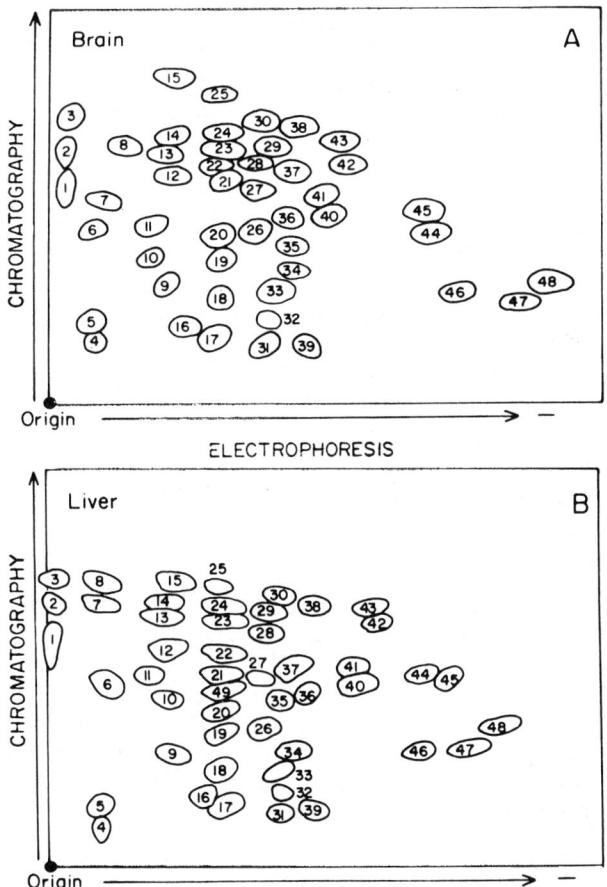

Fig. 5. Tryptic peptide patterns of S-carboxymethyl argininosuccinase from brain and liver. Peptides are numbered in A and B according to location. In both A and B, histidine was found in peptides 2,3,4,8,19,31,39,40,41,42,43,44; tyrosine in 5, 12,18,21,23,29,33,38; methionine and carboxymethyl cysteine in 1,2,3,7,12,13,14,16,17,18,22,24,26,27,30,34,37 and in peptide 8 in the liver pattern. From ref. 10.

as that of the liver enzyme, and full activity was similarly restored by warming for 30 min at 38°. Inactivation was accompanied by dissociation into catalytically inactive dimeric subunits as a first order process ($E_4 \rightleftharpoons 2E_2$).

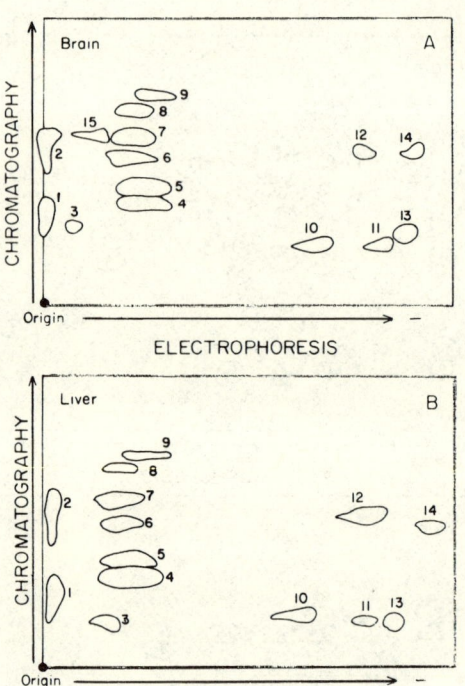

Fig. 6. Cyanogen bromide cleavage maps of S-carboxymethyl argininosuccinase from bovine brain and liver. In both A and B, peptides 4,5,7 and 13 stained deeply with ninhydrin; peptides 3,9,12 and 14 were faint; others stained moderately. From Murakami-Murofushi & Ratner[10].

Relation of SH Groups to Tetrameric Structure

Brain argininosuccinase contains four SH groups per mole of 200,000 which were found to interact with DTNB rapidly without any loss of catalytic activity (Table 4). On cold dissociation to dimeric subunits, 4 additional SH groups became available to DTNB (Fig. 8, Table 4), suggesting that the second 4 SH groups are located near or at the hydrophobic contact sites between the 2 dimers. In the presence of SDS, 8 further SH groups became available to the reagent (Table 4). The last 8 may be situated at or near contact sites between pairs of monomers. A total of 16 SH groups were found in agreement with titration values and amino acid analysis[7,8]. The tetrahedral model proposed for the quaternary structure possesses dihedral symmetry and takes into account the 2 types of subunit dissociation.

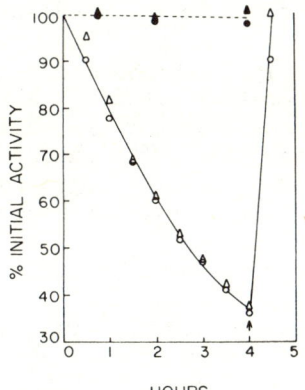

Fig. 7. Comparison of cold inactivation rates and thermal reactivation of brain and liver argininosuccinases. Enzyme solutions were incubated in 0.1 M Tris-HCl containing 2mM Na EDTA at 0°. After 4hr, the mixture was incubated at 38° for 30 min. Incubation at 0° : brain enzyme (o); liver enzyme (△). Incubation at 24.5° for controls : brain enzyme (●); liver enzyme (▲). From ref. 10.

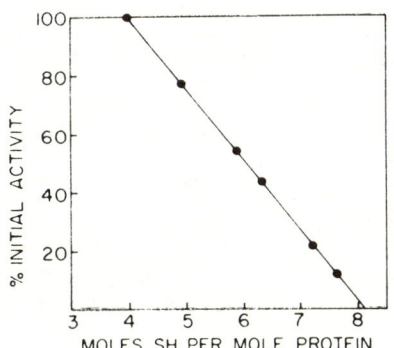

Fig. 8. The number of accessible SH groups in brain argininosuccinase related to the extent of cold inactivation. From ref. 10.

Table 4. Number of Accessibility of Sulfhydryl Groups[a]

Conditions	Oligomeric state	Moles of SH per mole (202,000 g)		
		Brain	Liver	Kidney
KPO_4, pH 7.5, 25°	Tetramer	3.97	3.88	–
Tris, pH 8.6, 0°	Dimer	8.13	8.16	–
0.1% SDS	Monomer	15.99	16.12	14.9

a) Sulfhydryl groups were determined with Ellman's reagent[11]. From Murakami-Murofushi & Ratner[10].

Cooperative Substrate Effects in the Regulation of Argininosuccinase

Double reciprocal plots of kinetic data exhibit a break in slope indicating negative cooperative substrate effects (Fig. 9). In 0.05 M Tris buffer, pH 7.5, linear double reciprocal plots were observed ($\underline{K_m}$ = 5.1 x 10^{-5} M), whereas in 0.06 M phosphate buffer, the curves were biphasic and gave $\underline{K_m}$ values for argininosuccinate of 6.4x10^{-5} and 1.8x10^{-4} M at low and high substrate concentrations, respectively.

Table 5 shows the comparison of kinetic data given for brain, liver and kidney enzymes. The capacity to regulate catalysis, presumably through conformational change, by maintaining appreciable activity over wide changes in substrate concentration has been attributed to this negative homotropic behavior[9].

Antigenic Properties

Antigenicity was studied by the Ouchterlony double diffusion technique with rabbit antisera prepared against the bovine liver and kidney enzymes. The brain enzyme formed similar precipitin bands whether antibody against liver or kidney enzyme was present as shown in Fig. 10. The bands indicated symmetrical fusion and no spur was present. It suggests the close similarity in antigenic properties of the enzymes from three different sources. To further examine the similarities in the antigenic properties, the inhibitory effects of

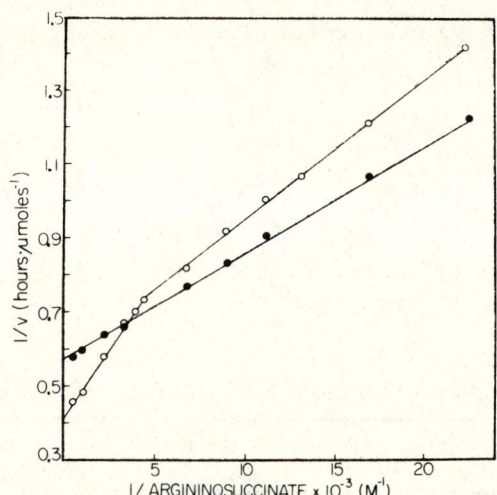

Fig. 9. Effects of phosphate and Tris-HCl on substrate dependence of brain argininosuccinase activity. Potassium phosphate buffer (o); Tris-HCl buffer (•). From Murakami-Murofushi & Ratner[10].

ISOLATION OF ARGININOSUCCINASE

Table 5. Comparison of Catalytic Properties of Bovine Brain, Liver and Kidney Argininosuccinases[a]

Property	Enzyme source		
	Brain	Liver	Kidney
$\underline{K_m}$ (M)			
KPO_4[b], low ASA concentration	6.4×10^{-5}	4.5×10^{-5}	–
high ASA concentration	1.8×10^{-4}	1.7×10^{-4}	1.4×10
Tris[c]	5.1×10^{-5}	3.7×10^{-5}	–
Inhibition by antibody	c.r.[d]	c.r.	c.r.
Optimal pH	7.5	7.5	7.5

[a]Modified from Ratner et al.[6,10].

[b]Ionic strength $\Gamma/2 = 0.15$

[c]Ionic strength $\Gamma/2 = 0.05$

[d]cross reactive

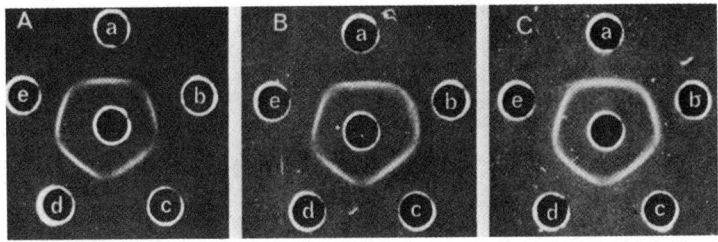

Fig. 10. Comparison of the cross-reactivity of brain argininosuccinase as antigen with antibody to argininosuccinases from liver and kidney. A: To wells of (a) and (b), the liver enzyme at 0.26 and 0.13 mg/ml were added, and to (c), (d) and (e), the brain enzyme at 0.185, 0.139 and 0.093 mg/ml were added, respectively. To the center well was added the rabbit antiserum to the liver enzyme. B : (a) and (b) contained the kidney enzyme at 0.230 and 0.115 mg/ml; (c), (d) and (e) contained the brain enzyme at the same concentration as in A. Center well contained liver enzyme antiserum. C : The outer wells contained the same amounts of kidney and brain enzymes as in B, and center well contained the rabbit antiserum to the kidney argininosuccinase. From ref. 10.

liver enzyme antibody on the activity of brain enzyme was investigated. Figure 11 shows that the inhibition of activity of brain and liver enzymes were increased to the same extent in proportion

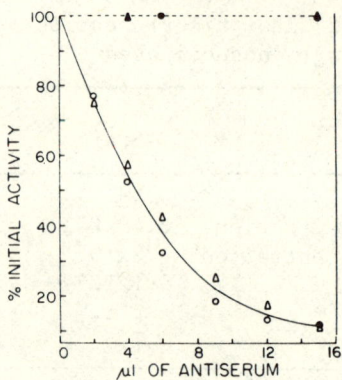

Fig. 11. Inhibition of catalytic activity of brain and liver argininosuccinases by rabbit antibody to liver enzyme. Brain enzyme (o); liver enzyme (△). From Murakami-Murofushi & Ratner[10].

to the amount of antibody present. Almost the same extent of inhibition by antibody of liver enzyme on the activity of kidney enzyme has been observed[8].

Genetic Consideration

The resemblance in kinetic, structural and immunological properties of the enzymes from different sources, and the close similarity in amino acid sequence of the four monomeric polypeptide chains, have suggested that argininosuccinases from brain, liver and kidney might be coded for by the same structural gene. The possibility that each is controlled by different regulatory genes has not been excluded.

REFERENCES

1. Ratner, S., Morell, H. and Carvalho, E., Arch Biochem. Biophys., 91 : 280, (1960).
2. Allan, J.D., Cusworth, D.C., Dent, C.E., and Wilson, V.K., Lancet, 1 : 182; (1958).
3. Shih, V.E., and Efron, M.L., in : "The Metabolic Basis of Inherited Disease", Stanbury, J.B., Wyngaarden, J.B., and Frederickson, D.S., eds., 3rd ed., pp. 370, McGraw-Hill, New York, (1972).
4. Tomilson, S. and Westall, R.G., Nature (London), 188 : 235 (1960).
5. Havir,E.A., Tamir, H., Ratner, S., and Watner, R.C., J. Biol. Chem., 240 : 3079, (1965).

6. Schulze, I.T., Lusty, C.J., and Ratner, S., J. Biol. Chem., 245 : 4534, (1970).
7. Lusty, C.J., and Ratner, S., J. Biol. Chem., 247 : 7010, (1972).
8. Bray, R.C., and Ratner, S., Arch. Biochem. Biophys., 146 : 531, (1971).
9. Rochovansky, O., J. Biol. Chem., 250 : 7225, (1975).
10. Murakami-Murofushi, K., and Ratner, S., Anal. Biochem., 95 : 139, (1979).
11. Ellman, G., Arch. Biochem. Biophys., 82 : 70, (1959).

ON THE MECHANISM OF THE ALTERATIONS OF RAT KIDNEY TRANSAMIDINASE

ACTIVITIES BY DIET AND HORMONES

John F. Van Pilsum, Denise M. McGuire and Howard Towle

Department of Biochemistry
The Medical School
Minneapolis, Minnesota, 55455, USA

The first reaction in the biosynthesis of creatine is the transfer of the amidine group of arginine to the amino group of glycine to form ornithine and guanidinoacetic acid. This reaction is catalyzed by the enzyme L-arginine : glycine amidinotransferase, EC 2.1.4.1, commonly called transamidinase. Transamidinase was discovered by Borsook and Dubnoff in 1941[1] and the only tissues from the rat that have significant transamidinase activities are kidney and pancreas [1-5]. The second reaction in the biosynthesis of creatine is the methylation of guanidinoacetic acid in the liver by S-adenosylmethionine catalyzed by the enzyme S-adenosylmethionine : guanidinoacetate N-methyltransferase, EC 2.1.1.2. Hormonal induced alterations in this enzyme activity have not been found[6]. Livers from rats fed complete diets supplemented with creatine had similar guanidinoacetate methyltransferase activities as from rats fed the complete diet without creatine[7]. Rat kidney transamidinase activities have been found to be altered greatly in a variety of dietary and hormonal states. Thus transamidination is thought to be the control step in creatine biosynthesis. Since ornithine is a product of this enzyme reaction, an interrelationship may exist between the control of creatine and urea synthesis.

The first evidence of an alteration of rat kidney transamidinase by diet was that weanling rats fed a protein-free diet for 12 days had 24% of the kidney transamidinase activities as rats fed a complete diet[8]. Total carcass creatine and phosphocreatine were determined in rats fed a complete diet and a protein-free diet for 16 days[9]. No difference in the amount of creatine or phosphocreatine per g carcass in the two groups of rats was found. However, the body weights of the rats fed the complete diet were about 3 times greater than those fed the protein-free diets. Therefore, the total creatine and phosphocreatine per rat was 3 fold greater in the rats fed the complete diet than in those fed the protein-free diet. In other words, a correlation between the transamidinase activities and the total amounts of creatine in the rats seemed to exist.

Fitch et al.[10] and Walker[11] reported that rats fed complete diets supplemented with 1-3g creatine/100 g diet for about one month had ~30% of the kidney transamidinase activities found in rats fed the unsupplemented diet. The weight gains of the two groups of rats were similar. Walker[12] reported that kidney transamidinase activities returned to normal levels 4 days after creatine was removed from the diet. No data were presented in these reports on the effect of creatine feeding on the amounts of creatine in the carcasses of the animals or on the amounts of creatine and creatinine excreted in the urine. Thus, there was no evidence that the low kidney transamidinase activities in the creatine-fed rats was accompanied by a decrease in the biosynthesis of creatine. Both Fitch et al.[10] and Walker[11] did find that the addition of creatine to the transamidinase assay mixture was without any effect on the enzyme activities of kidney homogenate from normal rats. Therefore, depression of transamidinase activities by dietary creatine did not involve an inhibition of the enzyme by creatine itself.

In 1966 Ungar and Van Pilsum[13] demonstrated that rat kidney transamidinase activities were regulated by growth hormone. Kidneys from hypophysectomized rats had ~20% of the transamidinase activities as kidneys from sham-hypophysectomized rats. Adrenalectomy, thyroidectomy, or castration of the male rat were without any effect on the enzyme activity. Hypophysectomized male weanling rats given 7 daily doses of 20 µg bovine growth hormone had two times the kidney transamidinase activities as hypophysectomized rats. The administration of a number of other hormones to hypophysectomized rats was without any effect on the low transamidinase activities.

Rats that were given large doses of either thyroxine or adrenal steroid hormones had low kidney transamidinase activities[14]. Male rats have kidney transamidinase activities two times greater than female rats[15,16]. Castrated female rats had kidney transamidinase activities similar to those of male rats. Male rats given injections of the female sex hormone had lower kidney transamidinase activities than the males given no injections. Female rats injected with testosterone had slightly higher enzyme activities than found in females given no injections.

In 1970 we reported that thyroidectomized rats had ~20% of the kidney transamidinase activities as intact rats[17]. The failure to find any effect of thyroidectomy in the earlier report[13] is attributed to incomplete surgical removal of the thyroid gland. Thyroidectomized rats injected with 0.5 µg thyroxine/day for 14 days had transamidinase activities similar to those found in intact rats. Injections of growth hormone into thyroidectomized rats or injections of thyroxine into hypophysectomized rats were without any effect on the low transamidinase activities. We have also found that thyroxine injected into hypophysectomized rats that were receiving small amounts of growth hormone (5-10 µg/100g body wt/day) was without any effect on the transamidinase activities. However, hypophysectomized rats that were receiving large amounts of growth hormone (100-200 µg/100 g body wt/day) and that were also injected with thyroxine had higher transamidinase activities than the rats injected only with growth hormone. Others have reported that small amounts of thyroxine are secreted by the thyroid gland in the absence of the pituitary[18]. It is also well established by others that thyroxine is necessary for the production of growth hormone by the pituitary[19]. Therefore, both growth hormone and thyroxine were concluded to be necessary for maintenance of rat kidney transamidinase activities. The small amount of thyroxine secreted by the thyroid gland in the absence of the pituitary is sufficient to permit stimulation of transamidinase activities by growth hormone in the hypophysectomized rat.

It was believed that conclusive evidence had been obtained that rat kidney transamidinase was indeed under dietary and hormonal control. Furthermore, the possibility existed for interrelationships between the dietary and hormonal control mechanisms. The requirement of growth hormone and thyroxine for transamidinase activities was considered of interest in view of the lack of knowledge of the action of these two hormones at the molecular level. The low transamidinase activities after feeding creatine supplemented diets was also of interest because this was one of the first demonstrations of metabolic control of a mammalian enzyme activity by the end product of the pathway. The isolation of rat kidney transamidinase in a homogeneous state was essential to further work in this area. Attempts made in our laboratory to isolate the rat enzyme by the procedure of Conconi and Grazi[20] for hog kidney transamidinase were not successful. A method for isolating rat kidney transamidinase in a homogeneous state was reported by McGuire et al. in 1980[21]. The procedure consisted of chromatography of a supernatant from rat kidney homogenate on DEAE-cellulose followed by chromatography on phenyl-Sepharose. Two fractions of enzyme, designated as α and β transamidinase were obtained with chromatography on DEAE-cellulose. The two fractions were individually purified to homogeneity as judged by their migration as a single band in native and sodium dodecyl sulfate gel electrophoresis and by sedimentation equilibrium experiments. No significant differences in the properties of the α and β fractions have been found. The minimum molecular weights, determined by sodium dodecyl sulfate gel electrophoresis, were 42,000 and 44,000. The apparent molecular weights, determined by sedimentation equilibrium were 82,600 and 83,300. The amino acid compositions and specific activities of the two fractions were similar. The pI values were 6.9 to 7.3. The Km values were 2.8 and 2.4 mM for arginine and 3.0 and 3.1 for glycine. The immunological properties of the α and β fractions were indistinguishable. At the present time no explanation can be given for the separation of transamidinase into α and β fractions on DEAE-cellulose.

Monospecific antibodies to the purified transamidinase were prepared and the amounts of transamidinase protein in kidneys from three groups of experimental rats were determined by immunotitration experiments. The purpose of the immunotitration experiments was to determine if the alterations of transamidinase activities after removal of the pituitary gland were due to a decrease in the amount of enzyme protein or a modification of the catalytic capacity of the enzyme. Hypophysectomized rats had 30 and 33%, respectively, of the activities and the amounts of immunoprecipitable transamidinase protein found in kidneys of sham-hypophysectomized rats. Hypophysectomized rats that had received nine daily injections of growth hormone and thyroxine had similar amounts of transamidinase activity and transamidinase protein as sham-hypophysectomized rats. Growth hormone and thyroxine were concluded to be necessary for maintaining the levels of transamidinase protein in rat kidney.

The amounts of transamidinase protein in kidneys from intact rats fed a complete diet and from rats fed a complete diet supplemented with creatine are being determined in our laboratories by immunotitration experiments. Preliminary evidence is interpreted to indicate that the major portion of the decrease in enzyme activities that occurs in creatine feeding can be accounted for by a decrease in the amount of transamidinase protein.

The low amounts of transamidinase protein in hypophysectomised rats or in intact rats fed a creatine-supplemented diet could be the result of a slower than normal rate of its synthesis or a faster than normal rate of its degradation. Therefore, some experiments have been done in an attempt to determine the relative rate of synthesis of transamidinase protein in hypophysectomized rats maintained with or without injections of growth hormone plus thyroxine and in intact rats fed diets with and without added creatine. The in vivo relative rate of synthesis of kidney transamidinase was expressed as the incorporation of $[^3H]$-leucine into immunoprecipitable kidney transamidinase relative to its incorporation into total cytosolic protein following a 60 minute labeling period. On the basis of a limited number of experiments it appears that the relative rate of synthesis of transamidinase is lower in hypophysectomized rats than it is in hypophysectomized rats that were given injections of growth hormone plus thyroxine. The relative rate of synthesis of transamidinase also appears to be lower in intact rats fed a creatine supplemented diet than it is in rats fed creatine-free diet.

Thyroxine and growth hormone have been reported to be necessary for the production of α_{2u}-globulin, a protein of unknown function found in the urine of male rats[22]. Growth hormone has been reported to be induced by thyroid hormone in rat pituitary-derived cell lines[23,24]. Albumin[25] and malic enzyme[26] in liver have been reported to be induced by growth hormone and thyroxine, respectively. In all of these examples the hormones have been shown to increase the synthesis of the specific protein and the mRNA levels for the protein[27].

A procedure has been developed in our laboratories for the synthesis of transamidinase, in vitro, using a lysate of rabbit reticulocytes and mRNA isolated from rat kidney. The amount of transamidinase is determined by precipitation with its monospecific antibody followed by gel electrophoresis. The amount of radioactivity localized in the portion of the gel containing the transamidinase is assumed to be proportional to the amount of mRNA for the enzyme in the kidney. The amounts of transamidinase mRNA in kidneys from rats in various physiological states are now being determined. In other words, we should be able to determine if the low transamidinase activities found in creatine-fed rats or hypophysectomized rats can be explained by low amounts of transamidinase mRNA.

REFERENCES

1. H. Borsook and J.W. Dubnoff, The formation of glycocyamine in animal tissues, J. Biol. Chem., 138 : 389 (1941).
2. J.B. Walker, Role for pancreas in biosynthesis of creatine, Proc. Soc. Exp. Biol. and Med., 98 : 7 (1958).
3. J.B. Walker and M.S. Walker, Formation of creatine from guanidinoacetate in pancreas, Proc. Soc. Exp. Biol. and Med., 101 : 807 (1959).
4. J.F. Van Pilsum, B. Olsen, D. Taylor, T. Rozycki, and J.C. Pierce, Transamidinase activities, in vitro, of tissues from various mammals and from rats fed protein-free, creatine-supplemented and normal diets, Arch. Biochem. Biophys., 100 : 520 (1963).
5. J.F. Van Pilsum, G.C. Stephens, and D. Taylor, Distribution of creatine, guanidinoacetate and the enzymes for their biosynthesis in the animal kingdom, Biochem. J., 126 : 325 (1972).
6. M. Carlson and J.F. Van Pilsum, S-adenosylmethyionine : guanidinoacetate N-methyltransferase activities in livers from rats with hormonal deficiencies or excesses, Pro. Soc. Exp. Biol. and Med., 143 : 1256 (1973).
7. J.F. Van Pilsum and M. Carlson, Assay for S-adenosylmethionine : guanidinoacetate N-methyltransferase activities of rat liver homogenates, Anal. Biochemistry, 35 : 424 (1970).
8. J.F. Van Pilsum, D.A. Berman, and E.A. Wolin, Assay and some properties of kidney transamidinase, Proc. Soc. Exp. Biol.

and Med., 95 : 96 (1957).
9. J.F. Van Pilsum, Creatine and creatine phosphate in normal and protein-depleted rats, J. Biol. Chem., 228 : 145 (1957).
10. C.D. Fitch, C. Hsu, and J.S. Dinning, Some factors affecting kidney transamidinase activity in rats, J. Biol. Chem.; 235 : 2362 (1960).
11. J.B. Walker, Metabolic control of creatine biosynthesis, J. Biol. Chem., 235 : 2357 (1960).
12. J.B. Walker, Repression of arginine-glycine transamidinase activity by dietary creatine, Biochim. Biophys. Acta, 36 : 574 (1959).
13. F. Ungar and J.F. Van Pilsum, Hormonal regulation of rat kidney transamidinase; Effect of growth hormone in the hypophysectomized rat, Endocrinology, 78 : 1238 (1966).
14. F. Ungar and J.F. Van Pilsum, Effects of adrenal steroids and thyroid hormone on creatine and kidney transamidinase in the rat, Endocrinology, 79 : 1143 (1966).
15. J.F. Van Pilsum and F. Ungar, Effect of castration and steroid sex hormones on rat kidney transamidinase, Arch. Biochem. Biophys., 124 : 372 (1968).
16. I. Krisko and J.B. Walker, Influences of sex hormones on amidinotransferase levels. Metabolic control of creatine biosynthesis, Acta Endocrinol., 53 : 655 (1966).
17. J.F. Van Pilsum, M. Carlson, J.R. Boen, D. Taylor, and B. Zakis, A bioassay for thyroxine based on rat kidney transamidinase activities, Endocrinology, 87 : 1237 (1970).
18. R.O. Scow, Effect of growth hormone on muscle and skin collagen in neonatal thyroidectomized rats, Endocrinology, 49 : 641 (1951).
19. J. Solomon and R.O. Greep, The effects of alterations in thyroid function on the pituitary growth hormone content and acidophil cytology, Endocrinology, 65 : 158 (1959).
20. F. Conconi and E. Grazi, Transamidinase of hog kidney. Purification and properties, J. Bio. Chem., 240 : 2461 (1965).
21. D.M. McGuire, C.D. Tormanen, I.S. Segal, and J.F. Van Pilsum, The effect of growth hormone and thyroxine on the amount of L-arginine : glycine amidinotransferase in kidneys of hypophysectomized rats. Purification and some properties of rat kidney transamidinase, J. Biol. Chem., 255 : 1152 (1980).
22. A.K. Roy and D.J. Dowbenko, Role of growth hormone in the multihormonal regulation of messenger RNA for $\alpha_{2\mu}$ globulin in the liver of hypophysectomized rats, Biochemistry, 16 : 3918(1977).
23. H. Seo, G. Vassart, H. Brocas, and S. Refetoff, Triiodothyronine stimulates specifically growth hormone mRNA in rat pituitary tumor cells, Proc. Natl. Acad. Sci. USA, 74 : 2054 (1977).
24. J.A. Martial, J.D. Baxter, H.M. Goodman, and P.H. Seeburg, Regulation of growth hormone messenger RNA by thyroid and glucocorticoid hormones, Proc. Natl. Acad. Sci. USA, 74 : 1816 (1977).

25. G.H. Keller and J.M. Taylor, Effect of hypophysectomy and growth hormone treatment on albumin mRNA levels in the rat liver, J. Biol. Chem., 254 : 276 (1979).
26. F. Isohashi, K. Shibayama, E. Maruyama, Y. Aoki, and F. Wada, Immunochemical studies on malate dehydrogenase (decarboxylating) (NADP), Biochim. Biophys. Acta., 250 : 14 (1971).
27. H.C. Towle, C.N. Mariash, and J.H. Oppenheimer, Changes in the hepatic levels of messenger ribonucleic acid for malic enzyme during induction by thyroid hormone or diet, Biochemistry, 19 : 579 (1980).

SCIATECTOMIC STIMULATION OF MUSCLE

ARGINASE AND ITS IMPLICATIONS

V. Mohanachari, P. Neeraja, K. Indira and K.S. Swami

Department of Zoology,
Sri Venkateswara University,
Tirupati - 517 502, India

Denervated muscle is characterized by high turnover of proteins, hyperammonemia, degeneration and regeneration of muscle fibers (Shahzad, 1977; Goldspink 1978; Chetty et al, 1980). In view of these conditions an enzyme like arginase, which cleaves L-arginine to L-ornithine and urea was investigated mainly because, its detoxification role in ammoniotoxemia is well established and its role in extra hepatic tissues, where other urea cycle enzymes are absent, still remains to be elucidated.

Unilateral sciatic denervation was performed always on right side in healthy frogs, Rana hexadactyla as described by Dass and Swami (1972). One month after sciatectomy the animals were examined for marked atrophy in gastrocnemius muscle. After ensuring it the animals were sacrificed, the control and denervated muscles were excised with minimal loss of time from six animals and chilled to 0°C followed by immediate homogenization to 20% (wt/vol) in ice cold 0.1% (wt/vol) cetyltrimethylammonium bromide (CTB) in a glass homogenizer embedded in ice. The homogenate was centrifuged at 2000 x g for 15 min to eliminate the cell debris and the clear crude supernatant was employed as the enzyme source after dialysis against 0.1% CTB.

Arginase activity was assayed by the method of Campbell (1961) in a medium containing 100 m moles of glycine-sodium hydroxide buffer (pH 9.5), varied concentrations of L-arginine adjusted to pH 9.5 and 5 m moles of manganese chloride in a final volume of 0.7 ml. The reaction mixture was incubated at 37°C for 30 min by adding 0.3 ml of enzyme source and then stopped with 2 ml of 1.0 M perchloric acid. The urea formed in the aliquots was measured spectrophotometrically by diacetyl monoxime method. All the enzyme velocities were corrected with non enzymatic hydrolytic velocities. The protein in the enzyme source was measured according to the procedure of Lowry et al (1951) using bovine serum albumin as the standard.

The results are summarized in the table 1.

The Vmax values of arginase in denervated muscle showed increased velocity of the enzyme indicating the increased rate of hydrolysis of L-arginine. The Km of arginase was found unaffected in denervated muscle as compared with the control muscle. Despite a reported 100% increase in L-arginine in denervated muscle, apparently no change in Km was witnessed in the present investigation. This fact suggests that the affinity between the substrate and enzyme molecule was not affected. The increase in Vmax could be due to substrate induced activation of the enzyme in denervated muscle. A similar induction of arginase by the addition of arginine in mammalian cell culture was reported by Klein (1960).

Table 1. Substrate dependent kinetic analysis of arginase activity in control and denervated amphibian gastrocnemius muscles.

(The Vmax values are per mg protein per hour. The Vmax and Km were derived from the Lineweaver Burke double reciprocal plots, using least squares as the best fit).

Sample	Kinetic parameters	
	Vmax	Km
Control muscle	0.0540×10^{-6} M	5×10^{-3} M
Denervated muscle	0.0796×10^{-6} M	5×10^{-3} M
Percent change	47.4	-

The biochemical necessity of intensified arginase activity during denervation appears to be three dimensionally oriented. First, the increased ammonia could be fixed into urea cycle and the ammonia toxicity can be avoided. Secondly, during the process of elimination of ammonia toxicity through arginase activity, urea is produced and the presence of urea within the tissues has been reported to play a vital chemohomeostasis role, as is the case with marine elasmobranchs which are dependent at least on some minimal urea concentration in tissues (Alexander et al, 1968), besides its functioning in enzymatic stabilization during certain physiological conditions (Baxter, 1976). Thirdly, one of the main functions of urea includes its indispensability in certain amphibians for muscle contraction (Thesleff and Schmidt-Nielsen, 1962). Since sciatectomized muscle is primarily deprived of neural regulation for muscular contraction, the high urea levels may contribute to the limited contractile activity of the muscle as in the case of crab-eating-frog, Rana cancrivora, in which the presence of urea is an exclusive necessity for muscle contraction (Thesleff and Schmidt-Nielsen, 1962). The stepped up arginase activity in denervated muscle is well in line with high urea (Chetty et al, 1980) and ornithine levels (Narayanareddy and Swami, 1975).

The other alternative role of arginase could possibly be the production of ornithine which is a precursor for polyamine synthesis via the ornithine decarboxylase reaction. These polyamines might be of utmost importance in the regeneration process of denervated muscle, since polyamines are required in the growth process. Though arginase activity is witnessed in the muscle accounting for the urea and ornithine levels, it is still not clearly known whether all the enzymes of urea cycle are effectively operating in the muscle. The presence or absence of complement of urea cycle enzymes and the role of ornithine via polyamine synthesis in the regeneration of denervated muscle are currently under investigation in this laboratory.

REFERENCES

Alexander, M.D., Haslewood, E.S., Haslewood, G.A.D., Watts, D.C. and Watts, R.L., 1968, Comp. Biochem. Physiol., 26: 971-978.
Baxter, C., 1976, Abstr. Amer. Soc. Neurochem. No.74: 100-101.
Campbell, J.W., 1961, Arch. Biochem. Biophys., 93: 448-455.
Chetty, C.S., Naidu, R.C., and Swami, K.S., 1980, J. Biol. Sci., 2: 135-138.
Dass, P.M.K., and Swami, K.S., 1972, Enzymologia, 42: 235-244.
Goldspink, D.F., 1978, Comp. Biochem. Physiol., 61B: 37-41.
Klein, E., 1960, Exptl. Cell. Res., 21: 421-429.
Lowry, O.H., Rosebrough, N.J., Farr, A.L. and Randall, R.J., 1951, J. Biol. Chem., 193: 265-275.
Narayanareddy, K., and Swami, K.S., 1975, Indian. J. Exp. Biol., 13: 343-345.
Shahzad, M.A., 1977, J. Neurol. Sci., 33: 251-266.
Thesleff, S. and Schmidt-Nielsen, K., 1962, J. Cell. Comp. Physiol. 59:31-34.

ENZYMES OF ARGININE METABOLISM IN THE LIZARD CALOTES VERSICOLOR

T. G. Baby and S. Raghupathi Rami Reddy

Department of Zoology
University of Poona
Pune 411 007, India

INTRODUCTION

Evolutionary transition from ureotelism to uricotelism in the animal kingdom is associated with the loss of a functional urea cycle and hence the ability to synthesize arginine (Campbell, 1973). Thus, while arginase is ubiquitously distributed in animals, one or more of the other enzymes of the urea cycle are absent in all the uricotelic animals that have been investigated (Brown and Cohen, 1960; Tamir and Ratner, 1963; Mora et al., 1965; Inokuchi et al., 1969; Kameyama and Miura, 1970; Reddy and Campbell, 1977). Arginase is usually regarded as a ureogenic enzyme, but ureogenesis can be a major function of arginase only in the ureotelic animals and not in the uricotelic animals. Knox and Greengard (1965) suggested that arginase functions in roles other than urea formation under situations where it exists without the other enzymes of the urea cycle.

In the uricotelic animals, ornithine formed by the action of arginase on exogenous arginine cannot be recycled in the absence of ornithine carbamoyltransferase. Presumably, arginase in these animals serves a catabolic role making ornithine available for conversion to proline and glutamate via the intermediate formation of 1-pyrroline-5-carboxylate (Adams, 1970) or for the decarboxylation to putrescine from which the polyamines, spermidine and spermine, are synthesized (Tabor and Tabor, 1972). The non-ureogenic functions of arginase in uricotelic animals are not well understood except that in insects it provides ornithine as a precursor for the synthesis of proline required in flight muscle metabolism (Reddy and Campbell, 1969). We present here enzymatic evidence for the metabolic conversion of ornithine formed in the arginase reaction to proline glutamate and putrescine in the lizard Calotes versicolor.

MATERIALS AND METHODS

The collection of lizards, their maintenance in the laboratory and the assay of tissues for the urea cycle enzymes, arginase, ornithine carbamoyltransferase (OCTase) and arginine synthetase (ASase; the combined activity of argininosuccinate synthetase and argininosuccinate lyase), have been described earlier (Baby et al., 1976; Baby and Reddy, 1980). Ornithine metabolizing enzymes were assayed as follows: ornithine aminotransferase (OATase) by the method of Peraino and Pitot (1963); 1-pyrroline-5-carboxylate reductase (P-5-C reductase), 1-pyrroline-5-carboxylate dehydrogenase (P-5-C dehydrogenase) and proline oxidase according to the method of Herzfeld et al.(1977) and ornithine decarboxylase by the method of Janne and Williams-Ashman (1971). DL-1-pyrroline-5-carboxylate was prepared by the periodate oxidation of DL-hydroxylysine as described by Williams and Frank (1975).

RESULTS

Urea Cycle Enzymes

In the liver, kidney and brain tissues of the lizard, OCTase and ASase activities could not be detected (Table 1). More concentrated homogenates and longer incubation periods were employed to ascertain the absence of these enzymes in the lizard tissues. Arginase activity,

Table 1. Urea Cycle Enzymes in Frog, Rat and Lizard Tissues[a]

Tissue	OCTase	ASase	Arginase
Frog Liver	160.60	0.11	126.80
Rat Liver	78.60	0.19	68.40
Lizard Liver	B.L.D.	B.L.D.	15.20 $\pm$ 9.90
Lizard Kidney	B.L.D.	B.L.D.	37.60 $\pm$ 16.80
Lizard Brain	B.L.D.	B.L.D.	1.00 $\pm$ 0.30

[a] Enzyme activities are expressed as µmoles of product formed per hour per mg protein; B.L.D.: Below the level of detection. Data taken from Baby and Reddy, (1980).

on the other hand, is present in all the three tissues of the lizard. Activity of this enzyme in lizard kidney and liver is comparable to that reported by Brown and Cohen (1960) in the liver of some ureotelic species. The activities of the three urea cycle enzymes assayed in the livers of frog (Rana tigrina) and rat (Rattus norvegicus) are also included in Table 1 for comparison.

Enzymes of Ornithine Metabolism:

The activities of OATase which catalyzes the conversion of ornithine to pyrroline-5-carboxylate in the presence of 2-oxoglutarate, P-5-C- dehydrogenase which oxidizes pyrroline-5-carboxylate to glutamate in the presence of NAD, P-5-C reductase which reduces pyrroline-carboxylate to proline in the presence of NADH, proline oxidase which oxidizes proline to pyrroline-5-carboxylate and ornithine decarboxylase which decarbodylates ornithine to putrescine are demonstrable in the tissues of the lizard (Fig. 1).

DISCUSSION

Uric acid, ammonia and urea constitute about 89%, 0.85% and 0.45%, respectively, of the urinary nitrogen in C. versicolor (Baby and Reddy, 1977). The apparent absence of OCTase and ASase in the tissues of this lizard is in agreement with the hypothesis that evolution of uricotelism is associated with the loss of a functional urea cycle (Campbell, 1973). The absence of these two enzymes and presence of arginase in C. versicolor is supported by in vivo experiments on the chameleon in which injected arginine is converted to ornithine, but arginine is not synthesized from ornithine and citrulline (Coulson and Hernandez, 1974). Maximum conversion of the injected arginine into ornithine occurs in the kidney of the chameleon (Herbert, 1973) supporting the observation that kidney has the highest arginase activity in the lizard.

The presence of arginase, OATase, P-5-C dehydrogenase, P-5-C reductase, proline oxidase and ornithine decarboxylase in the lizard suggests the metabolic conversion of arginine into glutamate, proline and polyamines by the pathway outlined in Fig. 2. Demonstration of enzymes can only indicate the metabolic potential and conclusive evidence for the operation of the pathway in lizards should await isotopic studies. However, the elevation of glutamate and glutamine levels in the tissues of chameleon injected with arginine and ornithine (Herbert and Coulson, 1972; Herbert, 1973) provide partial support to the operation of the pathway (Fig. 1). Glutamate is the most concentrated of the free amino acids in the liver, kidney, brain and heart of the chameleon (Herbert, 1973), while proline is abundant in connective tissue proteins like collagen (Adams, 1970) and polyamines are present in the tissues of C. versicolor at considerable concentrations (Baby and Reddy, 1981).

Studies on chicken and insects also suggest the catabolic function of arginase in uricotelic animals. In chicken liver, arginase and OATase activities increase in response to the stimulation of gluconeogenesis by hydrocortisone administration suggesting that

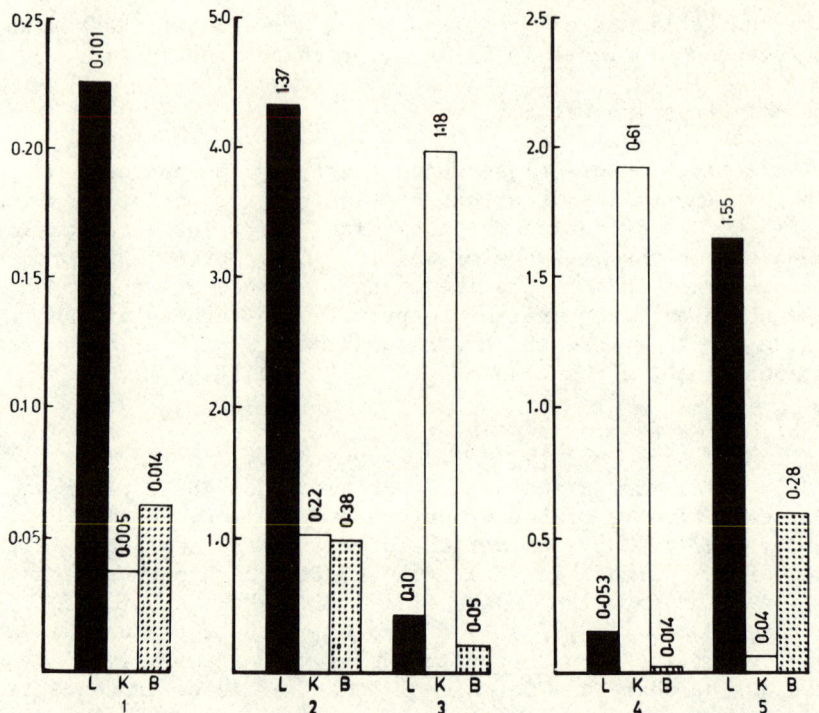

Fig. 1. Distribution of enzymes of ornithine metabolism (1. OATase, 2. P-5-C reductase, 3. P-5-C dehydrogenase, 4. proline oxidase, 5. ornithine decarboxylase) in lizard tissues (L=liver, K=kidney, B=brain). The activity units on the ordinate are μmoles product formed/hr/mg protein except in the case of ornithine decarboxylase where the units are nmoles CO_2 formed per hr per mg protein. Values plotted are the means of 6-8 assays. Standard deviations are given at the top of each column.

arginase in this uricotelic species functions as a catabolic enzyme (Vecchio and Kalman, 1968). In the silkmoth, Hyalophora gloveri, which lacks a functional urea cycle, arginase activity increases sharply at the time of emergence of the winged adult during pupal-adult development (Reddy and Campbell, 1969; 1977). Enzymatic and isotopic evidence suggests that arginase in the flying moth functions in the conversion of arginine to proline (Reddy and Campbell, 1969) which is a major substrate for insect flight muscle mitochondria (Sacktor and Childress, 1967; Hansford and Sacktor, 1970).

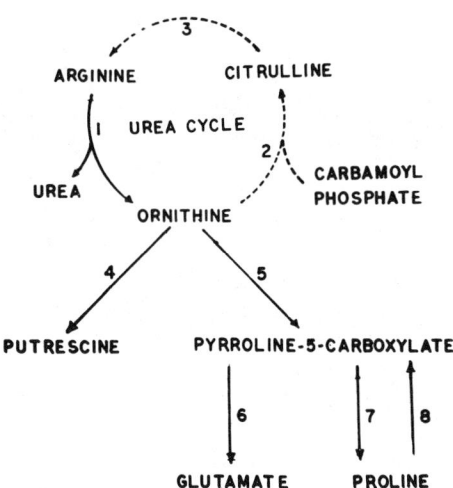

Fig.2. Pathway of arginine metabolism in lizard tissues.
1. Arginase, 2. OCTase, 3. ASase, 4. Ornithine decarboxylase, 5. OATase, 6. P-5-C dehydrogenase, 7. P-5-C reductase, 8. Proline oxidase. Enzymatic reactions shown in broken arrows could not be demonstrated.

Further evidence for the role of arginase outside the urea cycle comes from studies with the non-ureogenic tissues of ureotelic animals. In mammalian pancreas and submaxillary gland (Herzfeld and Raper, 1976) as well as liver cells in culture (Eliasson and Strecker, 1966), all of which lack OCTase, arginase appears to function mainly as a catabolic enzyme. Lactating mammary gland, which also lacks a functional urea cycle (Yip and Knox, 1972; Clark et al., 1975) takes up from the blood sigficantly more arginine and less proline and glutamate than it puts out in milk (Mepham and Linzell, 1966, 1967). The activities of OATase, P-5-C reductase and P-5-C dehydrogenase increase coordinately in the lactating mammary gland (Yip and Knox, 1972; Mezl and Knox, 1977) and isotopic studies are consistent with the conversion of arginine to proline and glutamate required for milk protein synthesis (Roets et al., 1974; Clark et al., 1975; Mezl and Knox, 1977). There is also evidence that arginase in lactating mammary gland functions to provide ornithine as a precursor for the formation of spermidine which stimulates the synthesis of casein and α-lactalbumin (Oka and Perry, 1974). A similar function has been suggested for epidermal arginase (Cotton and Mier ,1974). This is in accordance with the suggestion of Pegg and Williams-Ashman (1968) that provision of ornithine for ornithine decarboxylase reaction is perhaps the reason for the widespread occurrence of arginase in tissues where other enzymes of the urea cycle are absent.

Thus from a survey of the literature, it can be concluded that in cells and animals where a functional urea cycle is absent, arginase provides ornithine for the synthesis of proline, glutamate and polyamines. Considering the involvement of polyamines in the control of such fundamental processes as cell division and macromolecular synthesis (Tabor and Tabor, 1972; Russell, 1973), arginase is an important enzyme in the physiology of cells. Generation of ornithine appears to be a more basic function of arginase while ureogenesis is a more specialized function of arginase associated with the need to detoxify ammonia in certain animal groups. It therefore appears more appropriate to designate arginase as an ornithinogenic enzyme. Its function is to provide ornithine as a precursor either for the synthesis of citrulline in the urea cycle or for the formation of proline, glutamate and polyamines outside the urea cycle.

ACKNOWLEDGEMENTS

We thank the Head of the Department of Zoology for facilities and encouragement, the University of Poona for a grant-in-aid of research, and University Grants Commission for the award of Junior Research Fellowship to T.G.B.

REFERENCES

Adams, E., 1970, Metabolism of proline and hydroxyproline, Int. Rev. Connect.Tissue Res., 5: 1.

Baby, T.G., Goel, S.C. and Reddy, S.R.R., 1976, A comparative study of arginase activity in lizards, Physiol.Zool., 49:286.

Baby, T.G. and Reddy, S.R.R., 1977, Nitrogenous constitutents in the urinary deposits of the lizard Calotes versicolor, Brit.J. Herpetol., 5: 649.

Baby, T.G. and Reddy, S.R.R., 1980, Arginine metabolism in purinotelic animals. 1. Characterization of arginase and absence of other urea cycle enzymes in the lizard Calotes versicolor, J.Comp.Physiol., 140:261.

Baby, T.G. and Reddy, S.R.R., 1981, Distribution of polyamines and ornithine decarboxylase activity in the tissues of the lizard Calotes versicolor, J.Exp.Zool., 216: 37.

Brown, G.W. and Cohen, P.P., 1960, Comparative biochemistry of urea synthesis. 3. Activities of urea cycle enzymes in various higher and lower vertebrates, Biochem.J., 75: 82.

Campbell, J.W., 1973, Nitrogen excretion, in:"Comparative Animal Physiology",C.L. Proser, ed.,W.B. Saunders Co., Philadelphia.

Clark J.H., Derrig, R.G. and Dairs, C.L., 1975, Metabolism of arginine and ornithine in the cow and rabbit mammary tissue, J.Dairy Sci 58: 1808.

Cotton, D.W.K. and Mier, P.D., 1974, Role of arginase in epidermis, Nature, 252: 616.

Coulson, R.A. and Hernandez, T., 1974, Intermediary metabolism of reptiles, in: "Chemical Zoology", M.Florkin and B.T. Scheer, Eds., vol.9, Academic Press, New York.
Eliasson, E.E. and Strecker, H.J., 1966, Arginase activity during the growth cycle of Chang's liver cells, J.Biol.Chem.,241:5757.
Hansford, R.G. and Sacktor, B., 1970, The control of oxidation of proline by isolated flight muscle mitochondria, J.Biol.Chem., 245: 991.
Herbert, J.D., 1973, Amino acid metabolism in chameleon tissues, Comp. Biochem.Physiol.,46B: 229.
Herbert, J.D. and Coulson, R.A., 1972, Amino acid metabolism in Chameleon liver, Comp.Biochem.Physiol., 42B:463.
Herzfeld, A., Mezl, V.A. and Knox, W.E., 1977, Enzymes metabolizing Δ'-pyrroline-5-carboxylate in rat tissues, Biochem.J.,166: 95.
Herzfeld, A. and Raper, S.M., 1976, Amino acid metabolizing enzymes in rat submaxillary gland, normal or neoplastic, and in pancreas, Enzyme, 21: 471.
Inokuchi, T., Horie, V. and Ito, T., 1969, Urea cycle in silkworm Bombyx mori, Biochem.Biophys.Res.Commun., 35: 783.
Janne, J. and Williams-Ashman, H.G., 1971, On the purification of L-ornithine decarboxylase from rat prostate and effects of thiol compounds on the enzyme, J.Biol.Chem. 246: 1725.
Kameyama, A. and Miura, K., 1970, Changes in activities of carbamoyl phosphate synthetase and aspartate carbamoyltransferase in the life cycle of the blowfly Aldrichina grahami, Arch.Int. Physiol.Biochim. , 78: 435.
Mepham, T.B. and Linzell, J.L., 1966, A quantitative assessment of the contribution of individual plasma amino acids to the synthesis of milk proteins by the goat mammary gland, Biochem. J., 101: 76.
Mepham, T.B. and Linzell, J.L., 1967, Urea formation by the lactating goat mammary gland, Nature, 214: 507.
Mezl, V.A. and Knox, W.E., 1977, Metabolism of arginine in lactating rat mammary gland, Biochem.J., 166: 105.
Mora,J., Martuscelli, J., Ortiz-Pineda, J. and Soberon, G., 1965, The regulation of urea biosynthesis enzymes in vertebrates, Biochem.J., 96: 28.
Oka,T. and Perry, J.W., 1974, Arginase affects lactogenesis through its influence on the biosynthesis of spermidine, Nature, 250: 660.
Pegg, A.E. and Williams-Ashman,H.G., 1968, Biosynthesis of putrescine in the prostate gland of the rat, Biochem.J., 108: 533.
Peraino,C. and Pitot, H.C., 1963, Ornithine δ-transaminase in the rat.1. Assay and some general properties, Biochim.Biophys.Acta, 73: 222.
Reddy, S.R.R. and Campbell, J.W., 1969, Arginine metabolism in insects. Role of arginase in proline formation during silkmoth development, Biochem.J., 115: 495.

Reddy, S.R.R. and Campbell, J.W., 1977, Enzymic basis for the nutritional requirement of arginine in insects. Experientia, 33: 160.
Roets, E., Verbeke, R., Massart-Leen, A.M. and Peeters, G., 1974, Metabolism of (^{14}C) Citrulline in the perfused sheep and goat udder, Biochem.J., 144: 435.
Russel, D.H., 1973, The roles of the polyamines putrescine, spermidine and spermine in normal and malignant tissues. Life Sci. 13: 1635.
Sacktor, B. and Childress, C.C., 1967, Metabolism of proline in insect flight muscle and its significance in stimulating the oxidation of pyruvate, Arch.Biochem.Biophys., 120: 583.
Tabor, H. and Tabor, C.W., 1972, Biosynthesis and metabolism of 1,4-diaminobutane, spermidine, spermine and related amines, Adv.Enzymol., 36: 203.
Tamir, H. and Ratner, S., 1963, A study of ornithine, citrulline and arginine synthesis in growing chicks, Arch.Biochem.Biophys., 102: 259.
Vecchio, D.A. and Kalman, S.M., 1968, Ornithine transaminase in the liver of the chick embryo and in the young chick. Arch.Biochem.Biophys., 127: 376.
Williams, I. and Frank, L., 1975, Improved chemical synthesis and enzymatic assay of Δ'-pyrroline-5-carboxylic acid, Analyt. Biochem., 64: 85.
Yip, M.C.M. and Knox, W.E., 1972, Function of arginase in lactating mammary gland, Biochem.J., 127: 893.

BASIC BIOCHEMISTRY

3) Relation of Urea Cycle to Organic Acid Metabolism

OROTIC ACID IN URINE AND HYPERAMMONEMIA

C. Bachmann and J. P. Colombo

Dept. Clinical Chemistry
Inselspital
University of Berne, Switzerland

Orotoc acid (OA) an intermediary metabolite of the pyrimidine synthesis pathway is formed from carbamylphosphate (CP) and aspartate (Fig. 1). In the liver these precursors are synthetised mainly in the mitochondrium[1,2,3]. Natale and Tremblay[4] showed that CP formed by carbamylphosphate synthetase (CPS I) in the mitochondria crosses the mitochondrial membrane into the cytosol where the condensation with aspartate occurs. In addition especially in extrahepatic, fetal or regenerating tissues a cytosolic glutamine dependant carbamylphosphate synthetase (CPS II) exists[5,6]. Tremblay et al. showed that the synthesis of CP in the mitochondria accounts for a major part of the CP utilized for pyrimidine synthesis by rat liver slices.

Urinary OA has been shown to be overexcreted in ornithine transcarbamylase (OTC) deficiency[7]. In the last decade we have measured OA excretion in most of the hyperammonemic disorders described. We will try to summarize our experience. Because the colorimetric method is not reliable for measuring low concentrations[8] we developed a simple, rapid, accurate and precise anion exchange method[9]. The results we obtained fit those published with more cumbersome HPLC methods[10].

OA accumulates if its synthesis is increased or its metabolism impeded. This latter is found in primary orotic aciduria, azauridine treatment and conditions with PRPP depletion. We will focus on hyperammonemic disorders.

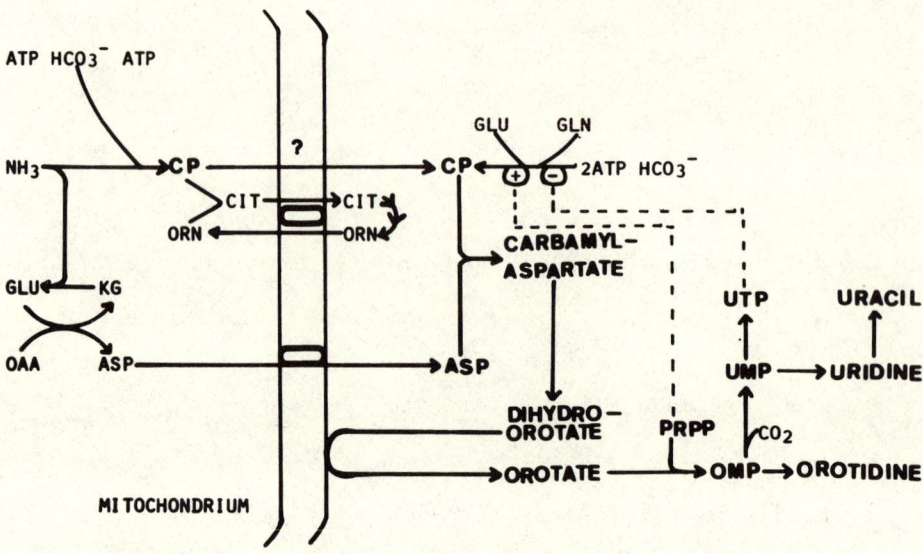

Fig. 1. Metabolic pathway of pyrimidine synthesis (CP : carbamylphosphate ; ORN : ornithine ; GLU : glutamate ;
GLN : glutamine ; KG : 2-oxoglutarate ; ASP : aspartate ;
OAA : oxaloacetate ; CIT : citrulline ; PRPP : phospho-
ribosylpyrophosphate

RESULTS

OA as well as uracil and uridine are increased in OTC deficiency (Fig. 2). Even if plasma concentrations of ammonia are normalized by a low protein diet, urinary OA, is still elevated. The wide range found in females illustrates the Lyonisation in this x-chromosomal defect. For asymptomatic mothers without hyperammonemia a better discrimination is obtained if OA is measured in a six hours urine after a protein load (1g/kg). OA is further elevated in citrullinemia (except benign variant). We did not find an increase in a patient with citrullinemia with pyruvate carboxylase deficiency[11]. OA is barely elevated in argininosuccinic aciduria but extremely high in argininemia. An elevation is also found in untreated lysinuric protein intolerance.

In contrast we found no increase of OA in CPS deficiency, N-acetylglutamate synthetase (NAGS) deficiency, or organic acidurias with hyperammonemia. It is not elevated in the transient hyperammonemia of the premature newborn (Table 1).

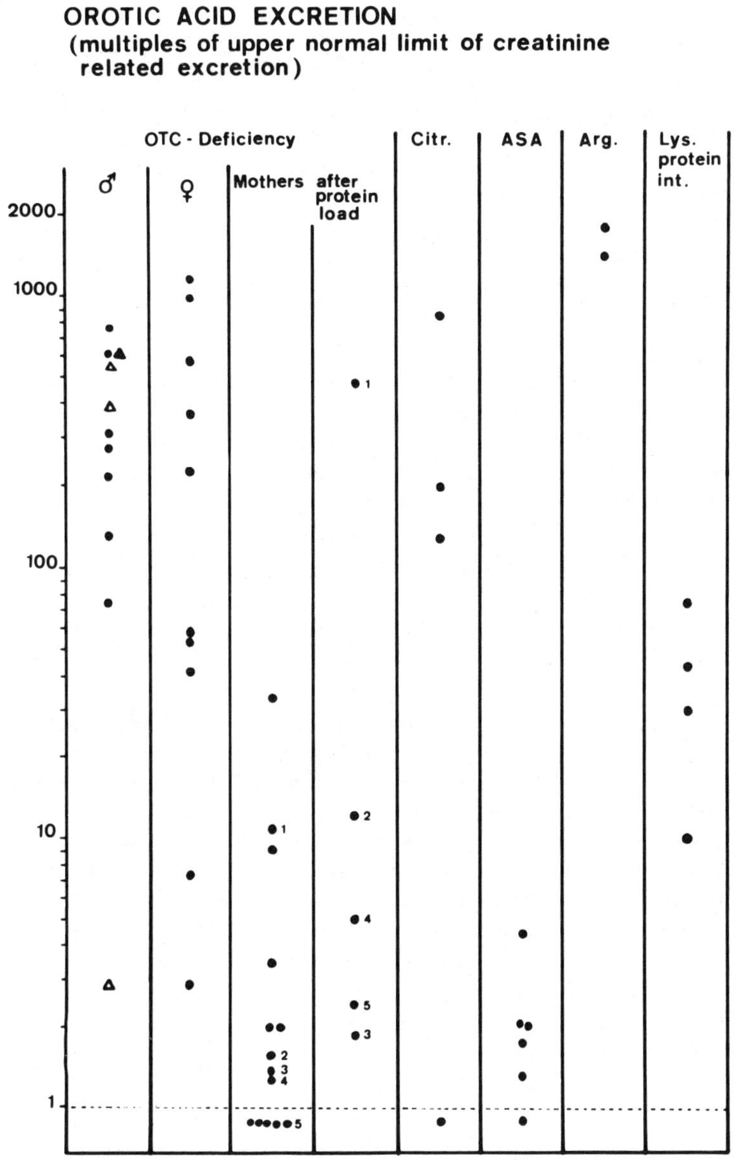
OROTIC ACID EXCRETION
(multiples of upper normal limit of creatinine related excretion)
OTC - Deficiency
Citr.
ASA
Arg.
Lys. protein int.
♂
♀
Mothers
after protein load
2000
1000
100
10
1
1
2
3
4
5

Fig. 2.

Table 1. Hyperammonemic patients without elevation of orotic acid excretion.

	Orotic acid μmol/g Creat.
Carbamylphosphate synthetase deficiency	12.9*
N-acetylglutamate synthetase deficiency	8*
Propionic acidemias	0.10*/0.06**/0.2**/11.4*/9.44**
Methylmalonic acidemia	0.2***
Biotin dependant multiple carboxylase deficiency	2.6**
3-methyl-3-hydroxyglutaric aciduria	0.1**
Transient hyperammonemia (premature)	2*/31*

Reference range *6-29, **2-34, ***0.7-4 μmol/g Creat.

DISCUSSION

We assume that if OA is not increased carbamylphosphate does not accumulate. Based on the findings of Shigesada and Tatibana[12] later confirmed by Coudé et al.[13] we have speculated[14] that hyperammonemia in propionic or methylmalonic aciduria is due either to inhibition of N-acetylglutamate synthetase or by formation of propionylglutamate. A recent publication of Stewart et al.[15] supports our hypothesis. It is at present not clear if the low activity of CPS is due to low N-acetyl-glutamate-be it by impeded NAG synthesis or low substrates (acetyl CoA or glutamate)- or to insufficient activation by propionylglutamate. Furthermore Cathelineau et al.[16] have given indications that ATP concentration might be rate limiting for CPS activity in organic acidurias. We suggest that if hyperammonia is due to insufficient activation of CPS, carbamylglutamate could be used for treatment in hyperammonemic organic acidurias. This would not be effective if ATP was limiting CPS activity in vivo.

We interpret an increased OA excretion in hyperammonemic disorders as an indicator of carbamylphosphate accumulation. Despite orotic acid overexcitation orotidine is not increased indicating that OMP-decarboxylase is not rate limiting. Whether a PRPP depletion is also operative in causing orotate accumulation is not clear. Uric acid excretion is not reduced, but its rate of de novo synthesis has not been measured to our knowledge in hyperammonemic disorders. We have not encountered patients with OTC deficiency who were at the same time hyperammonemic and had a normal OA excretion. One should however point out that in some patients the orotic aciduria was not very marked and could have been overlooked by a less specific method. A concomittant low activity of CPS could explain such findings[1,17,18].

The relatively low values found in argininosuccinic (ASA) aciduria are of interest especially when compared to those at least 500x higher in hyperargininemia. Plasma ammonia concentrations in those two patient groups were in the same range (100-250 μmol/l). In our opinion this illustrates the feedback mechanism of arginine on CPS through NAGS and NAG : While in argininosuccinic aciduria the product of the reaction - arginine-, is low it is obviously increased in argininemia. A high arginine in the mitochondrium will stimulate N-acetylglutamate synthesis and thus enhance the formation of CP. It is thus not surprising that patients with normal plasma ammonia and high OA were described[19].

An additional factor might be considered. The increased arginine in arginase deficiency will lead to a decreased tubular reabsorption of ornithine and thus could cause ornithine depletion. A deficiency of this substrate of OTC could in turn cause decreased utilisation of CP as postulated in rats on an arginine deficient diet[20]. Such a mechanism could also explain the OA overexcretion in lysinuric protein intolerance. It is of interest that despite a decreased arginine in ASA-uria a decrease of ornithine, if present, does not lead to that marked an orotic aciduria as in argininemia. Apparently due to the intact arginase ornithine is still formed and transported into the mitochondrium.

Our interpretation certainly simplifies the biochemical situation. We did not take into account the action of OKT or the transport mechanisms through the mitochondrial membrane which e.g. are certainly not clear for arginine and complex for ornithine.

Still, OA measurement has proven its usefulness as an indicator of carbamylphosphate accumulation. Thus in rats subjected to portacaval shunt[21] OA overexcretion is normalised by addition of branched chain ketoacids to the diet, but not by branched chain amino acids. Doubling the amount of calories or arginine (keeping the diet isonitrogenous) barely reduces orotate excretion. Doubling both calories and arginine increases orotate. Those rats further had an increased uric acid excretion which correlated with orotate significantly but not with weight loss. The uricosuric action of orotate has been described in primary orotic aciduria[22].

CONCLUSION

OA measurement in urine is useful in the differential diagnosis of hyperammonemia[23]. It is a good discriminatory parameter perhaps because of its high renal clearance (230 ml min^{-1}/1.73 m^2; 24). Done alone it will not allow to detect all the urea cycle disorders e.g. for screening purposes.

In hyperammonemia one can diagnose by aminoacid analysis of plasma and urine, citrullinemia, argininosuccinic aciduria, argininemia, lysinuric protein intolerance and hyperornithinemia with homocitrullinuria. If aminoacid changes are nonspecific an elevated OA will allow to diagnose OTC deficiency. If OA is normal gas chromatography/mass spectrometry on urine should be done to rule out an organic aciduria. Enzyme assays in liver are necessary for establishing a CPS deficiency or NAGS deficiency. Normal activities of these enzymes as well as the beneficial effect of treatment (peritoneal dialysis) without relapse on protein loading support the diagnosis of transient hyperammonemia in premature newborn, the cause of which is unknown.

ACKNOWLEDGEMENT

This work was supported by the Swiss National Research Foundation, Grant 3.551-0.79.

REFERENCES

1. G. C. Tremblay, D. E. Crandall, C. E. Knott and M. Alfant, Arch. Biochem. Biophys. 178:264 (1977).
2. A. Barton and N. J. Hoogenraad, Eur. J. Biochem. 116:131 (1981).
3. A. J. Meijer, J. A. Gimpel, G. Deleeuw, M. E. Tischler, J. Tager and J. R. Williamson, J. Biol. Chem. 253:2308 (1978).
4. P. J. Natale and G. C. Tremblay, Biochem. Biophys. Res. Comm. 37:512 (1969).
5. M. Mori, H. Ishida and M. Tatibana, Biochemistry 14:2622 (1975).
6. M. Tatibana and K. Ito, J. Biol. Chem. 244:5403 (1969).
7. A. Russell, B. Levin, V. G. Oberholzer and L. Sinclair, Lancet II:699 (1962).
8. L. Kesner, F. L. Aronson, M. Silverman and P. C. Chan, Clin. Chem. 21:353 (1975).
9. C. Bachmann and J. P. Colombo, J. Clin. Chem. Clin. Biochem. 18:293 (1980).
10. A. Van Gennep, E. J. Van Bree-Blom, J. Grift, P. K. De Brie and S. K. Wadman, Clin. Chim. Acta 104:227 (1980).
11. F. X. Coudé, H. Ogier, C. Marsac, A. Munnich, C. Charpentier and J. M. Sandubray, this volume.

12. K. Shigesada and M. Tatibana, Biochem. Biophys. Res. Comm. 44:1117 (1971).
13. F. X. Coudé, L. Sweetman and W. L. Nyhan, J. Clin. Invest. 64:1544 (1979).
14. C. Bachmann, Urea Cycle in "Heritable Disorders of Amino Acid Metabolism", W.L. Nyhan, ed. Wiley, New York (1974).
15. D. M. Stewart and M. Walser, J. Clin. Invest. 66:484 (1980).
16. L. Cathelineau, F. P. Petit, F. Coudé and P. P. Kamoun, Biochem. Biophys. Res. Comm. 90:327 (1979).
17. J. M. Saudubray, L. Cathelineau, J. M. Laugier, C. Charpentier, J. A. Lejeune and P. Mozziconacci, Acta Paed. Scand. 64:464 (1975).
18. S. Arashima and L. Matsuda, Tohoku, J. Exp. Med. 107:143 (1972).
19. S. D. Cederbaum, K. N. F. Shaw and M. Valente, Am. J. Hum. Genet. 25:20a (1973).
20. A. S. Hassan and J. A. Milner, Arch. Biochem. Biophys. 194:24 (1979).
21. J. P. Colombo and C. Bachmann, Enzyme 25:297 (1980).
22. H. A. Simmonds, D. R. Webster, D. M. O. Becroft and C. F. Polter, Eur. J. Clin. Invest. 10:333 (1980).
23. C. Bachmann and J. P. Colombo, Eur. J. Pediatr. 134:109 (1980).
24. D. R. Webster, H. A. Simmonds, D. M. J. Barry and D. M. O. Becroft, J. Inherit. Metab. Dis. 4:27 (1981).

COMPARISON BETWEEN AMINO ACIDS AND OROTIC ACID ANALYSIS IN THE DETECTION OF UREA CYCLE DISORDERS IN THE QUEBEC URINARY SCREENING PROGRAM

B. Lemieux, Ch. Auray-Blais, R. Giguère

Centre Hospitalier Universitaire, Département de Pédiatrie, Sherbrooke, Québec, Canada, J1H 5N4

INTRODUCTION

The Provincial Network of Genetic Medicine started its operations in 1969. The various programs of the network are located in four different regional centers ; two are located in Montreal, the mass screening of blood is in Quebec City and the mass screening of urine is in Sherbrooke. The funds to permit such a system come from the Ministère des Affaires Sociales of the Province of Quebec. Our participation in the urinary screening program has allowed us to screen over 700,000 newborns for the detection of inborn errors of metabolism.

Our urinary screening program[1] began in 1971 as a research and developmental project and has since evolved, due to many changes over the years in order to improve the cost-benefit ratio, into a service to the population which, in return, has shown a very good response. One example of these changes is the time of collection of the urine sample. Earlier in the program, we received the samples collected on filter paper at 5 days of age ; throughout the major part of the program the urine was collected at two weeks and since June 1981, it has been collected at three weeks of age. This has permitted us to eliminate, as much as possible, the false-positives and diminish the number of repeats.

Parents participation is on a voluntary basis and, as of now, averages over 91 %. Other changes throughout the years involve the discontinuation of biochemical analysis of uric acid/creatinine, keto acids/creatinine, and sugars, as these were found unsatisfactory as regards to the cost-benefit ratio.

MATERIALS AND METHOD

A brief description of the thin layer chromatography (TLC) of urinary amino acids and methylmalonic acid (MMA), as presently performed, is mentioned in the following paragraphs.

The preparation of the TLC plates is done with an automatic spreading machine. The slurry is composed of 60 % of cellulose (28.0 gm), 40 % silicia gel (18.6 gm) and water (160 cc) ; this is sufficient for four, 20 x 40 cm plates. The thickness of the plate is 250 μ. The plates are stored overnight in dessicated boxes. 100 plates are prepared in 1 1/2 hour.

Every filter paper is looked at under an ultra-violet light to detect if enough urine is on the filter paper or if there is a contamination by feces or creams.

A disc of 2 inches in diameter is punched from the filter paper and an elution is performed by shaking the filter paper with 2 cc of NH_4OH 0.01M in glass vials for 10 minutes on a New Brunswick shaker.

The samples are applied on TLC with our automatic spotting machine : 20 samples and 3 standards are spotted on the same plates ; 500 samples are applied on the plates in 1 1/2 hour.

Two unidimensional migrations of 2 1/2 hours each are done in butanol-acetic acid-water, 13-3-5. The plates are dried in the oven at 50° C for twenty minutes in between the migrations.

The upper half of the plates are first sprayed (under a fume hood) with o-dianisidine for MMA. They are then heated for 10 minutes at 50° C.

Preparation of the spray :

0.5 g of o-dianisidine in 90 ml of ethanol, 10 ml of water and 4 ml of glacial acetic acid. Shake for 30 minutes.

Afterwards, the plates are sprayed with ninhydrin for the detection of amino acids.

Preparation of the spray :

ninhydrin 0.2 % in ethanol	95 cc
cadmium acetate 1 % in 1 M glacial acetic acid	5 cc

The plates are then heated at 100° C for 20 minutes.

After the plates are looked at for abnormalities of amino

acids, they are sprayed with Ehrlich's reagent for the main detection of citrulline.

Preparation of the spray :

a) p-dimethyl-aminobenzaldehyde : 10 % in concentrated HCl — 1 volume
b) acetone — 4 volumes

Mix a and b prior to use. Then add 0.2 % isatin.

After spraying the plates, they are heated at 100° C for seven minutes.

Different Steps implicated in our Screening Program.

The filter paper samples are received, numbered and processed as mentioned previously. If an abnormal sample is detected on the MMA or amino acids chromatography, specific sprays or spot tests (for sulfur amino acids) are used in order to identify precisely the abnormality. A second sample (filter paper or liquid urine) is then requested by phoning directly the parents and sending them a reply kit. The second analysis is performed by doing first, an alpha-amino nitrogen test and repeating TLC. If, again, the sample is abnormal, the case is referred to one of our four regional centers. It is important to mention that a repeat urine specimen is requested whenever the first sample is inadequate (no urine on the filter paper, filter paper contaminated by feces, diaper creams or other substances, etc...).

Orotic Acid Technique.

In 1979 a study was introduced in order to improve our screening program for the detection of urea cycle disorders. Orotic acid was chosen as a single test by which it was possible to detect multiple urea cycle and related disorders.

For mass screening purposes, this analysis was performed by a non-specific spectrophotometric technique, based upon Rogers' method [2], which was modified according to our needs. The rate of analysis was 160 samples/hour. Details have been previously published[3].

DISCUSSION

As mentioned in the previous reference, there is a very good

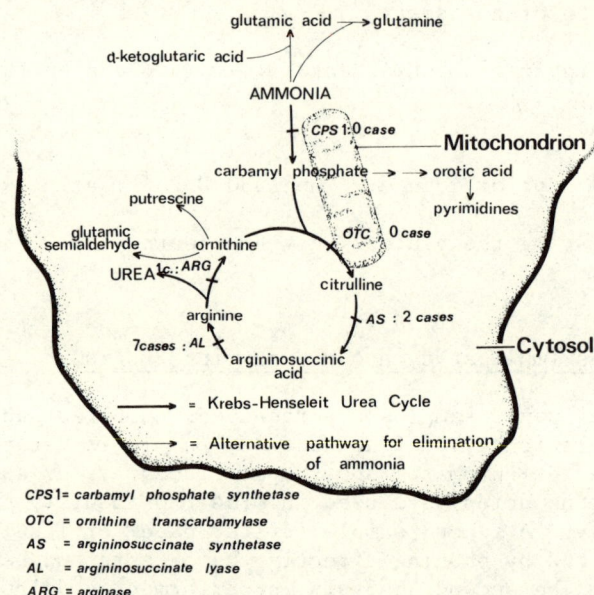

Fig. 1. Number of cases detected by the urinary screening program concerning urea cycle disorders.

linearity and reproducibility of the proposed method in the range of 1 to 50 μg/ml. The values of orotic acid were expressed as a ratio between the absorbance at 480 nm and 412 nm. This ratio takes into account the concentration of non-orotic acid urinary substances.

Nearly 90,000 analyses were performed in this manner over a period of one year. In the beginning, both programs were performed at the same time (for a few thousand samples) and then, the TLC program was temporarily discontinued.

Unfortunately, no cases were detected by the orotic acid technique even after re-checking second samples of elevated first analyses.

Although the pyrimidine technique used was somewhat sensitive enough to detect minute amounts of orotic acid in the urine and extremely useful for differentiating between the two mitochondrial enzyme deficiencies, it may not be a valuable single probe to detect certain forms of cytosol enzyme deficiency in urine speci-

Table 1. Comparison between Urinary Excretion of Amino Acids and Orotic Acid in the Genetic Enzyme Defects of the Urea Cycle Metabolism of Two to Three Weeks Old Babies

	MITOCHONDRIAL DEFECTS		CYTOSOL DEFECTS		
	CPS	OTC	AS	AL	ARG
amino acids (TLC)	−	−	++	+++	+
orotic acid (spectro. tech) a)	−	++	+++	±	+
b)			−	−	−

a) what should have been obtained (theoretically)
b) results obtained after the pilot study

CPS: carbamylphosphate synthetase, OTC: ornithine transcarbamylase, AS: argininosuccinate synthetase, AL: argininosuccinate lyase, ARG: arginase

Table 2. Comparison between two urinary screening programs in the detection of urea cycle disorders (cases per 100,000 analyses).

	QUEBEC	MASSACHUSETTS
argininosuccinic aciduria	1.3	1.2
citrullinemia	0.4	---
hyperargininemia	0.2	---

men collected at 2-3 weeks (table 1).

Moreover, previous cases detected by the TLC program did not show an increase in urinary orotic acid on the first urine sample, but only later, around 2-3 months of age.

Finally, in the last two cases, one argininosuccinate lyase (AL) and one argininosuccinate synthetase (AS) deficiency detected by the new TLC program this year had no increase of orotic acid in the first specimen collected at 3 weeks of age. Both colorimetric and HPLC methods were used to ascertain the orotic acid in the urine.

Figure 1 shows the total number of cases detected by the TLC program concerning the urea cycle disorders.

Table 2 shows the incidence of urea cycle disorders in our program compared to the Massachusetts Program and concerns exclusively the cytosol enzyme deficiencies. It is close to 1/50,000, two times less frequent than the "classical PKU" in our province.

CONCLUSION

In conclusion, the thin layer chromatography of amino acids on urine samples collected at three weeks of age seems more rewarding for early detection of cytoplasmic urea cycle enzyme deficiencies in an universal screening program (table 3).

Furthermore, it should be kept in mind that there is a variation in a few urinary amino acids in OTC cases : there is a possibility of a non-specific increase in urinary glutamine, alanine and sometimes lysine ; finally, there is an increase in orotic acid which is a typical secondary phenomenon. As for the higher risk screening or sick neonate, the investigation should

Table 3. Urea Cycle Disorders (Screening Programs)

	HIGH RISK SCREENING (sick neonate)	UNIVERSAL SCREENING (3 weeks neonate)
type of analyses	orotic acid, amino acids: urine, blood, ammonia (blood) etc...	urinary amino acids (+ blood) orotic acid
different forms	neonatal deficiency in: - NAGS - CPS - OTC - AS - AL - ARG (?)	subacute, chronic or transient deficiency in: - OTC (?) - AS - AL - ARG

NAGS: N-acetylglutamate synthetase, CPS: carbamylphosphate synthetase. OTC: ornithine transcarbamylase, AS: argininosuccinate synthetase. AL: argininosuccinate lyase, ARG: arginase be oriented towards the analysis of orotic acid, amino acids, ammonia, organic acids, etc... in order to detect which form of enzyme deficiency is implicated in the disease.

However, we feel that both TLC of amino acids and orotic acid determination are complementary and should be used. We do not have experience with blood screening using enzyme-multiple auxotroph assay as described by Naylor,[4] neither with TLC of blood amino acids on a large scale for possible detection of neonatal forms. These forms are probably detected more rapidly by each regional center as we are dealing usually with a high risk sick neonate[5].

We strongly believe that the urinary screening program in the province of Quebec fulfils the main criterias for mass screening ; it permits : 1. an effective early treatment for cases necessitating immediate medical intervention, (1/35,000) 2. a surveillance and follow-up of patients presenting transport disorders, (1/8,000) 3. an epidemiological research and enumeration, (1/3,600) 4. detection of heterozygotes (1/13,500) 5. and, finally, an appreciable incidence of abnormalities regarding the total number of cases detected per year, (1/1,600). Moreover, it permits suitable tests showing reliability, simplicity, reproducibility, few false-positives.

Finally, the urinary screening program is a unique mean of public education on genetic diseases as it was shown by an active participation and collaboration from the parents to a very high degree.

Thus, the urinary screening program should be viewed as a component of a comprehensive program, and as a mean to communicate directly with the parents (genetic education), and therefore, offering the possibility to detect in the near future other diseases encountered mainly in the urine.

REFERENCES

1. B. Lemieux, C. Auray-Blais, R. Giguère, Quebec Urinary Screening Program (in preparation).
2. L. E. Rogers and S. F. Porter, Hereditary Orotic Aciduria II. A urinary screening test, Ped. 42:423 (1968).
3. D. Paradis, R. Giguère, C. Auray-Blais, P. Draper, and B. Lemieux, An automated method for the determination of orotic acid in the urine of children being screened for metabolic disorders, Clin. Biochem. 13 (4):160, (1980).
4. T. D. Paul, E. W. Naylor, R. Guthrie, Urine screening for metabolic disease in newborn infants, J. Ped. 96 (4):653 (1980).
5. C. Bachmann and J. P. Colombo, Diagnostic value of orotic acid

excretion in heritable disorders of the urea cycle and in hyperammonemia due to organic acidurias, Eur. J. Pediatr. 134:109 (1980).

TRANSIENT HYPERAMMONEMIAS IN INFANTS WITH AND WITHOUT ORGANIC ACIDEMIA

W.L. Nyhan, V. Rubio, A. Jordá, S. Grisolia, F. Gutierez, and C. Canosa

University of California San Diego, La Jolla, Calif.
USA, Instituto de Investigaciones Citologicas, Valencia
Spain, and Hospital Infantil La Fe, Valencia, Spain

Transient hyperammonemia of the newborn was first described in 1978 in 5 preterm infants[1]. During that period this disorder appeared in our experience to be the most common form of symptomatic neonatal hyperammonemia. More recently, we have not studied a patient with this disorder in more than 2 years, while the number of patients referred to us for diagnosis of hyperammonemia has increased. Whether these observations reflect a real changing incidence of the disorder remains to be determined.

In each of the infants we studied, life threatening illness developed withing 48 hours of birth. Lethargy progressed rapidly to coma. Pupils were fixed and usually dilated. Each infant required intubation and mechanical ventilation. The severity of the process is indicated by the fact that the one infant in whom the respirator was turned off died at 84 hours of age.

Peak concentrations of ammonia in our 5 infants ranged from 977 to 7640 µg/dl. The median was 3400 µg/dl These concentrations are as great or greater than those observed in patients with defects in enzymes of the urea cycle. Plasma concentrations of glutamine were elevated at 16 and 31 mg/dl. in the two patients studied. The concentration of alanine was elevated to 8 mg/dl in one of them. The concentrations of citrulline were 0.25 and 2.3 mg/dl and those of arginine were 1.4 and 2.0 mg/dl We have detected citrulline in the plasma and an arginine concentration of 2.1 mg/dl in another patient with this syndrome. None of these patients had orotic aciduria, nor did they have organic aciduria. Thus it should be possible to distinguish these patients by metabolic investigation from those with other causes of hyperammonemia, except for those with carbamylphosphate synthetase deficiency or acetylglutamate synthetase deficiency. Furthermore, patients with carbamylphosphate synthetase deficiency in whom the defect was of such severity that the patient presented with massive hyperammonemia in the neonatal period should have no detectable citrulline in the plasma and very low or undetectable levels of arginine.

In three infants studied the activities of the classic enzymes of the urea cycle were normal. Two were studied by biopsy at 6 and 10 weeks of age. The third was studied postmortem at the height of illness. Therefore it seems likely that this disorder is not the result of delayed development of any of these 5 enzymes. This has focused attention on the acetylglutamate system. Bachmann and colleagues [2] have recently reported normal activity of acetylglutamate synthetase in biopsied liver of 2 infants with transient hyperammonemia of the newborn, but it is not clear that activity might not have been deficient earlier and developed by the time of biopsy.

Two of our patients were treated with exchange transfusion and peritoneal dialysis and two with exchange transfusion alone. In each the serum concentration of ammonia became normal at about 5 days of age. Clinical improvement was rapid, and each infant went on to develop normally. Thus the importance of rapid, definitive diagnosis is great. Without supportive treatment the disorder is certainly fatal. Furthermore, there is a tendency among neonatologists to discontinue mechanical ventilation and other support as heroic in an infant with a presumably fatal disorder of the metabolism of the urea cycle. Actually the availability of acylation therapy for instance with benzoate [3,4] and arginine replacement therapy [4,5] has brought into serious question the validity of the ethical argument for withdrawal of support even for the defects of urea cycle enzymes that in the past were uniformly fatal. Patients with transient hyperammonemia have not as yet been treated with benzoate, or with the keto acid analogues of essential amino acids[6]. It is also possible that the hyperammonemia might disappear at 5 days without therapy directed at the ammonia itself, but patients have not been treated with supportive measures alone. In the absence of data on this point treatment appears to be prudent.

Transient hyperammonemia is also seen in infants with a variety of disorders of organic acid metabolism and this may be symptomatic or even fatal[7]. It has been observed in propionic acidemia [7,8] methylmalonic acidemia,[9,10] and in glutaric aciduria type II[11]. In these patients hyperammonemia occurs at times when the basic metabolic abnormality is markedly out of control, when the organic acid intermediates which characterize these disorders are present in body fluids in very high concentrations, and there is severe acidosis and usually massive ketosis. These changes occur in episodi-

fashion and the patient may be desperately ill or even may die
in such an episode, which may be the infant's first, or later
may accompany a catabolic state induced by infection or surgery,
or may simply follow the gradual accumulation of organic acids
created by the regular ingestion of their amino acid precursors in
amounts much larger than those required for growth. In our experience there has been a distinct ontogeny in this,for episodes
of illness and extreme acidosis have been observed at any age,
while accompanying hyperammonemia has been observed only in very
young infants. This has now been studied systematically by
Cathelineau and colleagues[12] in 17 patients with methylmalonic
acidemia. Hyperammonemia was regularly encountered in the neonatal
period, occurring in all of 8 patients with neonatal onset of
symptoms and in none of the patients with a later onset, or even
in the same infants during episodes of ketoacidosis that occurred
later in life, despite similar levels of methylmalonic and propionic
acids. The concentration of ammonia observed in infants with organic acidemia may readily reach levels usually associated with
a defect in an enzyme of the urea cycle. In fact patients have
even been reported in abstract as having carbamylphosphate synthetase deficiency who later turned out to have organic acidemia.
Thus an investigation for the organic acids of the urine should
be carried out in any patient with hyperammonemia in which analysis
of the amino acids does not provide a clearcut diagnosis of a
urea cycle enzyme deficiency.

In infants with organic acidemia it is clear that hyperammonemia
occurs only under conditions of a plentiful supply of organic acids.
It has therefore been inferred that one or more of the organic
acids serves as an inhibitor of some aspects of the function of the
urea cycle. The absence of orotic aciduria in these infants
indicated that the inhibition was at an early step, prior to
ornithine transcarbamylase, even though propionic acid has been
shown to inhibit the overall synthesis of citrulline[13] and reduce
ATP content[14] in rat liver mitochondria. Propionyl CoA has been
found to inhibit the activity of carbamylphosphate synthetase
directly, that is even in the presence of acetylglutamate[15]. We
have studied the effects of a series of acyl CoA derivatives on
the activity of N-acetylglutamate synthetase in rat liver mitochondria[16]. Propionyl CoA was found to be a highly effective
competitive inhibitor of the synthetase. The inhibition constant
0.71 mM found is well within the range of concentrations of propionate found in the serum of infants with propionic acidemia and
methylmalonic acidemia. Methylmalonyl CoA was a less effective
inhibitor, yielding 28 % inhibition at 3 mM concentration, while
tiglyl CoA and isovaleryl CoA were 70 % inhibitory at 3 mM.
The latter two compounds are readily detoxified by the formation
of glycine conjugates which may account for the fact that hyperammonemia is not characteristic of infants with isovaleric acidemia
or β-ketothiolase deficiency. Propionyl CoA is also a substrate

for the acetylglutamate synthetase, and the propionylglutamate formed is a weak activator of carbamylphosphate synthetase. The activation constant approximated ten times that of acetylglutamate. Thus the decreased concentration of acetylglutamate that would result from the effects of propionyl CoA as an inhibitor should decrease the activity of carbamylphosphate synthetase and interfere with the clearance of ammonia, while the propionylglutamate formed would not be expected to stimulate ureagenesis. The synthesis of acetylglutamate has also been reported to be decreased in rat liver mitochondria by propionate[14].

The opportunity arose to study these issues in liver of an infant who died of a disorder of propionate metabolism after presenting with hyperammonemia. The patient, DDS, was a 2-day-old male infant transferred to the Hospital Infantil de la Seguridad Social La Fe, Valencia, Spain, because of hypoglycemia. He had been born following an uneventful pregnancy and delivery of a 28-year-old primipara. The father was 29 years old. The parents were healthy and unrelated. Birth weight was 3250 g and physical examination by a pediatrician on the first day of life was normal. Breast feeding was initiated.

The infant never began to suck well. At 11 hours of life he developed irritability and shortly thereafter he was found to be severely hypotonic and hypothermic. The rectal temperature was 35°C. Dextrostix was 0. He was treated with hypertonic glucose, sodium bicarbonate and calcium gluconate.

On admission following transfer he appeared gravely ill. The color was gray and the respirations appeared acidotic. The weight was 2560 g; height 50 cm; head circumference 35 cm. Temperature was 34.6°C.; pulse 180 and respirations 70 per minute. The liver was palpable 2 cm and the spleen 1 cm below the costal margin. He was weak, hypotonic and had no sucking reflex. Deep tendon reflexes were unobtainable. He appeared to be approximately 10% dehydrated.

Laboratory evaluation revealed a blood pH of 7.21, bicarbonate 5.8 mEq/l, and base excess -21. The blood urea nitrogen was 113 mg/dl and the glucose 199 mg/dl. The serum concentration of sodium was 153 and potassium 5 mEq/l. Urinalysis revealed a 2+ acetonuria. The concentration of ammonia was 482 µg/dl. The initial leukocyte count was 6300/cm with 66% polymorphonuclear forms, the hematocrit 51 and the platelet count 168,000/cmm, but on the second day the leukocyte count was 1100 and leukopenia was persistent. The hematocrit fell to 41 and in spite of multiple transfusions was 36 on day 4. The platelet count fell to 64,000 on day 2 and 35,000 on day 4. A lumbar puncture revealed xanthochromic fluid; its protein concentration was 144 mg/dl but it was otherwise unremarkable. Cultures were negative. The concentration of lactic acid in the blood was 42.9 mg/dl and that of pyruvic acid 0.57 mg/dl.

A few minutes after admission he began to have generalized tonic and clonic convulsions. The EEG was abnormal, and treatment was initiated with phenobarbital and dilantin. He was treated vigorously with parenteral fluids containing $NaHCO_3$ continuously, and in addition received frequent bolus injections of concentrated $NaHCO_3$ because of persistent or recurrent acidosis. Seven hours after admission apnea developed requiring intubation and mechanical ventilation. Peritoneal dialysis was instituted on day 2 when the results of the blood test for ammonia became known. By this time the EEG was flat. The blood urea nitrogen was brought to 35 mg/dl on this regimen and the acidosis largely controlled, but the infant never regained consciousness, or spontaneous respiration. When seen on day 4 he was flaccid, areflexic, and completely unresponsive to deep pain. He was given 1000 μg B_{12}. He died that night.

Analysis of the organic acids of the urine revealed a methylmalonic aciduria of 133,43 μEq/mg creatine. There was an enormous peak of lactic acid which could not be quantitated because of the size of the specimen available. The excretion of methylcitrate was 4.59, and that of hydroxypropionate 2.84 μEq/mg creatinine. Accordingly a diagnosis was made of methylmalonic acidemia.

At autopsy 10 hours postmortem, 5-10 g of liver were obtained, frozen immediately and kept at -70°C. for 49 days in a sealed 50 ml plastic container. The enzymes of the urea cycle were assayed using the methods of Nuzum and Snodgrass[17]. Carbamylphosphate synthetase and ornithine transcarbamylase were assayed in a 0.73 g sample. Argininosuccinic acid synthetase, argininosuccinase and arginase were assayed 2 days later in a 0.7 g sample. The concentration of N-acetylglutamate was assayed 3 weeks later using the method of Shigesada and Tatibana[18], in a 3 g sample. In each case fresh rat liver was assayed simultaneously. Protein was measured by the Lowry method, using bovine serum albumin (fraction V, Sigma) as standard.

The data on activity of the urea cycle enzymes are shown in Table 1. We have compared the results with the data of Nuzum and Snodgrass[17] as well. The results obtained on rat liver indicated general agreement between the assays in the two laboratories. The activities of carbamylphosphate synthetase, argininosuccinic synthetase and arginase were about 1.2 times higher in our hands, whereas ornithine transcarbamylase was 92% and argininosuccinase 70% of their values. Moderately elevated levels of carbamylphosphate synthetase and of arginase (both per g of tissue and per mg of protein) were observed in our human samples when compared with the results of Snodgrass on fresh human biopsied liver. Their data for each of the enzymes was lower in liver obtained at autopsy 3-14 hours after death. The levels of the other three enzymes in our human sample were of similar magnitude to those found in

Table 1. Activities of Urea-Cycle Enzymes.

Source of liver	Carbamyl Phosphate Synthetase	Ornithine Transcarbamylase	Argininosuccinate Synthetase	Arginino-succinase	Arginase	Protein
	u/g / u/mg P	u/g / u/mg P	u/g / u/mg P	u/g / u/mg P	u/g / u/mg P	mg/g
Autopsy, patient	354 / 2.94	3115 / 25.89	65.5 / 0.58	103.4 / 0.91	138240 / 1217	120;114
Rat liver (assayed simultaneously)	868 / 3.98	10591 / 48.58	383.6 / 1.92	259.2 / 1.3	145846 / 729	218;200
Rat liver (this laboratory) n = 3	912±46 / 4.39±0.39	13457±5040 / 64.1±28	333±45 / 1.65±0.24	238±23.3 / 1.18±0.10	181485±31600 / 895±146	206±13

Abbreviations: u = units of enzyme activity, expressed as micromoles of citrulline of urea formed per hour;
g = gram wet weight of liver;
mgP = milligram protein;
n = number of subjects.

The data in the bottom line represent the mean ± standard deviation.

Table 2. Concentration of acetylglutamate.

	(nmol/g liver)
Patient	34
Rat liver (assayed simultaneously)	48
Rat liver (n = 10) (assayed previously in this laboratory)	38 ± 8

necropsied liver by these investigators. From these data we concluded that the levels of urea cycle enzymes were normal or slightly elevated in this patient who died of methylmalonic acidemia.

The concentrations of acetylglutamate are shown in Table 2. Normal values for man are not available. The level observed in the patient was within the limits of normal for rat liver assayed in this laboratory.

Examination of the concentrations of the amino acids in this sample of liver revealed high concentrations of most of the amino acids except for serine which was decreased. The concentrations of glycine, glutamine and alanine were appreciably elevated, as was that of arginine. Elevated concentrations of glutamine and alanine are regular correlates of elevated concentrations of ammonia. The concentration of arginine suggests an adequate course of arginine synthesis via the urea cycle. Adequate amounts of arginine were also found in the urine and there was a peak of citrulline which was small, but more than usually seen in normal infants. It was concluded that the concentration of acetylglutamate was normal. However, ontogenetic changes may militate against the discovery of the pathogenesis by the time tissues are available for analysis.

REFERENCES

1. R.A. Ballard, B. Vinocur, J.W. Reynolds, R.P. Wennberg, A. Merritt, L. Sweetman, and W.L. Nyhan, Transient hyperammonemia of the preterm infant, New Eng. J. Med. 299 : 920 (1978).
2. C. Bachmann, G. Schubigger, K.H. Jaggi, S. Krähenbuhl, J.P. Colombo, and O. Tonz, N-acetylglutamate synthetase deficiency a new disorder of ammonia detoxification, New Eng. J. Med. 304 : 543 (1981).
3. S.W. Brusilow, J. Tinker, and M.L. Batshaw, Amino acid acylation : A mechanism of nitrogen excretion in inborn errors of urea synthesis. Science 207 : 659 (1980).
4. M.L. Batshaw, and S.W. Brusilow, Treatment of hyperammonemic coma caused by inborn errors of urea synthesis. J. Pediat. 97 : 893 (1980).

5. S. Brusilow, and M.L. Batshaw, Arginine therapy of argininosuccinase deficiency, Lancet 1 : 125 (1979).
6. J. Thoene, M. Batshaw, E. Spector, S. Kulovich, S. Brusilow, M. Walser, and W. Nyhan, Neonatal citrullinemia : Treatment with keto-analogues of essential amino acids. J. Pediat. 90 : 218 (1977).
7. T. Shafai, L. Sweetman, W. Weyler, S.I. Goodman, P.V. Fennessey, and W.L. Nyhan, Propionic acidemia with severe hyperammonemia and defective glycine metabolism, J. Pediat. 92 : 84 (1978).
8. L. Sweetman, W.L. Nyhan, J. Cravens, Y. Zomer, and D.C. Plunket, Propionic acidaemia presenting with pancytopaenia in infancy. J. Inher. Metab. Dis. 2 : 65 (1979).
9. S. Packman, M.J. Mahoney, K. Tanaka, and Y.E. Hsia, Severe hyperammonemia in a newborn infant with methylmalonyl-CoA mutase deficiency, J. Pediat. 92 : 769 (1978).
10. L. Shapiro, M.E. Bocian, L. Raijman, S.D. Cederbaum, and K.N.F. Shaw, Methylmalonyl-CoA mutase deficiency associated with severe neonatal hyperammonemia : Activity of urea cycle enzymes, J. Pediat. 93 : 986 (1978).
11. L. Sweetman, W.L. Nyhan, D.A. Trauner, T.A. Merritt, and M. Singh : Glutaric aciduria Type II, J. Pediat. 96 : 1020 (1980).
12. L. Cathelineau, P. Briand, H. Ogier, C. Carpentier, F.X. Coude, and J.M. Saudubray, Occurrence of hyperammonemia in the course of 17 cases of methylmalonic acidemia, J. Pediat. 99 : 279 (1981).
13. A.M. Glasgow, and H.P. Chase, Effect of pent-4-enoic acid and other short chain fatty acid on citrulline synthesis in rat liver mitochondria, Biochem. J. 156 : 301 (1976).
14. L. Cathelineau, F. Petit, F.X. Coude, and P. Kamoun, Effect of propionate and pyruvate on citrulline synthesis and ATP content in rat liver mitochondria, Biochem. Biophys. Res. Commun. 90 : 327 (1979).
15. J.A. Gruskay and L.E. Rosenberg, Inhibition of hepatic mitochondrial carbamylphosphate synthetase (CPSI) by acyl CoA esters. Possible mechanism of hyperammonemia in the organic acidemias. Pediat. Res. 13 : 475 (1979).
16. F.X. Coude, L. Sweetman, and W.L. Nyhan, Inhibition by propionyl-coenzyme A of N-acetylglutamate synthetase in rat liver mitochondria. A possible explanation for hyperammonemia in propionic and methylmalonic acidemia. J. Clin. Invest. 64 : 1544 (1979).
17. F. Nuzum and S. Snodgrass, in : "The Urea Cycle", pp. 325-349, S. Grisolia, R. Baguena, and F. Mayor, eds, J. Wiley & Sons, New York, (1976).
18. K. Shigesada and M. Tatibana, Role of acetylglutamate in ureotelism. I. Occurrence and biosynthesis of acetylglutamate in mouse and rat tissues. J. Biol. Chem. 246 : 5588 (1971).

MECHANISM OF HYPERAMMONEMIA IN AN EXPERIMENTAL MODEL OF PROPIONIC ACIDEMIA

P.M. Stewart and M. Walser

Johns Hopkins University School of Medicine, Baltimore Maryland, U.S.A.

SUMMARY

We have reported (J. Clin. Invest. 66:484, 1980) that rats injected with 10-20 mmol/kg of propionate develop severe hyperammonemia following amino acid loads, while acetate has no such effect. The mechanism was shown to be an impairment in the rise in N-acetylglutamate (AGA) and secondarily in carbamyl phosphate synthetase that normally follows amino acid loads. In order to determine whether this failure of AGA synthesis was caused (1) by competitive inhibition of AGA synthetase by propionyl CoA and/ or methylmalonyl CoA or (2) by depletion of acetyl CoA, one of the substrates for AGA synthesis, ethanol in doses of 0.1 to 5 mmoles/kg was given with propionate or acetate. The hyperammonemia was attenuated or prevented, and AGA levels were restored towards normal. Acetyl CoA levels were also restored towards normal, while unidentified medium chain thioesters changed little. Thus these results support mechanism (1) rather than (2) as the cause of hyperammonemia in propionic acidemia.

THE STUDY OF ORGANIC ACIDS METABOLISM IN A PATIENT WITH ORNITHINE

TRANSCARBAMYLASE (OTC) DEFICIENCY

Hiroko Kodama, Osamu Nose, Shintaro Okada and
Hyakuji Yabuuchi

Department of Pediatrics
Osaka University School of Medicine
Fukushima-ku, Osaka, Japan

INTRODUCTION

The formation of glutamate from ammonia and α-ketoglutarate (α-KG) is one of major pathways to remove excessive ammonia. This pathway is especially important in the urea cycle enzymopathies. It has been reported that the plasma α-KG concentration was increased in the patients with chronic hepatic dysfunctions, Reye's syndrome and fulminent hepatic failure[1,2]. On the other hand, Batshaw et al.[3] reported that plasma α-KG level was decreased in a hyperammonemic state in the patients with various urea cycle enzymopathies. In the patients with urea cycle enzymopathies, the citric acid cycle (TCA cycle) which is closely linked to the urea cycle, might be significantly influenced due to excessive ammonia. Thus it seems important to study the changes of the TCA cycle components in the urea cycle enzymopathy.

In this report, we describe the changes of several organic acids, especially α-KG and discuss the possible functions of α-KG in removing excessive ammonia in a patient with OTC deficiency.

CASE REPORT

The patient was a 9-year-old Japanese girl. She had several intermittent episodes of vomit, lethargy, anorexia and inappropriate behavior since 2 years of age. Her plasma ammonium level was 100-560 µg/dl (normal, < 50 µg/dl) and blood urea nitrogen level was 5-8 µg/dl. Transaminases (GOT, GPT) were usually within normal range, but sometimes were slightly increased (GOT, 120 I.U.; GPT,

Table 1. Urea cycle enzymes activities in the liver
(μ mole/hr/g)

	Patient	Control
carbamyl synthetase	384	457
ornithine transcarbamylase	188	5788
argininosuccinate synthetase	25	45
argininosuccinate lyase	135	168
arginase	58100	61400

86 I.U.). Her plasma glutamine level was increased to 1,450 µM in the intermittent period (normal, 196-344 µM), but other amino acids in the serum and the urine were within normal ranges. The urinary excretion of orotate was 20-150 µg/mg creatinine (normal, 4-8 µg/mg creatinine)[4]. The OTC activity in her biopsied liver was significantly decreased but the activities of the other urea cycle enzyme were within normal ranges (Table 1)[5].

METHODS

Identification and Quantitative Analysis of Plasma and Urinary Organic Acids

Filtered fresh urine and deproteinized plasma was applied to a column (3 x 1,000 mm SA-10A anion-exchange resin, Mitsubishi-Kasei, Tokyo, Japan) and eluted with 0.2 N HCl. The fraction which contained main organic acid was lyophilyzed and trimethyl-silylated with BSTFA (Pierce Co., Rockford Ill., U.S.A.). The individual organic acid was identified using GC-MS 9000 mass spectrometer (Shimadzu, Kyoto, Japan) with the conventional method.

Each organic acid was detected by the carboxylic acid analyser (Seishin Pharmaceutical Co., Tokyo, Japan), which equipped with the liquid chromatography and the specific colorimetric detection system for carboxylic acids[6]. Plasma α-KG level and ammonium level were measured by the enzymatic method[7]. Urinary contents of organic acids were expressed as mM/g creatinine. Creatinine was measured by the Folin-Wu method[8]. All blood samples were taken from cubital vein.

Oral Ammonium Chloride (NH₄Cl) Loading Test

After an overnight fasting, ammonium chloride (25 mg/kg) was orally given to the patient. Blood samples were taken 0, 30, 60, 120 and 180 min after the loading. Urine samples were collected every hour until 2 h after the loading.

Oral Citrate Loading Test

After an overnight fasting, citrate (150 µg/kg) was orally given to the patient and a control (healthy young woman). Blood and urine samples were taken as indicated in Fig. 5.

RESULTS

Contents of the Urinary Organic Acids

Fig. 1 shows typical chromatograms of urinary carboxylic acids from an age-matched healthy control (upper panel) and from the patient (lower panel). As shown in Fig. 2, the urinary excretion of citrate and α-KG was clearly increased and that of succinate was slightly increased in the patient in the early morning.

Plasma Ammonium Level vs Plasma α-KG Level in the Patient with OTC Deficiency

Fig. 3 shows the plasma concentration of ammonium and α-KG in our patient during the therapy of protein restriction. The plasma α-KG level in our patient was 9.5 - 20.2 µM (normal range, M+SD, 10.5 ± 4 µM). There was a negative linear correlation between plasma ammonium and α-KG levels ($r = -0.939$, $p < 0.001$) when plasma ammonium level was less than 200 µg/dl. When plasma ammonium level was beyond 200 µg/dl, plasma α-KG level tended to increase as plasma ammonium level was increased, but the positive correlation was not confirmed statistically.

NH₄Cl Loading Test

Plasma ammonium level was rapidly increased in the patient during the first 30 min after the administration of NH4Cl and gradually decreased afterwards. The changes of plasma α-KG and pyruvate were also similar to that of plasma ammonium (Fig. 4). Plasma citrate level was slightly increased at 1 h and then decreased. Plasma lactate level was decreased gradually.

Urinary excretion of α-KG, citrate, lactate and pyruvate

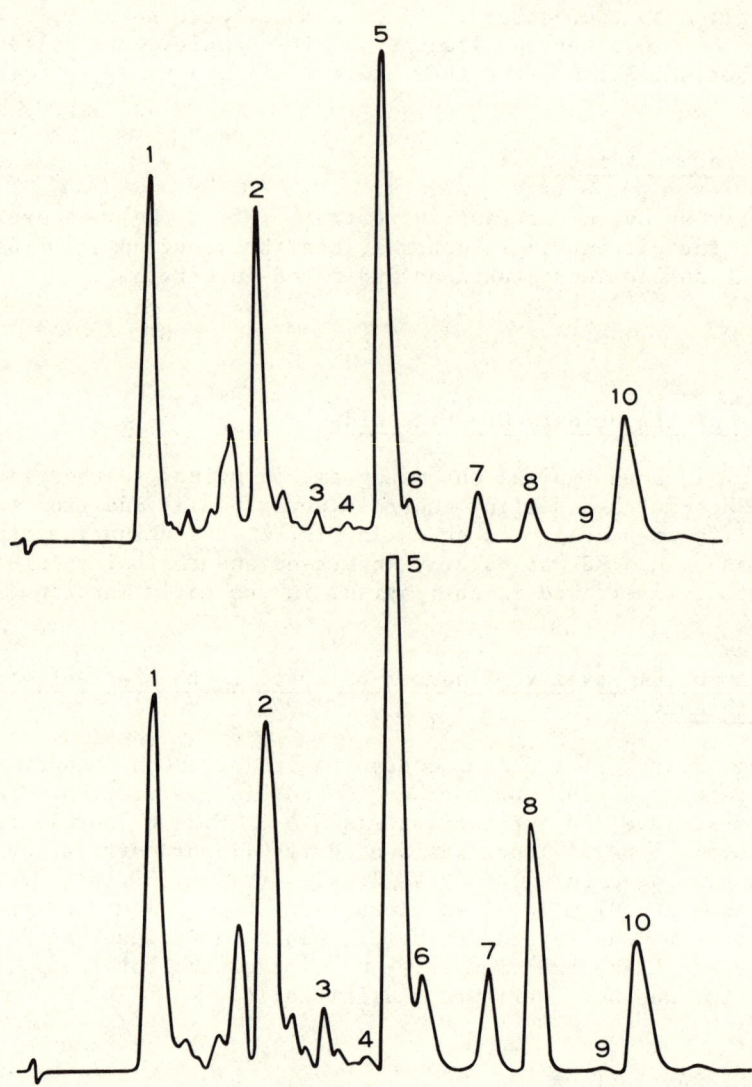

Fig. 1. Chromatogram of carboxylic acids present in urine from a healthy control (upper) and a patient with OTC deficiency (lower) 1, amino acids; 2, lactate; 3, pyruvate; 4, propionate; 5, citrate; 6, succinate; 7, isocitrate; 8, α-ketoglutarate; 9, adipate; 10, isovalerate (internal standard).

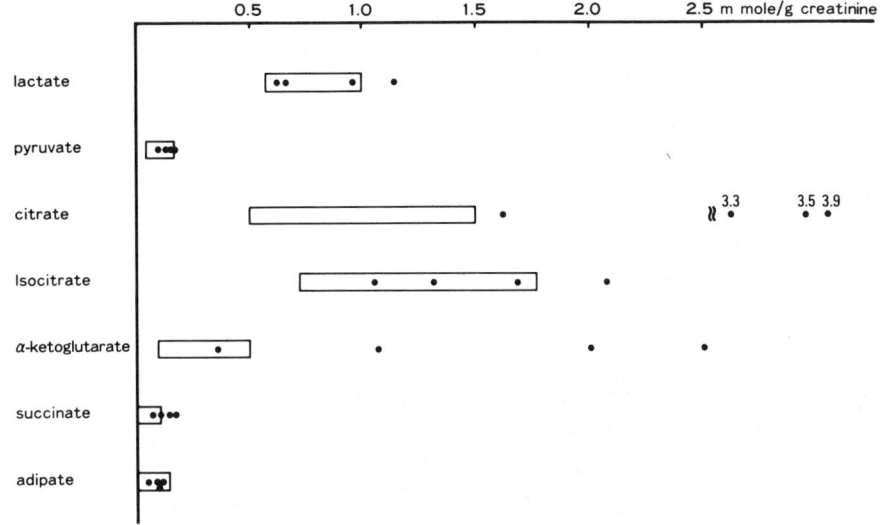

Fig. 2. Organic acid concentration in the urine from the patient with OTC deficiency and ▭ shows aged-matched normal subjects (n=9, - S.D. ~ +S.D.).

reflected their changes in plasma. That of succinate was slightly decreased after the loading.

Citrate Loading Test (Fig. 5)

Plasma α-KG was increased both in the patient and a control after the administration of citrate. Plasma ammonium level in the patient was decreased until 1 hr, but returned to the initial level at 2 hr, although plasma α-KG level was still high. Plasma and urinary pyruvate levels were slightly increased after the loading. The loading of citrate did not influence the lactate concentration of both plasma and urine.

DISCUSSION

The changes of plasma α-KG level in a hyperammonemia with various causes were reported by many authors[1-3, 9-11]. However, there are a few reports which were concerned with the changes of plasma α-KG level in OTC deficiency. Batshaw et al.[3] reported that there was a significant negative linear correlation between plasma ammonium and α-KG levels in patients with urea cycle enzymopathies including OTC deficiency when plasma ammonium level

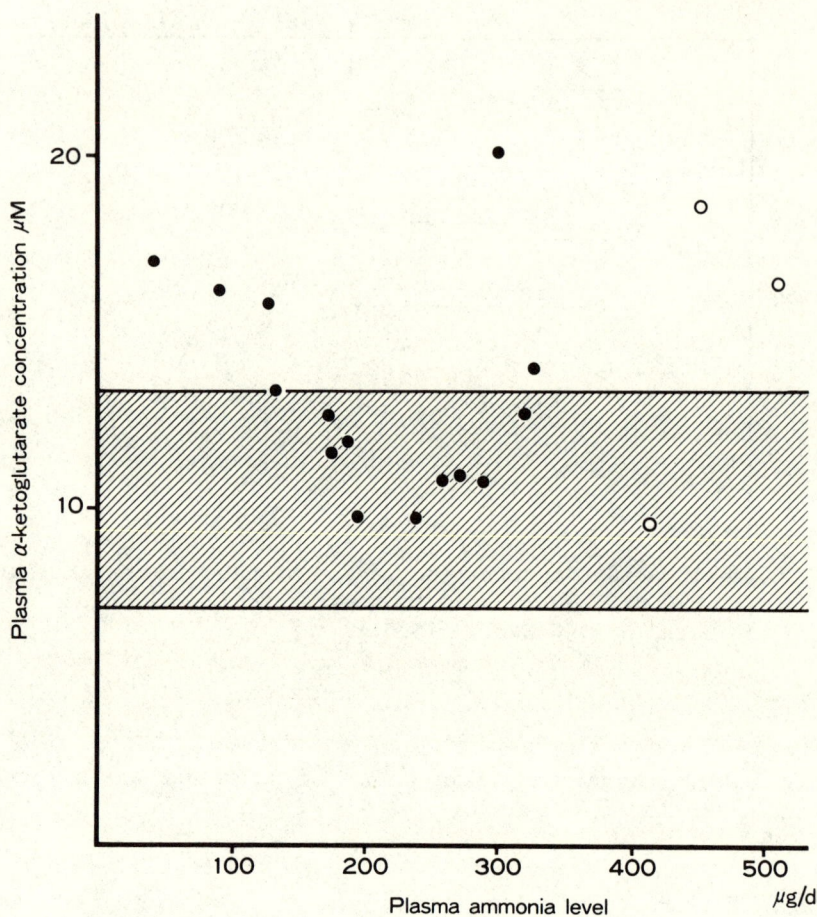

Fig. 3. The relation of the plasma α-ketoglutarate concentration to the plasma ammonia level in the patient with OTC deficiency, •, asymptomatic; ○, symptomatic. Normal range (n=9)(-S.D. ~ +S.D.).

was below 90 μM. In our patient, there was a significant negative linear correlation between plasma ammonium and α-KG levels (r= -0.939, $p < 0.001$), when plasma ammonium level was less than 200 μg/dl (Fig. 3). This result coincides with that of Batshaw et al.[3] In our data, the negative linear correlation between the levels of plasma ammonium and α-KG was not observed when plasma ammonium level was beyond 200 μg/dl. Furthermore, in the NH_4Cl loading test, the positive correlation between plasma α-KG and ammonium levels seemed to exist in hyperammonemic states (beyond 200 μg/dl) (Fig. 4).

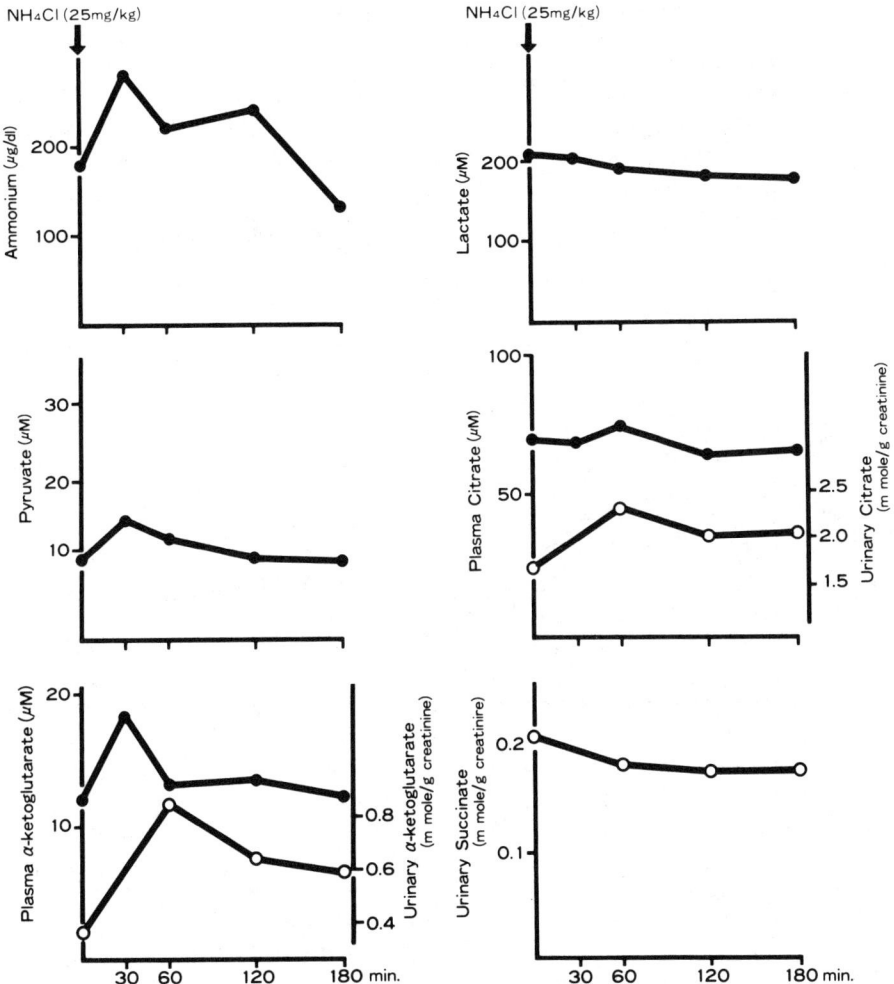

Fig. 4. The course of plasma levels of ammonium, lactate, pyruvate, α-ketoglutarate and citrate (●—●) and urinary excretion of α-ketoglutarate, citrate and succinate (○—○) in the patient with OTC deficiency after the administration of NH₄Cl.

The metabolism of α-KG is regulated by TCA cycle and glutamate dehydrogenase. The results of NH_4Cl loading test in the patient show that plasma and urinary citrate and α-KG level reached the maximum value within 60 min, while urinary excretion of succinate was slightly decreased after the administration of ammonium chloride (Fig. 4). These results suggest that α-KG dehydrogenase complex

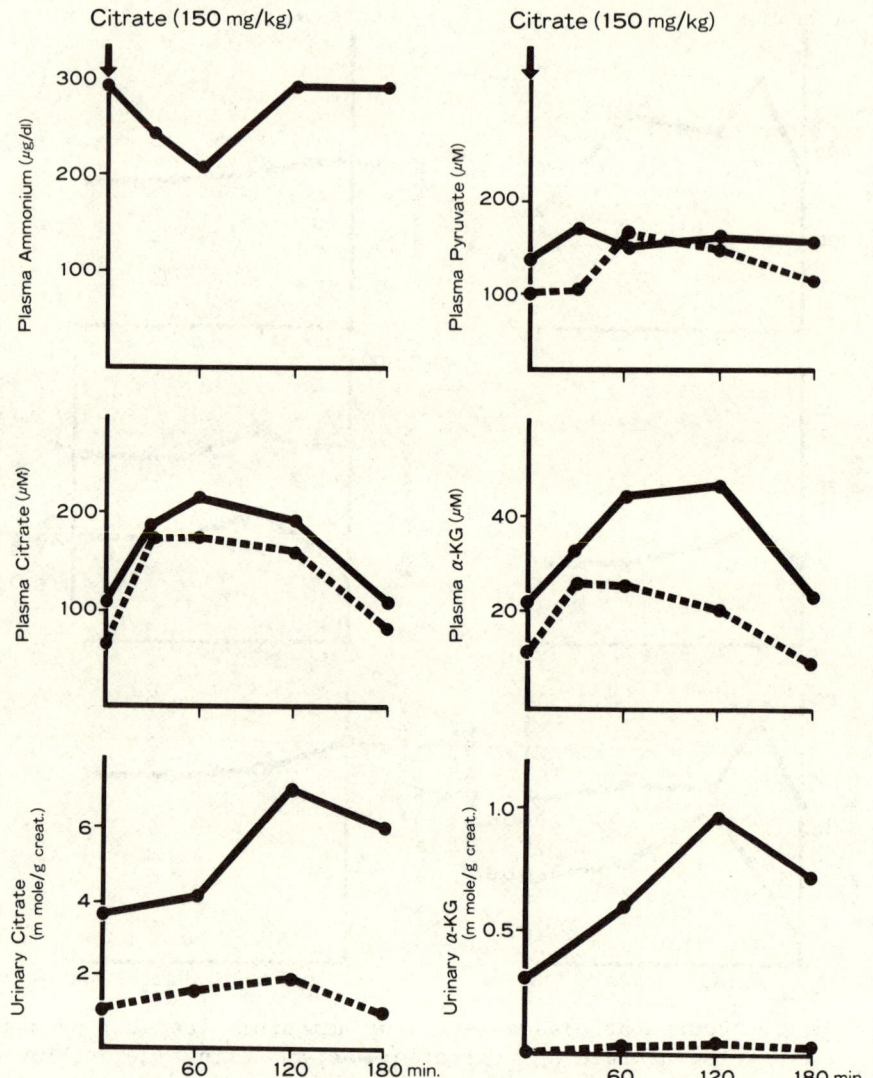

Fig. 5. The course of plasma levels of ammonium, pyruvate, citrate and α-ketoglutarate and urinary excretion of citrate and α-ketoglutarate in the patient with OTC deficiency (●—●) and a control (●---●) after the administration of citrate.

may be inhibited by the accumulation of ammonia because α-KG is metabolized to succinate by α-KG dehydrogenase complex. It was also shown that plasma pyruvate level increased in the same experiment (Fig. 4). The enzymatic mechanism of pyruvate dehydrogenase complex is known similar to that of α-KG dehydrogenase complex. Therefore, it is possible that these two dehydrogenases are suppressed by the high concentration of plasma ammonium.

Accumulated ammonia is also removed by the synthesis of glutamate. The reaction is controlled by glutamate dehydrogenase. Because increased glutamate and glutamine levels have frequently been described in hyperammonemic patients, this reaction may not be inhibited in hyperammonemic state. Bessman and Bessman[12] suggested that an increase of ammonia will concomitantly increase the rate of glutamate dehydrogenase in the direction of glutamate synthesis and higher concentration of ammonia would divert significantly amounts of α-KG from the TCA cycle to synthesize glutamate. However, from our results it is probable that the metabolic efficiency of glutamate synthesis by glutamate dehydrogenase does not seem to exceed the rate of α-KG accumulation due to the suppression of α-KG dehydrogenase.

There has been a therapy of citrate administration in patients with urea cycle enzymopathies[13-15]. The therapy has been performed by the possibility that citrate is converted to α-KG which can remove excessive ammonia. Our result of citrate loading test in the patient indicated that an increase in plasma α-KG level can decrease plasma ammonium level temporarily, but not continuously (Fig. 5). An increase in plasma α-KG level after the administration of citrate probably stimulates glutamate dehydrogenase system only temporarily. Therefore it is concluded that the administration of citrate or α-KG may not benefit in lowering the abnormally high plasma ammonium level in OTC deficiency.

REFERENCES

1. A.M. Dawson, J. Groote, W.S. Rosenthal and S. Sherlock, Blood pyruvic-acid and α-ketoglutaric acid levels in liver disease and hepatic coma, Lancet 1 : 392 (1957).
2. C.O. Record, R.A. Iles, R.D. Cohen, and R. Williams, Acid-base and metabolic disturbances in fulminant hepatic failure, Gut 16 : 144 (1975).
3. M.L. Batshaw, M. Walser, and S.W. Brusilow, Plasma α-ketoglutarate in urea cycle enzymopathies and its role as a harbinger of hyperammonemic coma, Pediatr. Res. 14 : 1316 (1980).
4. L.E. Rogers and S.F. Porter, Hereditary orotic aciduria. II A urinary screening test, Pediatrics 42 : 423 (1968).
5. C.T. Nuzum, and P.J. Snodgrass, Multiple assays of the five urea-cycle enzymes in human liver homogenates, in : "The urea cycle", S. Grisolia, R. Baguena, and F. Mayor, ed., Wiley & Sons, New York, (1976).
6. M. Nakajima, and Y. Ozawa, A highly efficient carboxylic acid analyser and its application, J. Chromatogr. 123 : 129 (1976).

7. H.U. Bergmeyer, and E. Bernt, 2-oxoglutarate, in : "Methods of enzymatic analysis", H.U. Bergmeyer ed., Academic Press, New York (1974).
8. R.W. Bonsnes, and H.H. Tausshey, On the colorimetric determination of creatinine by the Jaffe reaction, J. Biol. Chem. 158 : 581 (1945).
9. J.T. Brosnan, and D.H. Williamson, Mechanisms for the formation of alanine and aspartate on rat liver in vivo after administration of ammonium chloride, Biochem. J. 138 : 453 (1974).
10. B. Hindfelt, The effect of substained hyperammonemia upon the metabolic state of the brain, Scand. J. Clin. Lab. Invest. 30 : 245 (1972).
11. M.L. Batshaw and S.W. Brusilow, Asymptomatic hyperammonemia in low birthweight infants, Pediatr. Res. 12 : 221 (1978).
12. S.P. Bessman and A.N. Bessman, The cerebral and peripheral uptake of ammonia in liver disease with an hypothesis for the mechanism of hepatic coma, J. Clin. Invest. 34 : 622 (1955).
13. B. Levin, J.M. Abraham, V.G. Oberholzer and E.A. Burgess, Hyperammonemia. A deficiency of liver ornithine transcarbamylase, Arch. Dis. Childh. 44 : 152 (1969).
14. P. Sunshine, J.E. Lendenbaum, H.L. Levy and J.M. Freeman, Hyperammonemia due to a defect in hepatic ornithine transcarbamylase, Pediatrics 50 : 100 (1972).
15. C. Heiden, H.D. Bakker, J. Desplanque, M. Brink, P.K. de Bee, and S.K. Wadman, Attempted dietary treatment of a boy with hyperammonemia due to ornithine transcarbamylase deficiency, Eur. J. Pediatr. 128 : 261 (1978).

BASIC BIOCHEMISTRY

4) Relation of Urea Cycle to Proline Metabolism

HYPERORNITHINEMIA, GYRATE ATROPHY, AND

ORNITHINE KETOACID TRANSAMINASE

S. Hayasaka, T. Shiono, K. Mizuno,
T. Saito[*], K. Tada[*], T. Matsuzawa[**],
and I. Ishiguro[**]

Departments of Ophthalmology and Pediatrics[*]
Tohoku University School of Medicine
Sendai, Japan
and
Department of Biochemistry[**]
Fujita-Gakuen University School of Medicine
Toyoake, Japan

SUMMARY

We examined four cases of Japanese patients with gyrate atrophy of the choroid and retina. All cases had hyperornithinemia and deficiency of ornithine ketoacid transaminase (OKT). Two types of disease in responsiveness to vitamin B_6, the responsive and non-responsive, were found. The responsive type showed the in vivo reduction of serum ornithine level after oral vitamin B_6, and the OKT activity increased by high concentration of pyridoxal phosphate.

Among the bovine ocular tissues, the retinal pigment epithelium, ciliary body, iris, neuroretina revealed high activity of OKT. Arginase activity was high in the retina and uvea ; Δ'-pyrroline-5-carboxylate (P5C) reductase activity was high in the retina ; Δ'-pyrroline-5-carboxylate dehydrogenase activity was high in the uvea ; and proline oxidase activity was negligible in all ocular tissues. It is possible that ornithine in the retina may be converted into proline by OKT and P5C reductase, and that the toxicity of excess ornithine and/or the deficiency of proline in uveoretinal tissues may produce gyrate atrophy.

INTRODUCTION

Since Simell and Takki[12] found the association of gyrate atrophy of the choroid and retina with hyperornithinemia, much attention has been given to the significance of ornithine metabolism of the eye. The deficiency of ornithine ketoacid transaminase (OKT) was demonstrated recently in cultured fibroblasts and lymphocytes obtained from the affected patients[5,11].

We examined four cases of Japanese patients, and found different types of disease in responsiveness to vitamin B_6, the responsive and non-responsive. We further examined the role of OKT in the retina and choroid.

CASE REPORT

Case 1. A 5-year-old boy had yellow-white spots at the peripheral ocular fundus. He was the second son of consanguineous marriage. Pediatric examination was normal except for hyperornithinemia (Table 1). The electroretinogram was subnormal.

Case 2. A 8-year-old boy complained decreased vision and night blindness. He had myopia and typical chorioretinal atrophies with scallop margins. The ERG was subnormal. Hyperornithinemia was noted (Table 1).

Case 3. A 18-year-old man had a deterioration of vision with age. The parents were first cousins. Characteristic chorioretinal atrophy and hyperornithinemia (Table 1) were noted.

Case 4. A 23-year-old man suffered from night blindness. He had chorioretinal atrophy and hyperornithinemia (Table 1).

MATERIAL AND METHODS

Amino Acid Analysis

Serum amino acids were measured on a Hitachi amino acid analyzer (type 835).

Fibroblast Culture

Skin biopsy material was cut into small fragments and cultured in Eagle's medium supplemented with 10 % fetal calf serum and non-essential amino acids. Harvested cells were prepared by freezing and thawing and further sonication.

Bovine Ocular Tissues

About 40 bovine eyes were maintained at 4° C from the time of slaughter. The uveal and retinal tissues were dissected in ice-cold 250 mM sucrose containing 20 mM potassium phosphate buffer (pH 7.4), as described previously[3].

Enzyme Assays

Arginase activity was determined as described by Schimke[9]. Ornithine ketoacid transaminase activity was assayed by a modification of the method of Katunuma et al[6]. Δ'-Pyrroline-5-carboxylate reductase activity was determined as described previously[7]. Δ'-Pyrroline-5-carboxylate dehydrogenase activity was determined as described elsewhere[4]. Proline oxidase activity was determined as described elsewhere[4].

One unit was defined as the activity consuming 1 μ mole NADH or forming 1 μmol product per min at 37° C. Protein content was determined by the method of Bradford[2].

RESULTS

Serum Ornithine Level and OKT Activity in Patients with Gyrate Atrophy (Table 1).

Four cases had characteristic chorioretinal atrophy and hyperornithinemia, indicating gyrate atrophy of the choroid and retina. After 2 weeks of vitamin B_6 (600 mg per day) administration, serum ornithine level did not change in cases 1, 2, and 4. Serum ornithine level in case 3 decreased by vitamin B_6.

Ornithine ketoacid transaminase activity was negligible in all four cases, when assayed at 10^{-4} M pyridoxal phosphate. Fibroblasts from case 3 showed an increase in enzyme activity when concentrations of pyridoxal phosphate were increased in the assay medium.

Enzymes Metabolizing Ornithine-Proline Pathway in Bovine Uvea and Retina (Table 2).

OKT activity was high in the uvea and retina. Specific activity of the enzyme in the retinal pigment epithelium was about 10-fold higher than in liver. Moderate arginase activity was found in the retina and uvea. P5C reductase activity was higher in the retina. P5C dehydrogenase activity was higher in the iris and ciliary body. Proline oxidase activity was negligible in the retina and choroid.

Table 1. Serum Ornithine Level and Ornithine Ketoacid Transaminase Activity in Patients with Gyrate Atrophy

Cases	Serum Ornithine Level[a]		OKT Activity[c] in fibroblasts	
	before VB_6[b] administration	VB_6 (600 mg/day)	10^{-4}M PALP[d]	2×10^{-3}M PALP
Case 1	8.5	8.6	ND[d]	ND
Case 2	13.0	13.2	ND	ND
Case 3	6.9	2.5	ND	47
Case 4	17.6	17.4	ND	ND
Normal controls[f]	1.0 ± 0.6	1.0 ± 0.8	150 ± 25	148 ± 26

a : mg/dl
b : vitamin B_6
c : n moles pyrroline-5-carboxylate/30 min/ mg protein
d : pyridoxal phosphate
e : not detectable
f : mean $\pm$ S.D. (N = 5).

DISCUSSION

From the present study, several events and possibilities are represented. (1) There are different types of gyrate atrophy with hyperornithinemia in response to vitamine B6, the responsive (Case 3) and the non-responsive (Cases 1, 3, and 4). (2) In vivo responsiveness to vitamin B_6 correlates with in vitro enzyme activity. (3) OKT activity is high in the retinal pigment epithelium (4) It is possible that ornithine in the retina may be converted into proline by the action of OKT and P5C reductase.

OKT deficiency is also confirmed in Japanese patients with gyrate atrophy. Efficacy of vitamin B_6 is suggested for the responsive, but the sucessfull treatment for the non-responsive is still obscure. A low-arginine diet is tried recently, but the rebellion against the diet is serious[8]. Our present data suggest that the role of OKT in the retina is to convert ornithine into proline. It is likely that the pathogenesis of gyrate atrophy is due to the toxicity of excess ornithine and/or the deficiency of proline. Cases of hyperornithinemia without gyrate atrophy are

Table 2. Enzyme Metabolizing Ornithine-Proline Pathway in Bovine Uvea and Retina

	Arginase[a]	OKT[b]	P5C[a] Reductase	P5C[a] Dehydrogenase	Proline[a] Oxidase
Iris	5.58	96	0.44	1.63	0
Ciliary body	7.10	107	0.29	1.16	0
Neuroretina	10.86	71	14.74	0.07	0
RPE[c]	1.06	113	2.60	0.36	0
Choroid	12.5	14	0.18	0.30	0

a : m unit per mg protein
P5C : pyrroline-5-carboxylate
b : ornithine ketoacid transaminase
n moles pyrroline-5-carboxylate/30 min/mg protein
mean values from three different experiments
c, RPE : retinal pigment epithelium

reported[1,10]. Therefore, it is more likely that the deficiency of proline in uveoretinal tissues may produce atrophy.

ACKNOWLEDGMENT

The authors would like to thank Miss Y. Miura for secretarial assistance. The study was supported in part by a grant from The Ministry of Health and Welfare of Japan.

REFERENCES

1. H. Bickel, D. Feist, H. Müller, et al., Ornithinemie, Dtsch. Med. Wochenschrft, 93:2247 (1968)
2. M. M. Bradford, A rapid and sensitive method for the quantitation of microgram quantities of protein utilizing the principle of protein-dye binding, Anal. Biochem., 72:248 (1976)
3. S. Hayasaka, S. Hara, Y. Takaku and K. Mizuno, Distribution and some properties of cathepsin B in the bovine eyes, Exp. Eye Res., 26:57 (1978).
4. A. Merzfeld, V. A. Mezl and W. E. Knox, Enzyme metabolizing Δ'-pyrroline-5-carboxylate in rat tissues, Biochem. J., 166:95 (1977).
5. M. I. Kaiser-Kupfer, D. Vall and L. A. Del Valle, A specific enzyme defect in gyrate atrophy, Am. J. Ophthalmol.,

85:200 (1978).
6. N. Katunuma, Y. Matsuda and I. Tomino, Studies on ornithine-ketoacid transaminase, II, Purification and properties, J. Biochem., 56:499 (1964).
7. T. Matsuzawa and I. Ishiguro, Δ'-Pyrroline-5-carboxylate reductase from Baker's yeast, Biochem. Biophys. Acta., 613:318 (1980).
8. R. R. McInnes, S. A. Arshinoff, L. Bell, E. B. Marliss and J. C. McCulloch, Hyperornithinemia and gyrate atrophy of the retina, Improvement of vision during treatment with a low-arginine diet, Lancet, March, 513 (1981).
9. R. T. Schimke, Arginase, Methods in Enzymol., 17:313 (1970).
10. V. E. Shih, M. L. Efron and H. W. Moser, Hyperornithinemia, hyperammonemia and homocitrullinuria, Am. J. Dis. Child., 117:83 (1969).
11. V. E. Shih, E. L. Berson, R. Manadell and S. Y. Schmidt, Ornithine ketoacid transaminase deficiency in gyrate atrophy of the choroid and retina, Am. J. Hum. Genet., 30:174 (1978).
12. O. Simell and K. Takki, Raised plasma ornithine and gyrate atrophy of the choroid and retina, Lancet, 1:1031 (1973).

GYRATE ATROPHY OF THE CHOROID AND RETINA (GA) : TOXIC EFFECTS
OF ORNITHINE AND LONG-TERM THERAPY WITH AN ARGININE-RESTRICTED DIET

D. Valle, M. Walser, S. Brusilow and M. I. Kaiser-Kupfer

The Johns Hopkins University School of Medicine
Baltimore and National Eye Institute, Bethesda, USA

SUMMARY

GA is characterized by progressive chorioretinal degeneration and hyperornithinemia and is due to an inherited deficiency of ornithine-δ-aminotransferase (OAT). We have examined the ramifications of this enzymatic deficiency at the cellular and clinical levels. Unlike normal cells, intact GA fibroblasts do not convert radiolabeled ornithine to proline and glutamate. Although GA cells grow normally in standard medium, addition of ornithine (20 mM) kills > 98 % of the GA cells, while control cells grow at nearly normal rates. Addition of proline or other neutral amino acids (5 mM) blocks the ornithine toxicity ; whereas inhibitors of polyamine synthesis, guanidinoacetate, creatine, and the dibasic amino acids have no effect. These results suggest that high ornithine concentrations in association with OAT deficiency is detrimal to some cells and that reduction of ornithine may be beneficial in GA patients. We therefore placed 9 GA patients on an arginine-restricted diet for from 2-36 months. Within 40 days there was a 2-6 fold reduction in plasma ornithine which has been maintained for from 4-36 months in 4 of the patients. None of the patients has had progression of their ophthalmologic abnormalities while on the diet and as recently reported (Science 210:1128, 1980), one has had some improvement. At all plasma ornithine concentrations, total urine losses of arginine-derived carbon skeletons is less than predicted by arginine intake. This suggests that significant amounts of arginine (+/o ornithine) are metabolized in GA patients by either residual OAT activity, ornithine decarboxylase or by an as yet unelucidated pathway.

DISEASE OF ORNITHINE-PROLINE PATHWAY: A Δ^1-PYRROLINE-5-CARBOXYLATE REDUCTASE DEFICIENCY IN THE RETINA OF RETINAL DEGENERATION MICE

Takeo Matsuzawa, Koichi Iwasaki, Noriko Hiraiwa[*],
Etsuko Inagaki and Isao Ishiguro

Department of Biochemistry and Ophthalmology[*], School of Medicine, Fujita-Gakuen University, Toyoake, Aichi 470-11, Japan

INTRODUCTION

The discovery of a genetic defect in ornithine oxoacid aminotransferase (L-ornithine: 2-oxoacid aminotransferase, EC 2.6.1.13) in gyrate atrophy of the choroid and retina[1,2,3], prompted us to study the ornithine metabolism of the eye[4]. We soon found high activities of ornithine oxoacid aminotransferase and Δ^1-pyrroline-5-carboxylate reductase (L-proline:NAD(P)$^+$5-oxidoreductase, EC 1.5.1.2) (P5C reductase) in the retina and retinal pigment epithelium[4,5]. Gyrate atrophy of the choroid and retina is a disease related to the category of choroideremia and retinitis pigmentosa (primary, hereditary, pigmentary retinopathy), these are much frequent cause of the human hereditary blindness, however the biochemical pathogenesis of these diseases remains to be solved. In this paper we describe an intracellular proline synthetic pathway from ornithine uncovered in the bovine cornea and retinal outer layers(including the choroid) which presumably participates in the biosynthesis of proline-rich proteins such as collagen and glycoproteins, and the characteristic postnatal changes and the deficiency of this pathway in C3H retinal degeneration mice.

MATERIALS AND METHODS

1. Animals

Fresh bovine eyes were obtained from Holstein cows in a local slaughter-house. Ocular tissues were separated and frozen at -80°C until use. C3H/HeNCrj retinal degeneration mice and CRJ:B6C3F$_1$(C57 BL/6NCrj(♀) x C3H/HeNCrj(♂))mice(as the control) were supplied from Charles River Japan Inc.

2. Enzyme Assays

Ocular tissues and livers from cows and mice were homogenized with appropriate amounts of 0.25 M sucrose containing 0.01 M potassium phosphate buffer(PPB)(pH 7.5), freeze-thawed 3 times, centrifuged at 8500 x g for 10 min. The resultant supernatant was subjected to the assays of arginase, P5C reductase and ornithine oxoacid aminotransferase. Sonicated supernatant of bovine liver and ocular tissues was also applied to the assays of ornithine oxoacid aminotransferase, proline oxidase, P5C dehydrogenase and P5C syntase. All enzyme activities except P5C synthase activity were assayed similarly as previously described[4]. P5C synthase activity was determined according to the method of Smith et al.[6], but P5C formed was analyzed by high performance liquid chromatography[7]. The protein concentration was determined by the dye method[8].

3. Incorporation of ^{14}C-Ornithine into Proline

Bovine corneas and retinas were added one to two volumes of 0.25 M sucrose containing 0.01 M PPB(pH 7.5) and ground with quartz sand in a mortar at 4°C. Mice livers and their ocular posterior parts were homogenized as described above. The homogenates were centrifuged at 1200 x g for 10 min. The resulting supernatant was dialyzed against 0.01 M PPB(pH 7.5) overnight and the dialysate was used as the enzyme solution. The reaction mixture contained 40 µmol of PPB(pH 7.5), 5 µmol of 2-oxoglutarate, 1 µmol of L-ornithine, 0.2 µCi of L-[U-^{14}C]ornithine(200 mCi/mmol), 20 nmol of pyridoxal phosphate, 336 nmol of NADPH, 0.5 µmol of DL-isocitrate, 0.1 unit of isocitrate dehydrogenase(EC, 1.1.1.42), 10 µmol of glutamate and an appropriate amount of enzyme solution in a final volume of 1.0 ml. The mixture was incubated with constant shaking for 60 min at 37°C and terminated by the addition of 0.5 ml of 10% $HClO_4$. The blank flask was added 0.5 ml of 10% $HClO_4$ at 0 time, then incubated. Deproteinized mixture was neutralized with KOH, and proline fraction was separated by passing through a Dowex 50W(NH_3 form) column (1.0 x 1.2 cm), a AG 2-X10(formate form) column(1.0 x 4.0 cm) and again the Dowex 50W column as previously described[9]. An aliquot of this fraction was added to 10 ml of liquid scintillator(Scintisol-500, Wako Chemical Industries, Ltd., Osaka) and the radioactivity was counted.

RESULTS

1. Distribution of Enzyme Activities Relating to Ornithine Metabolism in Bovine Ocular Tissues

Intracellular proline synthesis is composed of ornithine- and glutamate-proline pathways as shown in Fig. 1. We determined the activities of ornithine transcarbamylase, arginase and 5 enzymes of the proline synthesis indicated in Fig. 1. In ocular tissue orni-

DISEASE OF ORNITHINE-PROLINE PATHWAY

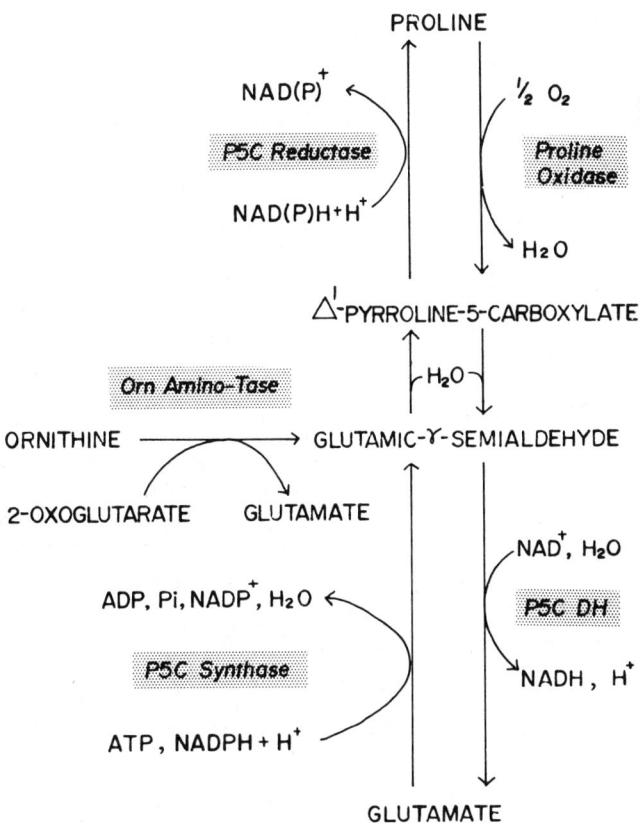

Fig. 1. Enzymatic reactions participating in ornithine- and glutamate-proline pathways.

Table 1. Distribution of Enzyme Activities Relating to Ornithine Metabolism in Bovine Ocular Tissues

	Arginase	OAT	P5C synthase	P5C-DH	P5C reductase	Proline oxidase
	(mUnit/mg protein) Mean value (n=3-4)					
Cornea	0.00	0.50	0	0.09	8.38	0
Lens	0.00	0.00	0	0.00	4.88	0
Iris	5.58	12.30		1.63	0.44	0
Ciliary Body	7.10			1.16	0.29	0
Retina	10.86	9.40	0	0.07	14.74	0
RPE	1.06	4.90		0.36	2.60	0
Choroid	12.50	0.50		0.30	0.18	0

P5C-DH, P5C dehydrogenase. Partially taken from Matsuzawa et al[4].

thine transcarbamylase activity was negligible, contrary arginase activity was high in the tissues derived from neural ectoderm, but not from epidermal ectoderm (Table 1). High activity of ornithine oxoacid aminotransferase was found in retina, retinal pigment epithelium(RPE), iris and ciliary body, but less activity found in cornea. Activity of P5C dehydrogenase was high in iris and ciliary body. High activity of P5C reductase, which catalyses the reduction of a common intermediate P5C to proline, was found in cornea, lens, retina and retinal pigment epithelium. P5C synthase and proline oxidase activities were almost nil. Thus the ornithine-proline pathway is presumed to be a main route for the intracellular proline synthesis of the eye(Table 1).

2. Incorporation of ^{14}C-Ornithine into Proline in Bovine Ocular Tissues

Table 2 shows the incorporation of ^{14}C-ornithine into proline by the two-steps reaction of ornithine-proline pathway in bovine cornea, lens and retina. Bovine cornea gave a higher specific activity than retina, however the total activity of the retina per eye was 3 times more than that of cornea. The incorporation observed in these tissues was completely inhibited by canaline, a specific inhibitor of ornithine oxoacid aminotransferase. The lens gave no incorporation in spite of its high content of P5C reductase.

3. Postnatal Changes of Enzyme Activities Involved in the Ornithine-Proline Pathway

The development of ocular tissues is known to be active in the

Table 2. Proline Formation from Ornithine with Bovine Ocular Tissue Homogenates

Tissue	^{14}C-Ornithine converted into proline	
	nmol/hr/mg protein	Total activity/eye
Bovine cornea	29.3 ± 3.3	208.5
+ canaline	0	
Bovine lens	0	0
Bovine retina	4.4 ± 1.5	310.9[a]
(including the choroid)		
+ canaline	0	

[a]Activity per ocular posterior half. Mean ± SD(n=3).

early fetal stages, but the enlargement of these tissues continues in the postnatal periods. Fig. 2 and 3 show the postnatal changes of P5C reductase and ornithine oxoacid aminotransferase in the ocular tissues, which were cut into two pieces: the posterior part including retina and retinal outer layers and the anterior part involving cornea and lens, and in livers of CRJ control and C3H retinal degeneration mice. In CRJ control mice P5C reductase of the posterior parts was already high at the early postnatal period, then it decreased to a basal activity at the weaning period. Ornithine oxoacid aminotransferase of the posterior parts and P5C reductase of the anterior parts were maintained throughout a constant level(Fig. 2A). Contrary in the livers of CRJ control mice P5C reductase showed a biphasic change: an early peak at the suckling period and the other at the weaning period. Ornithine oxoacid aminotransferase was induced after 10 days and reached a plateau by 20 days (Fig.2B). In the ocular tissues of C3H retinal degeneration mice P5C reductase of the posterior parts was clearly lower at the early period and decreased to a much lower level after 10 days. P5C reductase of the anterior parts was also weak throughout all the periods. However, ornithine oxoacid aminotransferase of the posterior parts was similarly constant as that of the control mice(Fig. 3A). In C3H mice the weaning was delayed, so that ornithine oxoacid aminotransferase activity in the livers was induced later after 15 days and plateaued by 25 days. The height of the early peak of P5C reductase of the ocular posterior parts in C3H mice was doubled on the 15th day, then decreased to the weaning level(Fig. 3B).

Furthermore, the incorporation of ^{14}C-ornithine into proline in CRJ and C3H mice showed almost an equal specific activity in their livers, but in the posterior parts of the eyes of C3H mice on the 19th day, it was about one third lesser than that of the control mice, in the specific and total activities (Table 3).

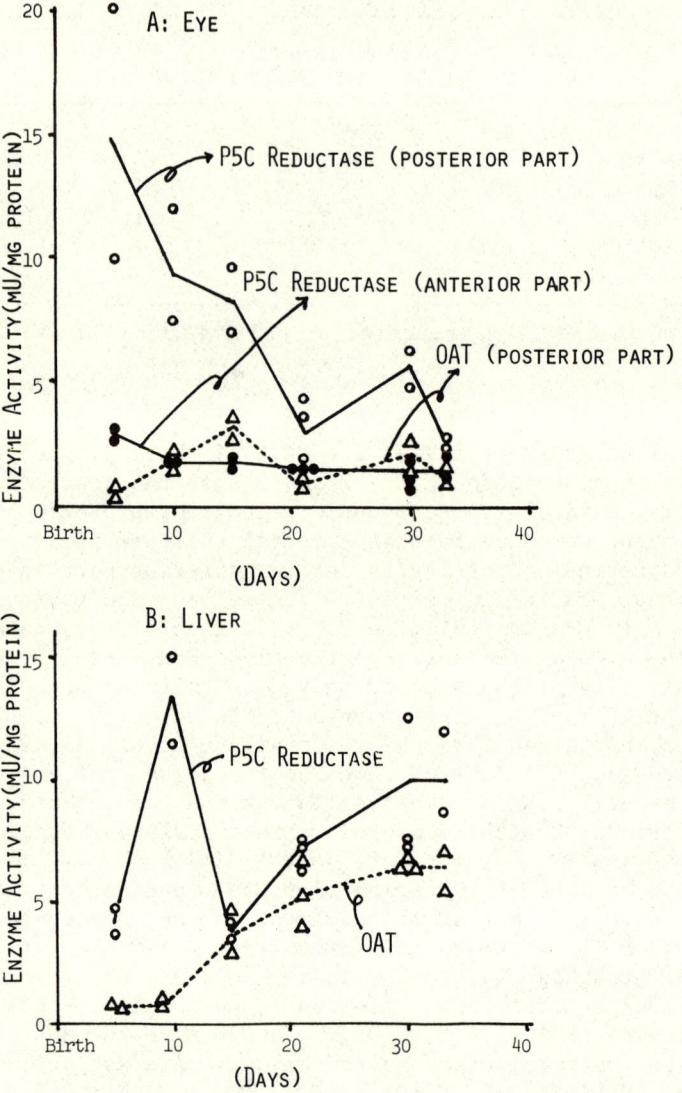

Fig. 2. Postnatal changes of P5C reductase and ornithine oxo-acid aminotransferase in CRJ control mice.
A: in the eye tissues; ○——○, P5C reductase of the posterior parts; ●——●, P5C reductase of the anterior parts; △---△, ornithine oxoacid aminotransferase of the posterior parts. B: in the livers.
Each value was obtained with 7-10 mice.

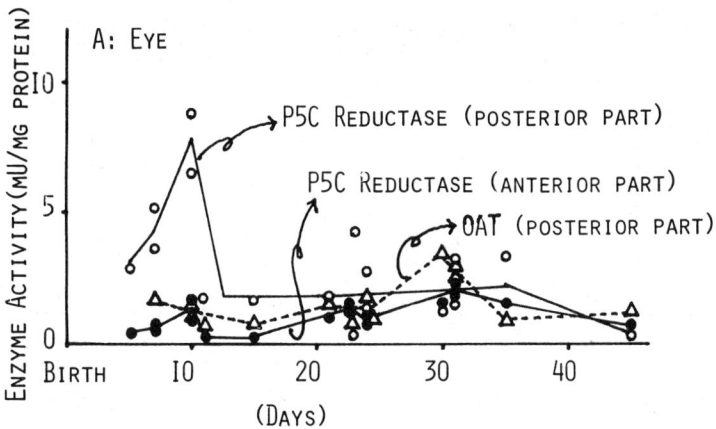
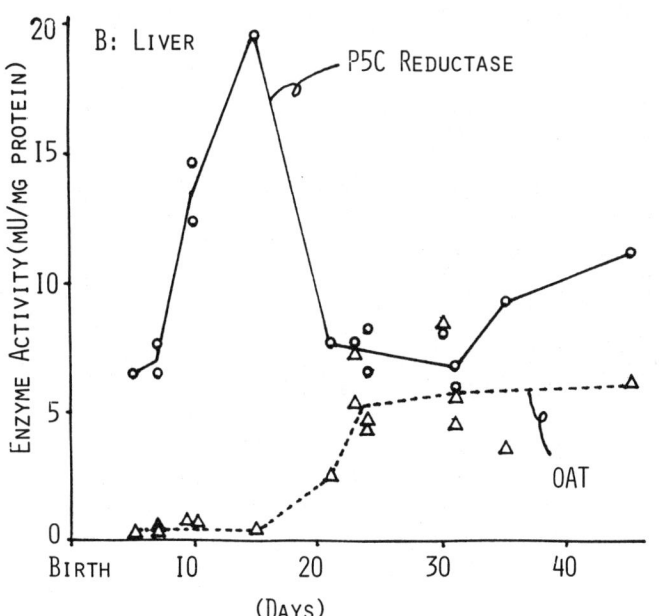

Fig. 3. Postnatal changes of P5C reductase and ornithine oxo-acid aminotransferase in C3H retinal degeneration mice. A: in the eye tissues; the symbols are similar as those in Fig. 2. B: in the livers.
Each value was obtained with 7-10 mice.

Table 3. Proline Formation from Ornithine with Mice Ocular Tissues and Livers Homogenates

Tissue	^{14}C-Ornithine converted into proline	
	nmol/hr/mg protein	Total activity/mouse
Ocular posterior parts:	Mean value(10 mice)	
CRJ mice,		
5 days-old	26.5	3.7
19 days-old	63.6	18.1
C3H mice,		
5 days-old	21.2	3.9
19 days-old	24.6	6.2
Livers:		
CRJ mice,		
5 days-old	1.1	12.2
19 days-old	3.0	115.5
C3H mice,		
5 days-old	1.3	10.8
19 days-old	3.6	221.1

DISCUSSION

The coenzyme requirement with highly purified bovine retinal P5C reductase was 1:1 for NADH and NADPH, respectively; i.e., the activity with NADH was almost equal to the activity with NADPH. The K_m value for NADH was 0.18 mM and it was larger than NADH level in the retina(0.067 mM), but the K_m value for NADPH was 7.1×10^{-6}M and it was small enough to meet the NADPH level in the retina(0.042 mM). Thus the NADPH-dependent activity may participate in the ornithine-proline pathway in the retinal outer layers. The present experiments showed that the intracellular proline synthesis in the ocular tissues is carried out by the ornithine-proline pathway. Since arginase activity was high in the retinal outer layers, the ornithine also can be derived from arginine in these tissues. Gyrate atrophy of the retina and choroid associated with hyperornithinemia is accompanied with muscular atrophy[10] and an abnormal collagen formation[11]. Collagen formation is known to be more active in child than in adult. A deficiency of proline formation through the ornithine-proline pathway is probably a major cause of this inborn error of ornithine metabolism. The present experiments showed evidence that the ornithine-proline pathway localizes in the photoreceptor cells or retinal pigment epithelium, those cells disappeared during the retinal degeneration process(Fig. 2 and 3)[12]. Bruch's membrane is rich in collagen and is assumed to fill a role of ultrafiltration membrane between the choriocapillaris and the retinal outer layers. High activity of the intracellular proline synthesis of the retinal outer layers seems to meet the proline requirement for the formation of Bruch's membrane. We would like to offer a newer understanding on "an abiotrophy of Bruch's membrane formation" in the retinal degeneration instead of Collins' classical view[13] from the biochemical standpoint of present experiments.

SUMMARY

In bovine ocular tissues, cornea and retinal outer layers provide the intracellular proline synthetic pathway from ornithine, but not from glutamate. In C3H retinal degeneration mice, P5C reductase activity in the retinal outer layers was decreased to about one third lesser than that of CRJ control mice in the specific activity, throughout suckling and weaning periods. The activity of an incorporation of ^{14}C-ornithine into proline in the retinal outer layers was also decreased to the same extent at the weaning period, in the retinal degeneration mice. Postnatal changes of P5C reductase and ornithine oxoacid aminotransferase activities in the livers of C3H mice showed a delayed weaning. Thus "an abiotrophy of Bruch's membrane formation" due to proline deficiency may contribute as a factor for the development of retinal degeneration.

ACKNOWLEDGEMENTS

We thank Professors Nobuhiko Katunuma and Katsuyoshi Mizuno, and Drs. Katsuhiko Niimi and Seiji Hayasaka for valuable discussion. We also thank Miss Kumiko Itoi, Sachie Fujii and Mr. Naofumi Sugimoto and Kazumi Mishima for their technical assistances. This investigation was supported by Fujita-Gakuen University Research Fund.

REFERENCES

1. J. M. F. Trijbels, R. C. A. Sengers, J. A. J. M. Bakkeren, A. F. M. DeKort and A. F. Deutman, L-Ornithine-ketoacid-transaminase deficiency in cultured fibroblasts of a patient with hyperornithinemia and gyrate atrophy of the choroid and retina, Clin. Chim. Acta 79:371 (1977).
2. M. I. Kaiser-Kupfer, D. Valle and L. A. Del Valle, A specific enzyme defect in gyrate atrophy, Am. J. Ophthal. 85:200 (1978).
3. S. Hayasaka, T. Shono, K. Mizuno, T. Saito and K. Tada, Ornithine ketoacid aminotransferase activity in gyrate atrophy patients and bovine eye, Invest. Ophthal. Visual Sci.(Suppl). ARVO abstract 19:185 (1980).
4. T. Matsuzawa, I. Ishiguro, S. Hayasaka, T. Shiono, H. Nakajima and K. Mizuno, Hyperornithinemia with gyrate atrophy and enzymes involved in ornithine metabolism of the eye, Biochem. International 1:179 (1980).
5. S. Hayasaka, T. Shono, Y. Takaku and K. Mizuno, Ornithine ketoacid aminotransferase in the bovine eye, Invest. Ophthal.

Visual Sci. 19:1457 (1980).
6. R. J. Smith, S. J. Downing, J. M. Phang, R. F. Lodato and T. T. Aoki, Pyrroline-5-carboxylate synthase activity in mammalian cells, Proc. Natl. Acad. Sci. USA 77:5221 (1980).
7. J. J. O'Donnell, R. P. Sandman and S. R. Martin, Assay of Ornithine aminotransferase by high-performance liquid chromatography, Anal. Biochem. 90:41 (1978).
8. M. M. Bradford, A rapid and sensitive method for the quantitation of microgram quantities of protein utilizing the principle of protein-dye binding, Anal. Biochem. 72:248 (1976).
9. T. Matsuzawa and I. Ishiguro, Ornithine metabolism in relation to stimulation of urea cycle, induced by high protein diet, Arch. Biochem. Biophys. 208:101 (1981).
10. C. McCulloch and E. B. Marliss, Gyrate atrophy of the choroid and retina with hyperornithinemia, Am. J. Ophthal. 80:1047 (1975).
11. K. Mizuno, Plenary lecture: Pathology and clinic of hereditary chorioretinal degeneration, Acta Soc. Ophthal. Jap. (Suppl). 85:3(Abstract) (1981).
12. K. Tansley, Hereditary degeneration of the mouse retina, Brit. J. Ophthal. 35:573 (1952).
13. T. Collins, Abiotrophy of the retinal neuroepithelium, or retinitis pigmentosa, Arch. Ophthal., N.Y. 48:517 (1919).

TOXIC EFFECTS OF ORNITHINE AND ITS RELATED COMPOUNDS ON THE RETINA

Yujiro Ishikawa, Toichiro Kuwabara
and Muriel I. Kaiser-Kupfer

Laboratory of Vision Research and Clinical Branch
National Eye Institute, National Institutes of Health
Bethesda, MD 20205, U.S.A.

INTRODUCTION

Recent studies revealed that intravitreous injection of L-ornithine causes severe damage in the pigment epithelium and photoreceptor cells of experimental animals and that the clinical and histological appearances of the damage are similar to those of gyrate atrophy, a hereditary blinding disease caused by hyperornithinemia due to the deficiency of ornithine-δ-aminotransferase[1]. In order to see cytologic toxicity, several agents which are structurally and metabolically related to ornithine were injected into the vitreous employing the technique that has been described earlier.

MATERIALS AND METHODS

Adult Sprague-Dawley and Evans black hooded rats were used in this experiment. The rats were anesthetized by an intraperitoneal injection of sodium pentobarbital. L-ornithine hydrochloride, N-α-acetyl-L-ornithine, L-arginine hydrochloride, L-citrulline, putrescin (1,4-diaminobutane) dihydrochloride, L-glutamic acid, L-proline and L-lysine monohydrochloride (Sigma Chemical Co., St. Louis, Missouri) and α-methylornithine hydrochloride (Vega Biochemicals, Tuscon, Arizona) were dissolved in physiologic saline to make 1 M solutions and the pH was adjusted to 7.2. After filtering through 0.22 µm millipore filters (Millipore Co., Bedford, MA.), 10 µl of the freshly prepared solution was injected into the vitreous via the upper equatorial region of the globe using a Hamilton microsyringe. The eyes which were injected with L-ornithine or

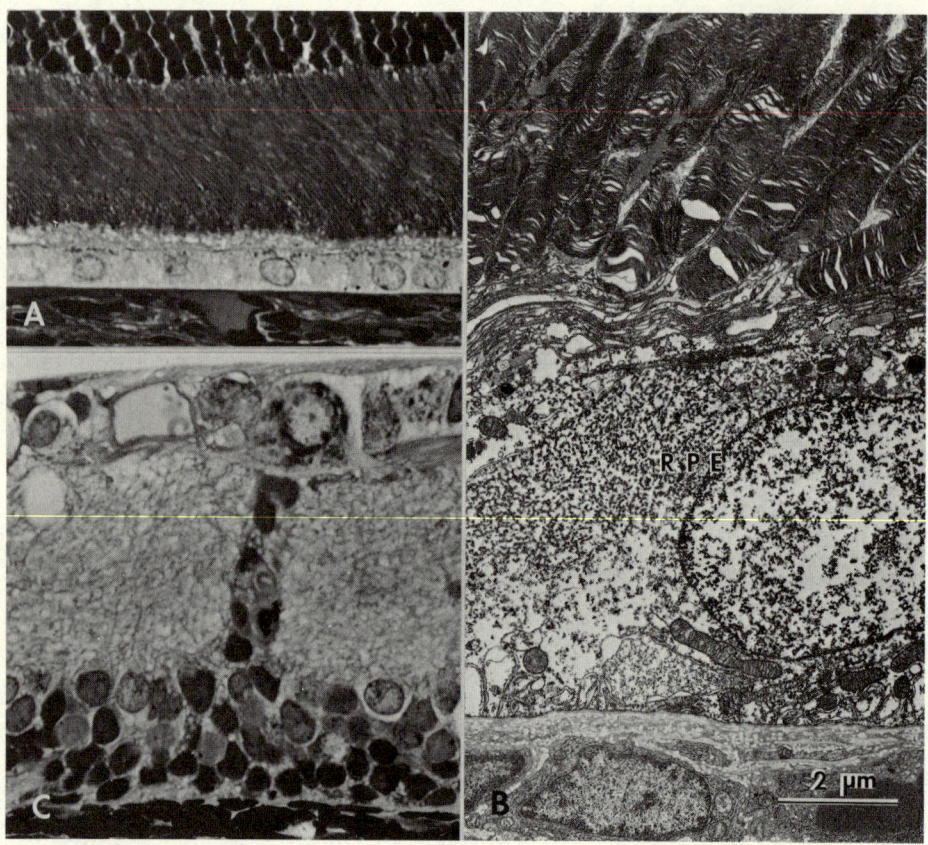

Fig. 1. Rat retinas following injection of 1 M solution of L-ornithine. A : 30 minutes after injection. RPE cells are markedly swollen, while photoreceptor cells remain normal. Toluidine blue stain, B : 30 minutes. Electron microscopy of the RPE reveals that swelling occurs in the cell matrix and microorganelles are not affected. C : 3 months. The RPE and photoreceptor cells are absent. Toluidine blue stain.

D-ornithine were examined at 30 minutes, 1, 3, 6, 12 and 24 hours, and 3, 7, 14, 30, 90 and 120 days, and the eyes which were injected with other compounds were studied at 3 hours and 14 days after the injection. Also, effects from diluted solutions of each agent were examined. Enucleated eyes were fixed in 4 % glutaraldehyde in 0.1 M phosphate buffer (pH 7.2) at room temperature. The chorioretinal tissues were embedded in glycol methacrylate for light microscopy and in an epoxy resin for electron microscopy.

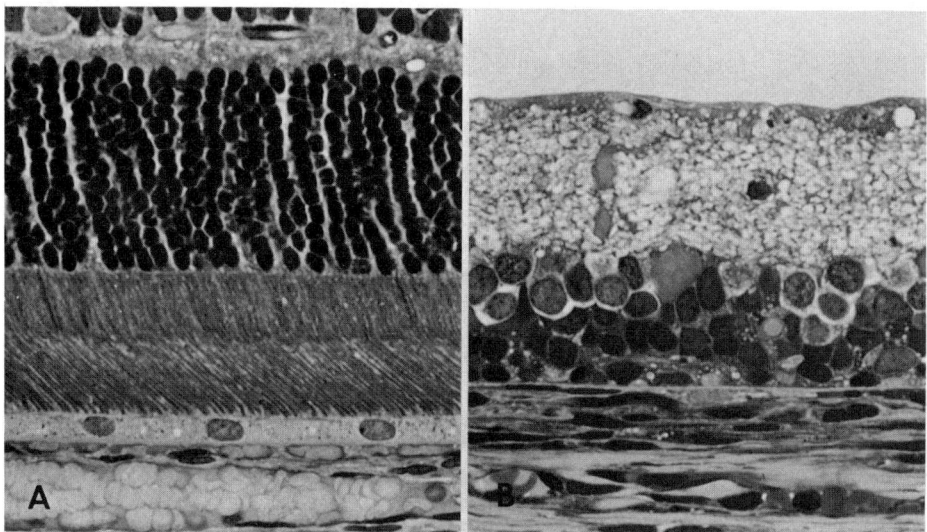

Fig. 2. Rat retinas following injection of 1 M solution of
D-ornithine. A : 3 hours. Swelling in RPE cells is mild.
B : 2 weeks. Outer layers of the retina have been
severely degenerated. Both toluidine blue stain.

RESULTS

Five minutes after the injection of 10 µl of 1 M L-ornithine
solution, the retinal pigment epithelium (RPE) began to show
swelling, which became apparent by 30 minutes (Fig. 1A, B). The
swelling of the RPE which occurred uniformly in all cells, progressed
until the fourth hour, and then started to subside in many cells.
By 24 hours after injection, destruction of certain areas of RPE
and disorganization of outer segments became apparent. Other parts
of the retina remained normal, although Müller cell components
showed mild swelling. In severely damaged eyes, the disappearance
of the RPE was followed by degeneration of outer retinal layers
at later stages (Fig. 1C). A 10 µl injection of 0.1 M solution
caused mild swelling in the RPE by 3 hours, but no permanent
effects were noted in the retina.

D-ornithine showed a somewhat different course in the damaging
process. Unlike L-ornithine, injection of 10 µl of the 1 M D-
ornithine solution did not cause acute swelling in the RPE at an
early stage (Fig. 2A). However, progressive degeneration began in
many RPE cells and in the photoreceptor layer at around the 12th
hour, and by the 2nd week, the cells in these layers had mostly
disappeared (Fig. 2B).

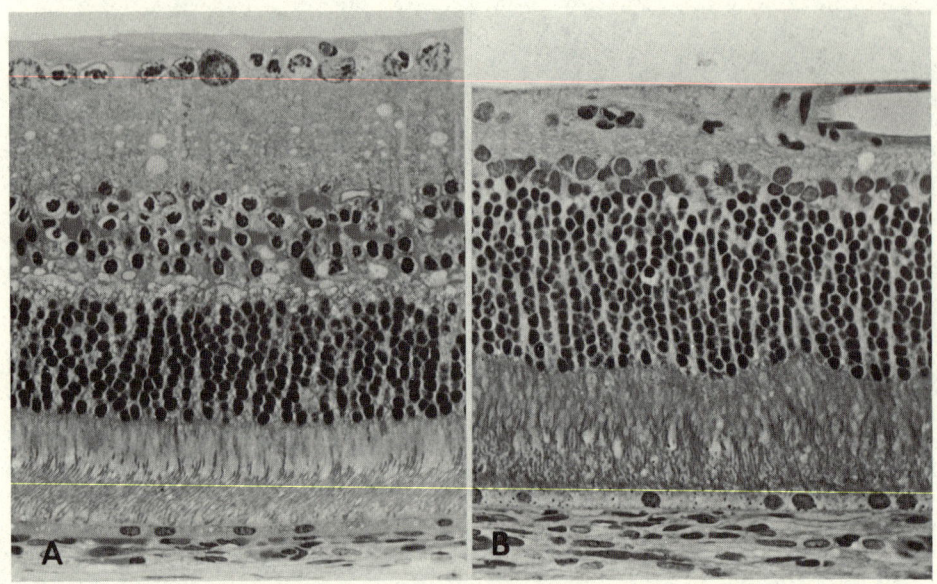

Fig. 3. Rat retinas following injection of 0.1 M solution of putrescin. A : 3 hours. Numerous cells in the ganglion cell and inner nuclear layers are pyknotic. B : 2 weeks. The inner layers of the retina become extremely thin, while the photoreceptor cell layer and the RPE remain normal. Both toluidine blue stain.

N-α-acetyl-L-ornithine or α-methylornithine did not induce any appreciable changes in the retina at 3 hours or at 2 weeks after the injection. Arginine, a main precursor of ornithine, caused slight swelling in the RPE at 3 hours, but no further damage was observed in the retina at 2 weeks after the injection. Since citrulline, a metabolite of ornithine was not soluble to make a 1 M concentration, a saturated solution was used. The solution did not produce any changes in the retina and the RPE.

Injection of putrescin, another metabolite of ornithine, caused marked changes in the retina. At 3 hours after the injection of 0.1 M solution, many ganglion cells and certain cells in the inner nuclear layer became pyknotic and vacuolated (Fig. 3A). Also, slight swelling was observed in Müller cell processes near the inner limiting membrane. The outer retina and RPE, did not show prominent changes. By the second week, gradual degeneration and thinning occurred in the inner layers of the retina, including the outer plexiform layer, while photoreceptor cells and RPE remained normal (Fig. 3B). Injection of 1 M solution of putrescin caused severe degeneration of the inner layers of the retina.

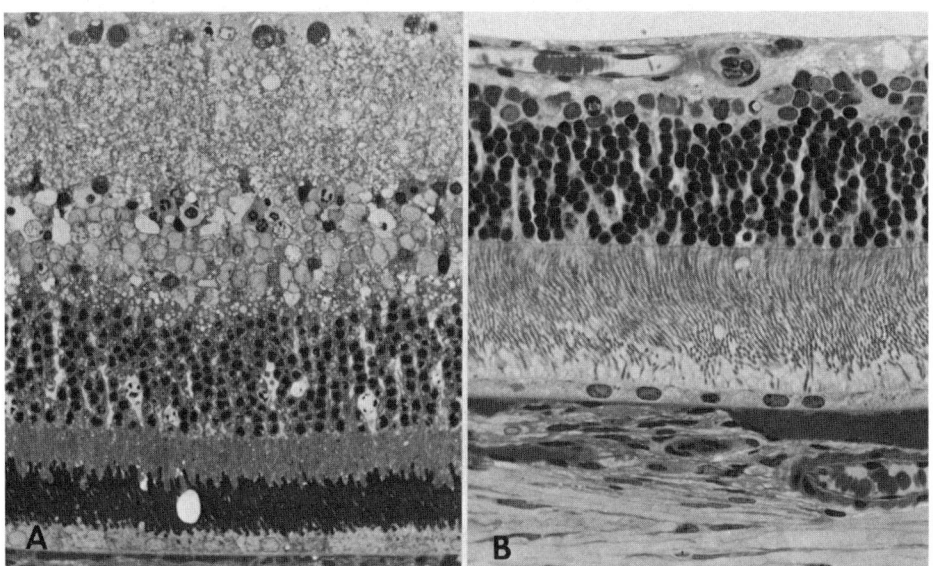

Fig. 4. Rat retinas following injection of 1 M solution of glutamic acid. A : 3 hours. Ganglion and amacrine cells are damaged. Also, cone cells are markedly swollen. B : 2 weeks. The inner layer including the outer plexiform layer becomes markedly atrophic. The number of the cone cells is decreased. Both toluidine blue stain.

Glutamic acid caused pyknotic changes in certain ganglion cells and amacrine cells 3 hours after injection (Fig. 4A). Noteworthy was that the cone cells became markedly edematous at an early stage of experiment. In addition, the inner and outer plexiform layers were finely vacuolated. Two weeks after the injection, the inner half of the retina became extremely thin and cone cells were markedly decreased in the photoreceptor cell layer, though the RPE layer appeared relatively normal (Fig. 4B).

Three hours after the injection of proline, a possible metabolite of ornithine, certain cells in the inner nuclear layer became markedly swollen (Fig. 5). In addition, inner plexiform layer was finely vacuolated. However, no permanent damage was observed at later stages.

Lysine, like ornithine a dibasic amino acid, caused slight swelling in Müller cells at 3 hours. The cells contained finely granular aggregations within the cytoplasm and nuclei. Two weeks after injection, it was found that groups of photoreceptor cells protruded into the subretinal space, and the outer nuclear layer

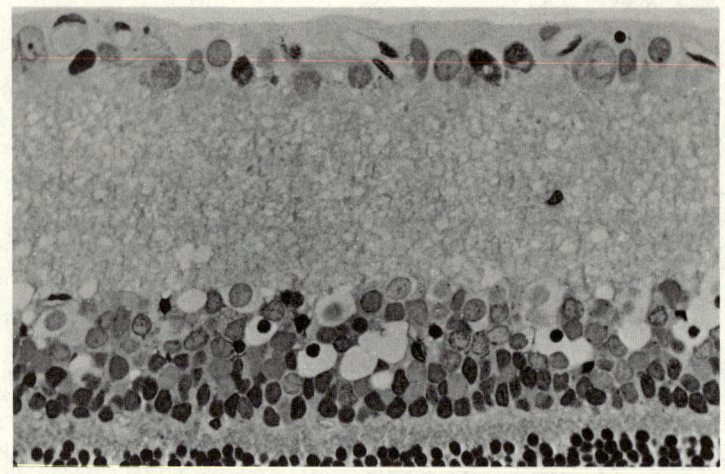

Fig. 5. Rat retina 3 hours after injection of proline. Certain cells in the inner nuclear layer are swollen and their nuclei are pyknotic. Small vacuoles are present in plexiform layers. Toluidine blue stain.

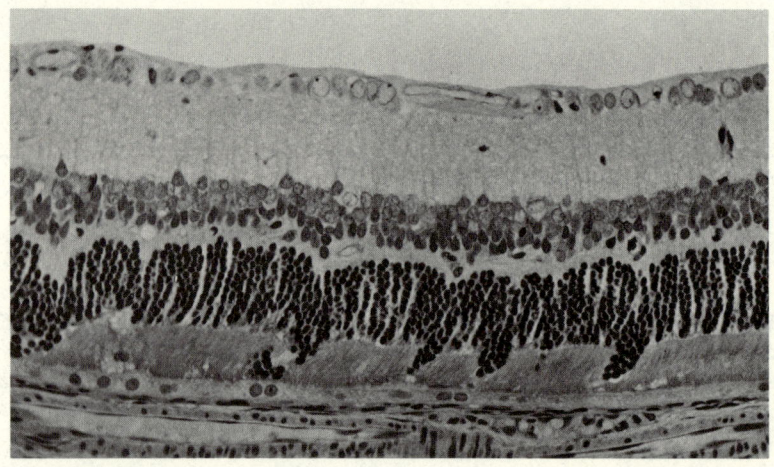

Fig. 6. Rat retina 2 weeks after injection of lysine. The arrangement of nuclei in the outer nuclear layer becomes markedly irregular. Toluidine blue stain.

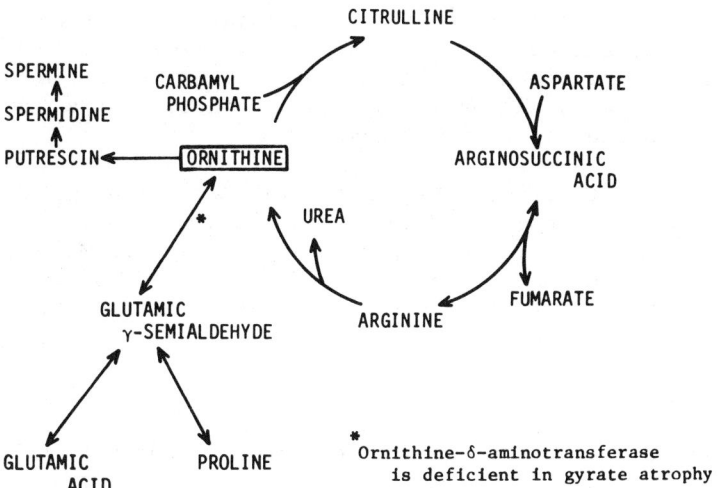

Fig. 7. Metabolism of ornithine.

became markedly irregular in arrangement (Fig. 6). The outer segments of dislocated photoreceptor cells gradually degenerated. No damage was noted in the RPE.

No compounds used in this study caused any histological changes in the anterior segments of the eye, the iris or the ciliary body. However, eyes severely damaged by injection of 1 M putrescin developed cataracts.

DISCUSSION

In the present study the retinal effects of various amino acids and derivatives related to ornithine metabolism were studied (Fig. 7). The intravitreal administration of ornithine, both L and D forms, is found to be specifically toxic to the pigment epithelium, and to cause severe chorioretinal degeneration secondarily. The pathologic process which occurs in patients with gyrate atrophy due to ornithine-δ-aminotransferase deficiency seems to be simulated in the animal model studied. No other compounds studied caused appreciable changes in the pigment epithelium. Arginine, citrulline, and acetylated and methylated forms of ornithine seem to be nontoxic to the retina. Glutamic acid caused degeneration in cells in the inner layers, as shown by earlier investigators[2,3]. The histologic appearance of the damaged retina at a late stage resembled that caused by arterial occlusion. However, no vascular damage was demonstrated immediately following the injection. Effects of glutamate on the cone will be discussed in detail elsewhere. Putrescin was found to be toxic to the retina without demonstrating affinity to any specific cells, though the damage caused by intravitreal injection of a low dose seemed to localize in the inner layer.

Lysine is not directly associated with ornithine metabolism, but is a dibasic amino acid. This amino acid was not toxic to the pigment epithelium, but caused progressive damage in Müller cells. Similar pathologic changes caused by administration of aminoadipic acid have been demonstrated by earlier investigators[3,4].

ACKNOWLEDGMENT

The authors thank Ms. Mary Alice Crawford for her technical assistance.

REFERENCES

1. T. Kuwabara, Y. Ishikawa, and M.I. Kaiser-Kupfer. Experimental model of gyrate atrophy in animals. Ophthalmology (Rochester). 88 : 331 (1981).
2. D.R. Lucas, and J.P. Newhouse, The toxic effect of sodium L-glutamate on the inner layers of the retina. Arch. Ophthalmol. 58 : 193 (1957).
3. J.W. Olney, Neurotoxicity of excitatory amino acids, In : "Kainic acid as a tool in neurobiology", E.G. McGeer, J.W. Olney, and P.L. McGeer, ed., Raven Press, New York (1978).
4. O.Ø. Pedersen, and R.L. Karlsen, Destruction of Müller cells in the adult rat by intravitreal injection of D, L-α-aminoadipic acid. An electron microscopic study, Exp. Eye Res. 28 : 569 (1979).

BASIC BIOCHEMISTRY

5) <u>Relation of Urea Cycle to Guanidino Compounds</u>

QUANTITATIVE DETERMINATION OF GUANIDINO COMPOUNDS:

THE EXCELLENT PREPARATION OF BIOLOGICAL SAMPLES

Akio Ando, Takeo Kikuchi, Hiroshi Mikami, Masamitsu
Fujii, Kazuo Yoshihara, Yoshimasa Orita, and Hiroshi Abe

The First Department of Medicine, Osaka University
Medical School
1-1-50 Fukushima, Fukushima-ku, Osaka, 553 Japan

INTRODUCTION

It is well known that the serum concentrations of some guanidino compounds, for example, methylguanidine and guanidinosuccinic acid, increase in the uremic state. Metabolic abnormality of guanidino compounds might be occurred by the aberrations of urea cycle in uremia. The analysis of guanidino compounds in biological fluids provides us with a great deal of useful information on the pathophysiology and treatment for uremia. Recently, the quantitative determination of various guanidino compounds has been performed with the use of high-pressure liquid chromatography and with the use of fluorometric detection of the 9,10-phenanthrenequinone derivative of guanidino compounds[1]. A rapid, sensitive and quantitative determination of various guanidino compounds with a small amount of sample became possible by this method, but the evaluation of this method for clinical application has not yet been completed.

The analysis of plasma guanidino compounds by the present method, with the same sample in different laboratories, revealed that the variance of measured values was great[2]. This variance seems to be mainly caused by the difference of deproteinizing methods. To evaluate our method of measuring guanidino compounds, it is necessary to establish the optimal deproteinization condition. In the present study, we have studied the optimal deproteinizing method for the present automatic analysis of guanidino compounds in biological samples.

MATERIAL AND METHODS

Analytical Principle

The analysis of guanidino compounds was done with a Guanidine Analyzer G-520 (Japan Spectroscopic Co., Hachioji, Tokyo, Japan). The analytical principle was shown in Figure 1. An eluent-selecting valve, which is controlled by an electronic monitor, was inserted between five buffer-chambers. The eluent was pumped through an automated sample-injection with a 200-μL sample loop. The 3.0 x 150 mm chromatographic column was packed with CM-10S (a styrene-divinylbenzene co-polymer), cross-linked at 10%, with a particle diameter of 11.5 μm. The column eluent was mixed with streams of 2 mol/L NaOH and 9,10-phenanthrenequinone reagent, and the guanidino compounds in the sample were converted at 60°C to fluorescent product, which was detected with the fluorescence photometer (excitation 365 nm, emission 495 nm).

Pretreatment of Biological Sample

(1) The concentrations of free guanidino compounds in plasma were determined by ultrafiltration. Plasma sample was centrifuged (800 x g, 30 min) in a Centriflo membrane CF-25 (cat.no.CB0437A; Amicon Corp., Lexington, MA02173, USA). Two aliquots of the ultrafiltrate, one of them was adjusted to pH 2.0, were used for the chromatographic analysis.

(2) The concentration of total guanidino compounds was determined by treatment with several concentrations (60,100,300 and 450 g/L) of trichloroacetic acid. One milliliter aliquots of plasma was deproteinized with half volume of trichloroacetic acid solution and the supernates were used in chromatographic analysis to determine the extraction efficiencies.

(3) The concentration of total guanidino compounds in erythrocytes were determined by pretreatment with trichloroacetic acid. A 1.0 ml aliquot from each sample of erythrocytes was hemolyzed and deproteinized with 5 ml of 40,60,80, and 120 g/L solutions of trichloroacetic acid. This mixture was homogenized in a glass homogenizer, the homogenate was centrifuged (1000 x g, 20 min), and the supernate was analyzed.

(4) The concentrations of total guanidino compounds in several tissues were also determined after the deproteinizing procedure by using trichloroacetic acid (Fig.2). Tissue sample was obtained from normal rats and uremic rats by using Platt's method[3].

In each biological sample, the optimal deproteinizing procedure

QUANTITATIVE DETERMINATION OF GUANIDINO COMPOUNDS

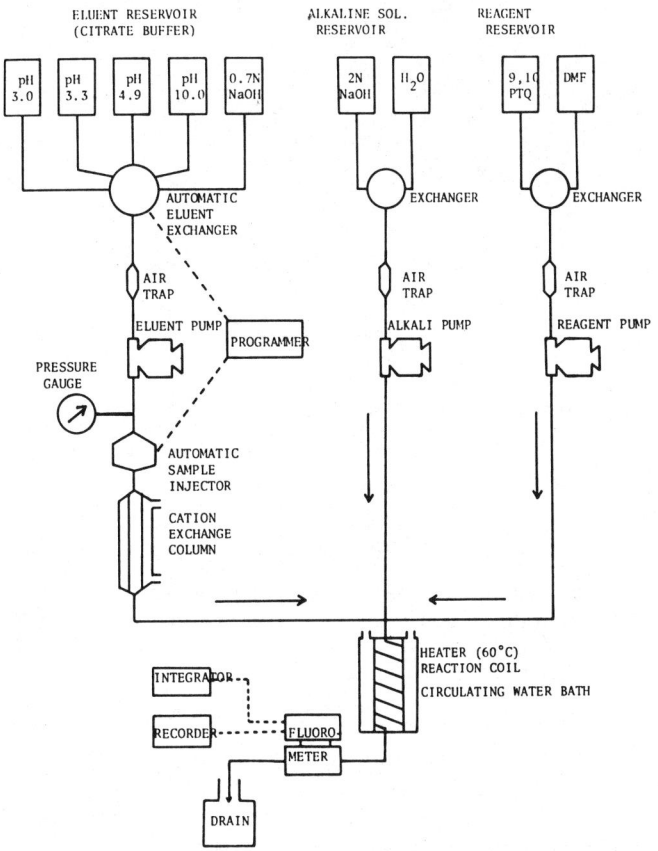

Fig.1 The analytical principle of Guanidine Analyzer G-520. Upper is an illustration of liquid chromatography and lower is the reaction formula of fluorescent derivative.

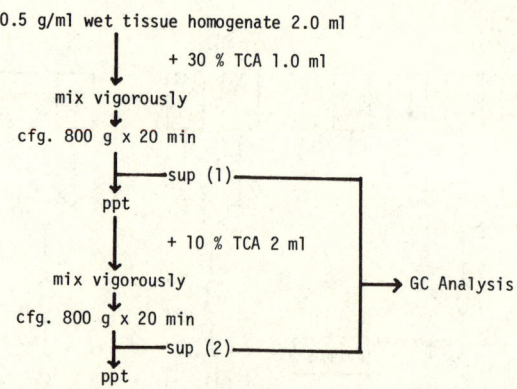

Fig.2 Guanidino compounds extraction procedure from tissue.

was determined by the percentage recovery, which was calculated by the analysis of standard solution, a sample solution, and a combined solution.

RESULTS

(1) Analytical recoveries of plasma guanidino compounds which were deproteinized by ultrafiltration were shown in Table 1. Separation and percentage recovery were good in chromatograms of samples with pH adjustment, but very poor in those with no pH adjustment, especially in taurocyamine, guanidinosuccinic acid, guanidinoacetic acid and guanidinopropionic acid.

(2) Analytical recoveries of plasma guanidino compounds treated with several different concentrations of trichloroacetic acid were shown in Table 2. The supernate of plasma treated with 100 g/L trichloroacetic acid solution (final concentration 3.3%) gave an abnormal high percentage recovery. The chromatogram of plasma supernate treated with 300 g/L trichloroacetic acid (the final concentration 10.0%) exhibited good reproducibility with no interferance from unidentified peaks. Trichloroacetic acid did not influence the chromatogram except taurocyamine eluted closely to the peak of trichloroacetic acid.

(3) Each guanidino compound in erythrocytes was extracted more efficiently and conveniently by the treatment with five volumes of 120 g/L trichloroacetic acid than by treatment with other concentration of trichloroacetic acid. The concentrations of each

Table 1. Percentage recovery of guanidino compounds in plasma filtrate.

	NON-TREATED PLASMA ULTRAFILTRATE	ACIDIFIED ULTRAFILTRATE PH 2.0
Tau	21.9 ± 39.9 (7)	92.5 ± 47.3 (7)
GSA	24.9 ± 10.2 (7)	112.7 ± 14.0 (7)
GAA	31.3 ± 3.9 (7)	106.5 ± 6.3 (7)
GPA	44.0 ± 8.3 (7)	101.7 ± 5.5 (7)
GBA	85.3 ± 12.8 (7)	101.5 ± 7.7 (7)
G	90.1 ± 6.4 (7)	97.7 ± 7.7 (7)
MG	92.0 ± 3.1 (7)	105.7 ± 8.0 (7)

(%), Mean ± S.D. (n)
Abbreviations used in Tables correspond, in order of elution, to compounds shown in Fig.3.

guanidino compound in plasma and erythrocytes were shown in Table 3. The plasma concentrations of guanidinosuccinic acid, creatine, guanidine and methylguanidine in patients with chronic renal failure were higher than those in normal subjects. In such patients, the concentrations of methylguanidine and guanidinosuccinic acid in erythrocytes were 1.5 times higher than those in plasma.

(4) Each guanidino compound in several tissues was extracted efficiently by re-extraction with trichloroacetic acid, and the percentage recovery of each guanidino compound was 85-100%. One

Table 2. Percentage recovery of plasma guanidino compounds treated with several different concentrations of trichloroacetic acid.

REAGENT CONCN (g/L)	100	150	300	450
FINAL CONCN	33	50	100	150
GSA	111.8	96.8	108.6	108.6
GAA	113.7	97.3	113.1	98.1
GPA	123.9	97.6	96.8	106.7
GBA	n.d.	n.d.	106.8	99.2
G	114.6	92.3	108.4	102.1
MG	108.7	95.6	102.6	103.3

Mean %, n=3, n.d.; not detectable, owing to interference from unidentified peaks.

Table 3. The plasma and erythrocyte concentration[a] of guanidino compounds in normal subjects and patients with chronic renal failure.

	NORMAL (n=6)			CHRONIC RENAL FAILURE (n=4)		
	PLASMA	ERYTHROCYTE	E/P RATIO	PLASMA	ERYTHROCYTE	E/P RATIO
GSA	0.4±0.2	0.2±0.5	-	20.1±1.6	30.6±11.6	1.56±0.71
GAA	2.8±0.9	2.3±0.9	0.81±0.20	2.4±0.7	3.7±0.9	1.75±1.03
Creatinine	82.6±26.2	100.3±23.8	1.25±0.20	440.7±44.2	537.5±87.0	1.23±0.24
GBA	0.0	0.0	-	0.0	1.0±1.9	-
Arg	99.0±22.8	15.1±2.2	0.16±0.07	101.6±39.2	42.5±12.4	0.45±0.13
G	0.0	0.0	-	3.1±1.1	8.1±0.9	3.03±1.63
MG	0.0	0.0	-	3.3±1.3	5.1±1.7	1.47±0.29

a) Concentration is in mol/L. Means ± S.D. are given. Concentration in erythrocytes refers to that in supernate analyzed.

case of chromatogram of the uremic rat liver was shown in Figure 3. The peak of methylguanidine which could not be detected in normal rats appeared in uremic rats, and the peak of guanidinosuccinic acid heightened in uremic rats. Some unidentified peaks appeared in severe uremic state.

DISCUSSION

The methods for quantitative determination of several guanidino compounds have been previously reported by Pfiffner[4], Stein et al.[5], Menichini et al.[6] and Baker and Marchall[7]. However, these analytical methods are time consuming and the use of colorimetric methods lacks sensitivity. Moreover, a large amount of sample is required in these methods; making them inadequate for clinical use.

Liquid chromatographic and fluorometric procedures enabled a rapid, sensitive, approximate and reproducible determination of various guanidino compounds. We tried to apply this method to clinical examination and determine the optimal condition of pretreatment for the analysis of guanidino compounds in biological fluids.

The ultrafiltrate technique is useful for analysis of free guanidino compounds in plasma. Separation and recovery were good in the case of chromatograms of samples with pH adjustment. But the component of samples without pH adjustment were not separated and there was substantial loss of guanidino compounds (e.g, taurocyamine, guanidinosuccinic acid, guanidinoacetic acid, and guanidinopropionic acid), possibly owing to nonelution because their strongly electronegative charge caused a strong retention by the

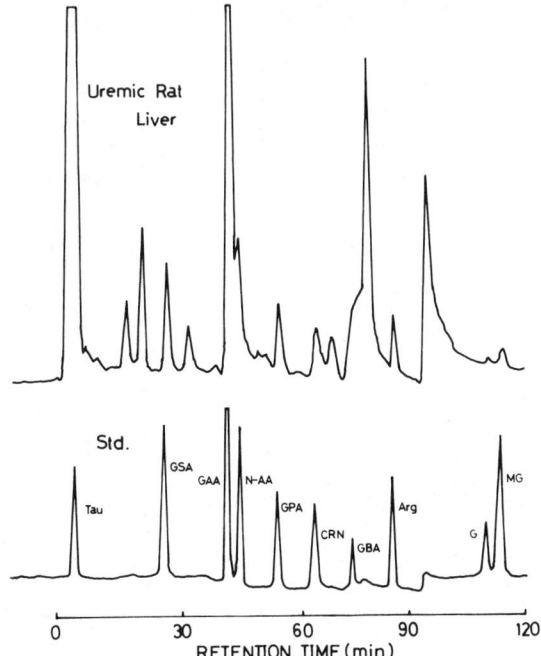

Fig. 3. Chromatogram of guanidino compounds in uremic rat liver.
lower; chromatogram of standard solution of guanidino compounds.

electropositive charge of the resin used.

Deproteinizing agents are useful for analysis of guanidino compounds, trichloroacetic acid has advantages over sulfosalicylic acid, perchloric acid, picric acid, and molybdenic tungstate[8]. By the treatment of uremic plasma with sulfosalicylic acid and perchloric acid, the recovery of guanidino compounds was about 100%, but the determination of guanidinobutyric acid was interfered with an unidentified peak. Taurocyamine and guanidinosuccinic acid were inseparable, because the elution peak of sulfosalicylic acid overlapped on each peak. By the treatment with picric acid the chromatogram of guanidino compounds was good, but the deproteinizing procedure was complicated and the sensitivity was decreased on account of increasing the relative amount of picric acid. These results suggest that trichloroacetic acid is a useful deproteinizing agent for the present automatic analysis of guanidino compounds (Table 4). Trichloroacetic acid was immediately eluted without being retained in the resin and did not significantly influence on the chromatogram. Each guanidino compound except

Table 4. Percentage recovery of guanidino compounds after the pretreatment with several deproteinizing agents.

Deproteinizing Agent	Mixture Ratio Plasma : Agent	Final Concentration	Percentage Recovery				
			GSA	GAA	GPA	GBA	MG
Trichloroacetic acid	1 : 0.5	10 %	108.7	113.1	96.8	106.8	102.6
Sulfosalicylic acid	1 : 0.5	10 %	102.8	96.7	99.7	?	101.1
Perchloric acid	1 : 1	5 %	104.0	96.0	-	-	103.2
Picric acid	1 : 5	0.2 %	85.0	81.0	87.0	92.0	93.0
Molybdenic tungstate	3 : 2	1.4 %	140.0	94.0	91.0	81.0	89.0

N-acetylarginine, was stable in trichloroacetic acid solution at room temperature. When we used 100 g/L final concentration of trichloroacetic acid for deproteinization, each guanidino compound exhibited good chromatographic reproducibility, with no interferance from unidentified peaks. The mixing ratio of deproteinizing agent influenced the determination of guanidino compounds in erythrocytes and tissues, but not in plasma. Increasing the relative amount of trichloroacetic acid was not helpful because sensitivity was decreased, and we sought optimal conditions for the analysis of each biological sample.

The present study confirms that the concentrations of methylguanidine and guanidinosuccinic acid in erythrocytes were higher than those in plasma. And the tissue concentrations of methylguanidine and guanidinosuccinic acid were also higher than those of plasma. Information on tissue or erythrocytes concentrations of guanidino compounds seems to be a more useful index in uremic state than that on plasma concentration, because methylguanidine and guanidinosuccinic acid act as an uremic toxin in the cells. In human subjects, erythrocytes concentration of guanidino compounds is useful for the clinical evaluation of uremic state in patients with chronic renal failure. For the present, muscle, liver, and kidney concentration of guanidino compounds is useful for experimental examination of uremic state, but may be meaningful from the clinical view point in the near future.

In all of the cases, our deproteinizing conditions of biological samples should be applied for the automated analysis of guanidino compounds.

SUMMARY

We described a procedure for quantitative determination of guanidino compounds in plasma, erythrocytes and some of the tissues. The guanidino compounds were separated using a high-pressure liquid chromatography and were detected with fluorometry. Analyzed mate-

rials were obtained from uremic patients and uremic rats.
As a deproteinizing agent, trichloroacetic acid had no influence, except on taurocyamine. Guanidino compounds can be efficiently extracted from plasma with 0.5 volume of 300 g/L trichloroacetic acid, from erythrocytes with 5 volumes of 120 g/L trichloroacetic acid, and from tissue by re-extraction with 300 and 100 g/L trichloroacetic acid. In all of the cases, our deproteinizing procedure could be applied to the automated analysis of guanidino compounds.

REFERENCES

1. Y.Yamamoto, A.Saito, T.Manji, H.Nishi, K.Ito, K.Maeda, K. Ohta, and K.Kobayashi, A new automated analytical method for guanidino compounds and their cerebrospinal fluid levels in uremia, Trans.Am.Soc.Artif.Intern.Organs, 24:618 (1978).
2. A.Mori, Y.Watanabe, Recovery rates for guanidino compound analysis and variance of measured values by different laboratories, Jpn.J.Clin.Chem., 9:217-231 (1980).
3. R. Platt, M.H. Roscoe, and F.W. Smith, Experimental renal failure, Clin. Sci, 11 : 217-231 (1952).
4. J.J. Pfiffner and V.C. Myers, Colorimetric estimation of methylguanidine in biological fluids. Proc. Soc. Exp. Biol. Med., 23 : 830-832 (1926).
5. I.M. Stein, G. Perez, R. Johnson, and N.B. Cummins, Serum levels and urinary excretion of methylguanidine in chronic failure, J. Lab. Clin. Med., 77 : 1020-1024 (1974).
6. G.C. Menichini, M. Gonella, G. Barsotti and S. Giovanetti, Determination of methylguanidine in serum and urine from normal and uremic subjects, Experientia, 27 : 1157-1158 (1971).
7. L.R.I. Baker, R.D. Marshall, A reinvestigation of methylguanidine concentration in sera from normal and uremic subjects., Clin. Sci., 41 : 563-568 (1971).
8. A. Ando, H. Mikami, T. Kikuchi, Y. Orita, and H. Abe, An automated analytical method for guanidino compounds; Determination of optimal deproteinization method, Jpn. J. Clin. Chem., 9 : 191-197 (1980).

RECOMMENDED DEPROTEINIZING METHODS FOR PLASMA GUANIDINO COMPOUND

ANALYSIS BY LIQUID CHROMATOGRAPHY

T. Hoshino

Pharmaceutical Institute, School of Medicine
Keio University, 35-Shinanomachi, Shinjukuku
Tokyo 160, Japan

INTRODUCTION

The quantitative determination of guanidino compounds in plasma has been performed in many laboratories working with liquid chromatography and fluorometry ; however, the obtained data were still dispersive for rational discussion. In the course of the determination, several procedures, sample storage, deproteinization, sample charging and so on were thought to cause such variance.

As a result of the cooperative examinations by twelve laboratories in Japan (the 2nd Symposium on the Analysis of Guanidino Compounds, Osaka, 1979), many of the pitfalls of the determination of the compounds were disclosed. Guanidinoacetic acid (GAA) and metylguanidine (MG) were remarkably increased when the plasma sample was left at room temperature for long time, but the increase was prevented by the storage at - 20° C immediately after the plasma separation[1]. The concentration of hydrogen ion of the charged sample had a great influence on the chromatographical elution behavior of guanidinosuccinic acid (GSA) and GAA, and their elution curves misfitted quantitatively when the concentration of hydrogen ion of the sample was not high enough. It was preferable to adjust the sample to pH 2.0 - pH 2.5 for accurate determination[1,2]. Guanidino compounds seemed to be adsorbed on the ultrafiltration membrane (CF-25, Amicon Corp.). The absorbing ratio was different according to the kinds of compounds and to the conditions used. In the case of GSA and MG, they were recovered 105.6 %, 92.6 % at pH 2.0 and 92.0 %, 77.3 % at pH 10.0, respectively, when they were filtrated without annexing to a plasma[3]. Guanidino compounds added to a plasma were preferably recovered by the deproteinization with the membrane under acidic conditions[1-4] ; if uremic plasmas were acidified and left for one hour at room temperature, and the plasma was determined for guanidino compounds via ultrafiltration, guanidine (G) and MG yield ten to thirty p.c. excess against non-acidified plasma[1]. It was suspected here that the compounds exist in plasma free and bound[4].

Five methods of chemical deproteinization were examined for appropriate pretreatment for the analysis of guanidino compounds Methods with trichloroacetic acid, sulfosalicylic acid, perchloric acid, picric acid and tungstate-molybdenic acid were tested and their distinctive characteristics were clarified and the optimal conditions for the analysis were established[2].

On the basis of the 2nd Symposium on the Analysis of Guanidino Compounds the re-experiment was performed in cooperation with fourteen laboratories. That report highlights the re-experiment to standardize the analytical procedures for the determination of guanidino compounds in plasma.

MATERIALS AND METHODS

For the establishment of the variances between laboratories, three kinds of ECUMs were used. Collected ECUM was ultrafiltrated and adjusted to pH 2.2 and divided every 3 ml in stoppered polypropylene tubes and frozen at twenty degrees below zero. Three kinds of plasma, one normal and two uremic, were for accuracy test of the obtained values measured by given deproteinizing method. Plasma was divided every 2 ml in stoppered polypropylene tubes immediately after the separation and frozen as mentioned.

Samples were distributed to every laboratory by dry ice packed cargo with the mixture of authentic guanidino compounds as a standard solution.

Storage condition was set to keep under twenty degrees below zero. Four deproteinizing methods and every procedures are also given here (Table 1). Other procedures for the determination of guanidino compounds without storage and deproteinization were left to the care of each laboratory. Duplicate or triplicate measurements for one sample were required to represent the value of each compound.

Reported data, observed values and representative values, were computed statistically on the Hoffmann's calculation.

TABLE 1. THE OPTIMAL CONDITIONS IN DEPROTEINIZNING PROCEDURE

I) TRICHLORO ACETIC ACID:**
(Plasma:30%TCA=1:0.5)

1ml Plasma+0.5ml 30%TCA
↓
  leave for 20min
  under ice cold
↓
  centrifuge,
  3,000rpm, for 15min
↓
Supernatant
↓
  adjust to pH2.2
↓
Sample for chromatography

II) PERCHLORIC ACID:**
(Plasma:10%PCA=1:1)

1ml Plasma+1ml 10%PCA
↓
  leave for 20min
  under ice cold
↓
  centrifuge,
  3,000rpm, for 15min
↓
Supernatant
↓
  adjust to pH2.2
↓
Sample for chromatography

III) TUNGSTATE-MOLYBDENIC:**
(Plasma:W-Mo=2:3)

2ml Plasma+3ml W-Mo*
↓
  leave for 20min
  under ice cold
↓
  centrifuge,
  3,000rpm, for 15min
↓
Supernatant
↓
  adjust to pH2.2
↓
Sample fo chromatography

IV) ULTRA FILTRATION:***
(CF-25)

CF-25 membrane
↓
  dehydrate 2,400rpm,
  for 10min, at 4°C
+2ml Plasma
↓
  centrifuge,
  800xG, for 120min,
  at 4°C
↓
Filtrate
↓
  adjust to pH2.2
↓
Sample for chromatography

*: Mix A and B (1:1) just before use.
A) 8g sodium tungstate/60ml H_2O+ 1g molybdenic acid/5ml 1N NaOH and make it 100ml with water.
B) 0.5N H_2SO_4

**: by A. Ando (2)
***: by T. Hoshino (1)

Table 2. Variance of Determined Values from different Facilities.

	Sample No1 n mol/ml	CV(%)	n/N	Sample No2 n mol/ml	CV(%)	n/N	Sample No3 n mol/ml	CV(%)	n/N
G-Tau	22.14	56.69	5/5	24.11	74.46	4/4	23.80	58.31	5/5
GSA ****	4.92	17.06*	11/12	10.39	5.23**	9/12	39.79	12.29*	11/12
GAA ******	1.40	6.83**	10/12	1.19	8.72**	10/12	1.61	9.33**	10/12
n-AA	0.88	31.70	3/3	0.81	57.54	3/3	0.54	41.00	3/3
GPA	3.65	188.95	4/4	0.67	162.10	5/5	0.53	203.67	6/6
CRN ***	377.23	6.39**	9/11	624.37	16.80*	9/1o	922.61	20.42	9/9
GBA	11.73	207.29	5/5	0.55	66.80	6/7	1.14	81.57	5/8
Arg ******	53.53	9.45**	10/11	96.41	8.94**	10/11	45.67	3.38**	9/11
G	0.19	47.78	8/9	2.22	66.05	9/9	2.12	53.74	9/9
MG ****	0.97	34.19	12/13	3.21	9.77**	10/13	4.79	7.19**	10/13

Representative data from each laboratory were computed statistically on the Hoffmann's calculation and listed. Data are mean value (n mol/ml), coefficient of variation (%) and number of reported representative value.
n/N = number of non-rejectable/total number of reported
* = under ten p.c.,   ** = ten to twenty p.c.

RESULTS AND DISCUSSION

Technical and Instrumental Variances

Data were reported from thirteen laboratories ; the number of representative data from the respective laboratories varied.

Computed values of each guanidino compound in the ECUMs (Table 2) are variant. In general, the coefficient of variation (CV) is high when the mean value is low. CVs of GSA, GAA, CRN, Arg and MG were fairly low, and that of GPA and GBA were excellently high.

G-Tau is eluted at the brake-through of the analytical column and the values are non-reliable. Reported values of GPA and GBA are still dispersive and cannot be rejected by computation. As it is possible to mistake other phenanthrenequinone positive substances as GPA and GBA and because of the size of the data, it is difficult to comment upon. The same applies to n-acetylarginine (n-AA).

In order to compare the data of this experiment with that of pre-experiment[1], CV value of each GSA, GAA, Arg and MG was reduced to below the half. It was remarkable in the cases of GSA and GAA, that CV of GSA decreased from 64.5 % to 11.5 % and that of GAA decreased from 55.5 % to 8.3 %. This successful reduction was thought to be performed by the standardization of the concentration of hydrogen ion when the sample was applied to the column. On the contrary, CV of G was not reduced. It might be caused by the level of G in the sample : the levels of this experiment were 0.2, 2,1 and 2.2, and that of the pre-experiment were 1.9, 3.6 and 6.1 (nmol/ml).

Comparison of Deproteinizing Methods

Guanidino compounds in the same plasma were determined using given deproteinizing methods mentioned above by fourteen laboratories. The results were listed in Table 3-A, 3 -I. The number of representative data from each laboratory were variant and the data were computed the same as the data of ECUM. As shown in table 3, the mean value of respective guanidino compounds from the same plasma was preferably high when the plasma was deproteinized by TCA and PCA, and was rather low by W-Mo and CF-25. It was curious that PCA gave remarkably high values of GBA and that W-Mo resulted about a half level of arginine value observed by another deproteinizing methods. The reason for this is not clear.

The CV value of each guanidino compound by the respective

G-Tau (taurocyamine)

Tabl.3-A

	Plasma No1			Plasma No2			Plasma No3		
	n mol/ml	CV(%)	n/N	n mol/ml	CV(%)	n/N	n mol/ml	CV(%)	n/N
TCA	4.40	-	2/3	11.37	-	2/4	13.19	19.99	3/4
PCA	11.52	43.52	3/4	20.18	17.23	4/5	24.66	20.13	3/4
W-Mo	15.82	-	2/2	17.66	-	2/2	22.38	-	2/3
CF-25	12.56	-	1/1	16.14	-	1/1	12.80	-	2/2

GSA (guanidinosuccinic acid)

Tabl.3-B

	Plasma No 1			Plasma No2			Plasma No3		
	n mol/ml	CV(%)	n/N	n mol/ml	CV(%)	n/N	n mol/ml	CV(%)	n/N
TCA	0.45	79.80	7/8	13.07	9.09	10/11	22.76	6.94	8/11
PCA	0.42	81.77	7/8	12.13	38.46	8/10	21.22	33.50	9/10
W-Mo	0.18	15.47	3/5	10.28	14.43	6/8	17.84	29.01	7/8
CF-25	0.20	52.35	3/3	10.23	19.68	4/5	25.26	11.26	4/5

GAA (guanidinoacetic acid)

Tabl.3-C

	Plasma No1			Plasma No2			Plasma No3		
	n mol/ml	CV(%)	n/N	n mol/ml	CV(%)	n/N	n mol/ml	CV(%)	n/N
TCA	3.95	4.47	8/10	2.44	7.35	10/11	4.21	7.23	9/11
PCA	3.97	10.36	6/9	2.44	14.14	7/9	4.32	7.44	7/10
W-Mo	3.37	13.88	5/8	2.08	16.13	6/8	3.70	20.08	6/8
CF-25	4.00	8.66	3/5	2.45	4.14	4/5	4.58	5.34	4/5

GPA (β-guanidinopropionic acid)

Tabl.3-D

	Plasma No1			Plasma No2			Plasma No3		
	n mol/ml	CV(%)	n/N	n mol/ml	CV(%)	n/N	n mol/ml	CV(%)	n/N
TCA	0.00	-	2/3	0.00	-	2/3	0.00	-	2/2
PCA	0.00	-	2/3	0.10	173.20	3/3	0.00	-	2/2
W-Mo	0.00	-	2/2	0.00	-	2/2	0.00	-	2/2
CF-25	-	-	/0	-	-	/0	-	-	/0

CRN (creacinine)

Tabl.3-E

	Plasma No1			Plasma No2			Plasma No3		
	n mol/ml	CV(%)	n/N	n mol/ml	CV(%)	n/N	n mol/ml	CV(%)	n/N
TCA	73.95	2.34	7/9	648.16	7.53	6/8	1,004.47	14.37	6/8
PCA	71.59	5.01	8/9	648.31	10.85	7/8	998.47	3.96	3/7
W-Mo	67.71	14.06	5/7	552.09	3.69	5/7	926.52	6.70	4/6
CF-25	72.70	5.93	4/5	498.14	12.33	3/4	855.66	17.47	4/4

GBA (γ-guanidinobuthyric acid)

Tabl.3-F

	Plasma No1			Plasma No2			Plasma No3		
	n mol/ml	CV(%)	n/N	n mol/ml	CV(%)	n/N	n mol/ml	CV(%)	n/N
TCA	0.00	0.00	4/4	0.00	-	2/4	0.13	173.20	3/4
PCA	39.86	-	2/4	162.81	-	2/3	210.88	-	2/5
W-Mo	2.68	173.20	3/3	0.00	-	2/2	0.00	-	2/3
CF-25	-	-	/0	0.71	-	1/1	1.35	-	1/1

Arg (arginine)

Tabl.3-G

	Plasma No1 n mol/ml	CV(%)	n/N	Plasma No2 n mol/ml	CV(%)	n/N	Plasma No3 n mol/ml	CV(%)	n/N
TCA	140.26	4.12	6/8	99.92	15.57	8/9	146.91	7.23	8/10
PCA	140.18	6.71	6/8	93.74	16.88	7/9	143.05	15.48	3/5
W-Mo	75.11	18.09	5/7	55.17	15.75	5/7	85.95	11.68	5/7
CF-25	122.78	3.00	3/5	85.13	2.94	3/5	131.75	6.39	4/5

G (guanidine)

Tabl.3-H

	Plasma No1 n mol/ml	CV(%)	n/N	Plasma No2 n mol/ml	CV(%)	n/N	Plasma No3 n mol/ml	CV(%)	n/N
TCA	0.00	0.00	3/4	1.76	15.24	5/8	2.40	25.83	6/8
PCA	0.00	0.00	3/4	1.57	23.90	3/4	2.11	29.32	3/5
W-Mo	0.00	-	2/2	1.27	14.48	3/5	1.31	11.82	3/5
CF-25	-	-	/0	1.86	-	2/3	1.45	84.28	3/3

MG (methylguanidne)

Tabl.3-I

	Plasma No1 n mol/ml	CV(%)	n/N	Plasma No2 n mol/ml	CV(%)	n/N	Plasma No3 n mol/ml	CV(%)	n/N
TCA	0.00	0.00	3/5	4.82	3.21	8/11	9.54	5.51	7/11
PCA	0.00	0.00	3/4	4.75	4.85	7/9	8.57	11.24	7/10
W-Mo	0.00	-	2/2	4.29	23.77	6/8	8.77	9.48	6/8
CF-25	-	-	/0	4.23	6.80	3/5	7.77	5.36	3/5

deproteinizing methods were very dispersive. However, CV value of the values which were reported by more than five laboratories showed preferably low levels. Among the data reported by more than five laboratories, CV value of the data was not so widely distributed and each mean of the CV from plasma No1, plasma No2 and plasma No3 was 17.6 %, 9.7 % and 11.2 % for TCA, and 26.0 %, 17.1 % and 16.8 % for PCA, and 15.4 %, 14.7 % and 14.8 % for W-Mo, and 5.9 %, 8.4 % and 7.1 % for CF-25.

The deproteinizing method by TCA was shown to yield fairly high levels of guanidino compounds i.e. good recoveries, and to be preferably reproducible. The method by PCA gave good recoveries but was not so reproducible, and contained disadvantage of misconception. Comparing other methods, deproteinization by W-Mo fell behind in recovery and reproduciblity. Recovery rate of the method by CF-25 did not seem to be good, but this method was reproducible.

In this experiment, CV value of the data proved to be reduced by the standardization of the procedures including storage, deproteinization and sample charging. And two deproteinizing methods, TCA and CF-25, were recommended for the analysis of plasma guanidino compounds by liquid chromatography.

ACKNOWLEDGMENTS

The author thanks Prof. A. Mori, President of the Symposium on the Analysis of Guanidino Compounds, and all the members of this symposium for their fruitful cooperation.

REFERENCES

1. T. Hoshino, M. Sakuma, E. Ohsawa, W. Yamayoshi, K. Sakakibara, T. Sato, H. Sakurai and S. Toyoshima, Effects of hydrogen ion concentration on determination of free guanidino compounds in plasma, Jap. J. Clin. Chem., 9:198 (1980).
2. A. Ando, U. Mikami, T. Kikuchi, Y. Orita and H. Abe, An automated analytical method for guanidino compound, Jap. J. Clin. Chem., 9:191 (1980).
3. M. Ariura, Z. Ito, A. Shiotsuki, Y. Takahashi, K. Sakurai and M. Yamagami, The problem of ultrafiltration used as pretreatment in analysis of guanidino compounds, Jap. J. Clin. Chem., 9:206 (1980).
4. A. Mori and Y. Watanabe, Recovery rates for guanidino compound analyses and variance of measured values by different laboratories, Jap. J. Clin. Chem., 9:183 (1980).

COOPERATED BY :
Akitane Mori and Yoko Watanabe
    Institute for Neurobiology, Okayama University Medical School

Akio Ando, Hiroshi Mikami, Takeo Kikuchi and Yoshimasa Orita
    The First Department of Medicine, Osaka University Medical School

Kenji Sakurai, Makoto Ariura, Yuichiro Takahashi and Zenichi Ito
    Kidney Center, Sagamidai Hospital

Tadao Akizawa and Ken Takahashi
    Internal Medicine, Fujigaoka Hospital, Showa University

Mitsuhiko Kuroda, Yohei Tohfuku and Tohru Kita
    Second Department of Internal Medicine, Faculty of Medicine
    Kanazawa University

Hikaru Koide and Chieko Ito
    Division of Nephrology, Department of Medicine, Juntendo
    University School of Medicine

Eiichi Chiba and Yoshinori Hatakeyama
    Kidney Center, Mikasa Municipal Hospital

Hiroaki Sugizaki
    Chyofu Hospital

Chozo Hayashi, Ichiro Furukawa, Kazuma Kohda, Hideo Hosotsubo and
Kazutaka Arisue
    Central Laboratory for Clinical Investigation, Osaka University
    Hospital

Kimiaki Kadono, Hideo Kushiro, Junzo Kodama and Makoto Satani
    Clinical Laboratory, National Cardiovascular Center

Makoto Ishizaki and Keiko Aoyama
    Kidney Center, Sendai Shakai-Hoken Hospital

Kazuo Isoda and Toshiyuki Nakao
    The Jikei University, School of Medicine, Department of the
    Second Internal Medicine

Kohichi Sakakibara, Setsuko Jitsukawa and Masaru Tawara
    Kumpuhkai Yamada Hospital

Wataru Yamayoshi and Etsuko Ohsawa
    Pharmaceutical Institute, School of Medicine, Keio University

EVOLUTIONARY RELATIONSHIPS BETWEEN ARGININE AND CREATINE IN MUSCLE

F.J.R. Hird, Siva Prasad Davuluri and R.M. McLean

Russell Grimwade School of Biochemistry
University of Melbourne
Parkville, Victoria 3052, Australia

In addition to being a constituent of proteins, arginine during evolution, has come to have important additional functions as a free amino acid. Thus, there have been three extensions to the metabolic pathway of its synthesis - phosphorylation to form phosphoarginine, hydrolysis to form urea, and transfer of its amidino group to certain amines to form a variety of guanidino compounds used as phosphagens in muscle. The present paper briefly refers to the special significance of arginine in the way it has permitted important evolutionary advances. It also raises the question of possible secondary effects of an arginine deficiency to the musculature and nervous system in the very young animal.

Evolutionary Significance of Arginine

The majority of species (excepting mammals) living in water solve the problem of eliminating surplus nitrogen (as ammonium ions) by a dilution into a large volume of surrounding water.[1a,b] They have evolved special exchange mechanisms and pumps to do this. Many of the invertebrates use phosphoarginine in their muscles as a reserve of ATP. The amounts of arginine - including phosphoarginine - reach surprisingly high values in muscles of invertebrates (up to 64 µmoles/g wet wt).[2] Phosphoarginine greatly exceeds ATP in amounts and in some animals its status determines the permissible duration of exercise.

The next step in evolution mediated by arginine was the use of urea for osmoregulation and this is seen clearly in the elasmobranchs where the blood levels reach over 350 mM.[1b] It is worth remembering that urea in these animals is not excreted - ammonia persists as the nitrogenous compound excreted in the 'fishes'.

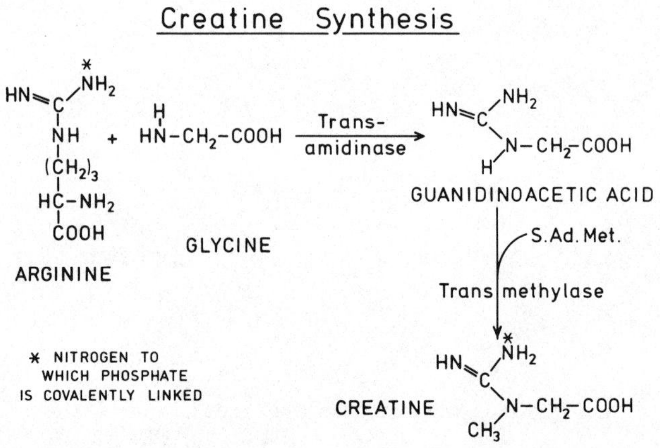

Creatine Synthesis

Fig. 1

A major advance can be seen in the amphibia where the relatively non-toxic urea, in the adult terrestrial species, is the main substance used as the means of excreting waste nitrogen.[1] This utilisation of the hydrolytic extension of the arginine pathway for excretion allowed vertebrates to leave the relative stability of the water and expose themselves to the highly varied selective pressures of life on the land - ultimately leading to the evolution of people discussing the special significance of this amino acid.

At the same time that arginine became used as a source of urea a competitive pressure was set up for another phosphagen as it would seem difficult for arginine in the presence of large amounts of arginase to fulfil these two major functions simultaneously. In the case of vertebrates, instead of the amidino group of arginine being hydrolysed to form urea, it was transferred to the amino group of glycine to form guanidinoacetate. Later this compound was methylated to form creatine (Fig. 1).

Functions of Phosphocreatine (and Phosphoarginine)

There are three clearly defined functions known for the phosphagens:

1. They act as a metabolic reserve of ATP and inorganic phosphate during muscle contraction (Fig. 2, 3) and as a reserve of ATP in nerve tissue.

2. Arginine, in invertebrate, and creatine in vertebrate muscle acts as a sink for inorganic phosphate when the muscle is at

Phosphocreatine Metabolism

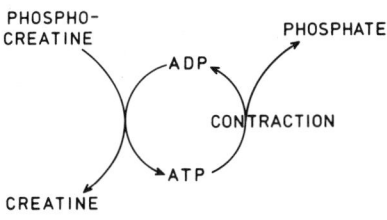

Fig. 2

rest[3] (Fig. 3). In this way glycogenolysis and glycolysis may be controlled by the removal of surplus inorganic phosphate from the myoplasm, i.e. over and above the amounts required or involved in the fluxes of ATP occurring in the resting state.

3. In muscle, rich in mitochondria e.g. heart and red skeletal muscle, creatine and phosphocreatine act, through random mixing and distribution, as intermediaries in the transport of high energy phosphate from mitochondria through the myoplasm to the contractile filaments[4] (Fig. 4). An isoenzyme of creatine phosphokinase is reported to be on the outside of the inner mitochondrial membrane[5,6] and is thought, as such, to present special vectorial advantages. Our work (to be published) with liver mitochondria synthesizing phosphocreatine and phosphoarginine shows that within one second of adding the appropriate kinase, synthesis of the phosphagen begins and this suggests strongly that the synthesis of phosphocreatine and phosphoarginine can be catalysed by the cytosolic enzyme. Similar results have been obtained with muscle mitochondria and arginine kinase.

Possible Implications of Arginine Deficiency

In the case of a serious deficiency of one of the enzymes involved in the synthesis of arginine this amino acid becomes essential. There will be three general demands on whatever arginine is available - hydrolysis to ornithine and urea, protein synthesis and creatine synthesis. A 70 Kg man contains about 120 g total creatine in muscle and nerve tissue and with excretion

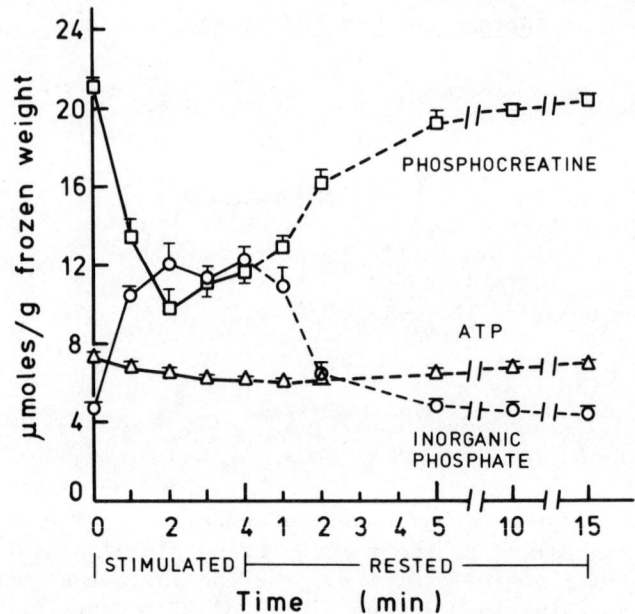

Fig. 3. Changes in levels of ATP, phosphocreatine and inorganic phosphate during electrical stimulation of rat gastrocnemius muscle through the sciatic nerve at 2 square wave pulses/sec at 5 volts for 1 msec duration. n = 5.

of creatinine there is a loss of about 2 g/day[7]. In the case of a young child unless milk contains creatine this loss must be replaced from dietary arginine. The requirements of a young, growing child to stock and service developing muscles, brain and other nerve tissue with creatine are unknown but with a half life of about 40 days in the human[8], the creatine status of the newborn may steadily deteriorate with time. The matter raised may have some importance as it has been shown in rat brain that 14-15 days after birth there is a marked increase in creatine phosphokinase over the next 10 days. This period is reported to coincide with increasing co-ordination of neuronal activity[9]. The levels of total creatine in 10 day old[10] and adult[11] mice and adult rats[12] is reported to be in the range 7.7 - 9.2 µmole creatine/g wet weight. There are therefore functional implications for both muscle and nerve tissue in relation to the amount of arginine synthesized and available. As the kidney, in the rat, is thought

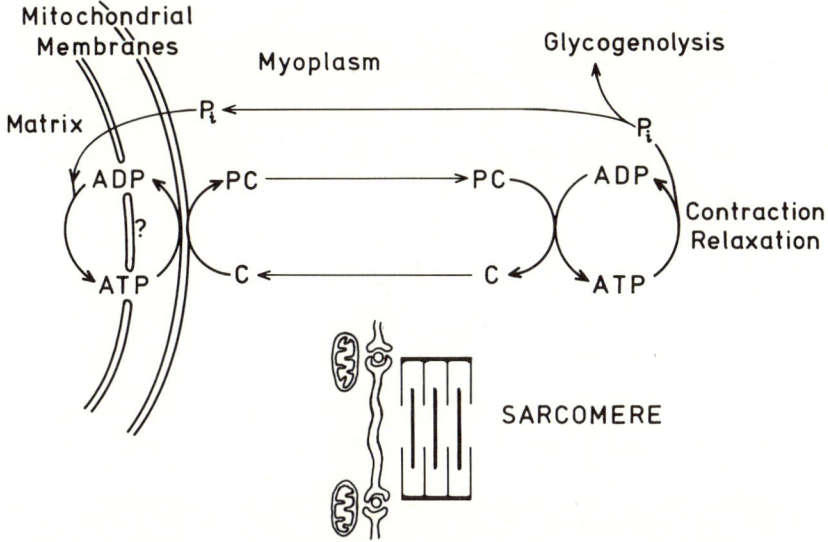

Fig. 4. Schematic representation of mitochondrial synthesis of ATP and phosphocreatine and its utilisation during the contraction - relaxation cycle in muscle. The inset indicates the spatial relationships involved. The contraction - relaxation processes serve to mix creatine and phosphocreatine throughout the myoplasm. Creatine in muscle greatly exceeds inorganic phosphate present in molar amounts and thus is always available at mitochondrial sites to be phosphorylated. The question mark is intended to simplify the diagram and not place doubt on the observation.

to be a more important source of arginine than the liver in providing this amino acid for protein synthesis in a number of tissues,[13] the conversion of citrulline to arginine in this organ is also of some significance to the present discussion.

ACKNOWLEDGEMENT

We wish to acknowledge financial support from the Australian Research Grants Committee. SPD holds an award under the Commonwealth Scholarship and Fellowship Plan.

REFERENCES

1. J. W. Campbell (ed.), "Comparative Biochemistry of Nitrogen Metabolism" (a) Vol. 1. "The Invertebrates". (b) Vol. 2. "The Vertebrates", (1970).
2. G. S. Sidhu, W. A. Montgomery, and M. A. Brown, Post mortem changes and spoilage in rock lobster muscle, J. Food. Technology 9:357 (1974).
3. S. P. Davuluri, F. J. R. Hird, and R. M. McLean, A re-appraisal of the function and synthesis of phosphoarginine and phosphocreatine in muscle, Comp. Biochem. and Physiol. 69(B):329 (1981).
4. S. P. Bessman and P. J. Geiger, Transport of energy in muscle: the phosphorylcreatine shuttle, Science 211:448 (1981).
5. W. E. Jacobus and A. L. Lehninger, Creatine kinase of rat heart mitochondria. Coupling of creatine phosphorylation to electron transport, J. Biol. Chem. 248:4803 (1973).
6. V. A. Saks, V. V. Kupriyanov, G. V. Elizarova, and W. E. Jacobus, Studies of energy transport in heart cells. The importance of creatine kinase localization for the coupling of mitochondrial phosphorylcreatine production to oxidative phosphorylation, J. Biol. Chem. 255:755 (1980).
7. J. B. Walker, Creatine: biosynthesis, regulation and function, Adv. Enzymol. 50:177 (1979).
8. C. D. Fitch, D. D. Lucy, J. H. Bornhofen, and G. V. Dalrymple, Creatine metabolism in skeletal muscle. II. Creatine kinetics in man, Neurology 18:32 (1968).
9. R. F. G. Booth and J. B. Clark, Studies on the microchondrially bound form of rat brain creatine kinase, Biochem. J. 170:145 (1978).
10. D. T. Woznicki and J. B. Walker, Utilization of cyclocreatine phosphate, an analogue of creatine phosphate, by mouse brain during ischemia and its sparing action on brain energy reserves, J. Neurochem. 34:1247 (1980).
11. O. H. Lowry, J. V. Passonneau, F. X. Hasselberger, and D. W. Schultz, Effect of ischemia on known substrates and cofactors of the glycolytic pathway in brain, J. Biol. Chem. 239:18 (1964).
12. R. L. Veech, R. L. Harris, D. Veloso, and E. H. Veech, Freeze blowing: a new technique for the study of brain in vivo. J. Neurochem. 20:183 (1973).
13. W. R. Featherston, Q. R. Rogers, and R. A. Freedland, Relative importance of kidney and liver in synthesis of arginine by the rat, Am. J. Physiol. 224:127 (1973).

METABOLISM OF ARGININE IN INVERTEBRATES : RELATION

TO UREA CYCLE AND TO OTHER GUANIDINE DERIVATIVES

Y. Robin

Laboratoire de Biochimie Marine
Ecole Pratique des Hautes Etudes (Collège de France)
Paris, 75005, France

INTRODUCTION

The metabolism of arginine in invertebrates, even when limited to its relation to urea cycle and to the biosynthesis of the other guanidine derivatives, is somewhat complex. This is due, in a large part, to the number and disparity of the invertebrate phyla, and to the resulting dispersion of the data.

In this short survey, we shall approach the following problems : (i) considering that arginine is universally present in invertebrates and that most of them lack a functional urea cycle, how do invertebrates get arginine : is it brought by the diet or biosynthesized in the tissues and, in that case, what is the contribution of the urea cycle in this process, (ii) what is the relation between the urea cycle and urea excretion in invertebrates, since even those species which possess a functional urea cycle usually do not utilize urea as their main nitrogen excretory product, and (iii) how can we schematize the metabolic interrelations between arginine and the other guanidino compounds present in the invertebrates, which seem unable to synthesize their amidine group *de novo* and take it from arginine through a variety of reactions that will be examined here.

We shall also try to point out the biological role of the urea cycle in non-ureotelic animals, and that of the different guanidine derivatives characterized in the invertebrates.

According to the general character of the survey, we shall refer when possible to the appropriate reviews.

ARGININE AND THE UREA CYCLE

Urea Cycle and Arginine Biosynthesis

Though the urea cycle had been established in vertebrates as soon as 1932[1] and arginine had been characterized in all invertebrate phyla, very little was known during the next two decades concerning its occurence in invertebrates. The discovery of phosphoarginine[2] in most invertebrates had focussed the attention upon the role of arginine in energy storage and the biosynthesis of arginine itself had received little attention. The general opinion, based on the scattered distribution of arginase in most invertebrate phyla, was that the urea cycle was not functional, arginine being most probably provided by the diet[3].

The biosynthesis of arginine from its amino precursor by *de novo* incorporation of CO_2 and NH_3 according to the classical pathway established in vertebrates :

$HCO_3^- + NH_4^+ + ATP \longrightarrow$ carbamylphosphate + ADP

Carbamylphosphate + L-ornithine $\longrightarrow$ L-citrulline + Pi

L-citrulline + L-aspartate + ATP $\longrightarrow$ L-argininosuccinate + AMP

L-argininosuccinate $\longrightarrow$ L-arginine + fumarate

L-arginine $\longrightarrow$ L-ornithine + urea

has now been demonstrated, in whole or in a part, in some groups of invertebrates, essentially worms and molluscs. The data are due, in a large part, to the extensive studies of Cohen, of Needham, of Campbell and of their coworkers, which have been collected in specialized reviews [4-8].

Among worms, only some steps of the ornithine-urea cycle could be established in the parasitic helminths[9], but a complete arginine biosynthetic pathway has been found in the land planarian *Bipalium kewense*[10] and in the earthworm *Lumbricus terrestris*[5,8]. In the earthworm, it was shown that arginine synthesis also provides protein arginine, so that this can not be an "essential" amino acid for the worm. It is also the pathway for urea synthesis, to be considered later. On an evolutionary point of view, the report of the ornithine cycle in the land planarian[10], supposed to bear some similarities to first metazoans, indicates that the arginine pathway was probably present in the most primitive invertebrates, thus supporting the idea that the pathway was continuous in its evolution from unicellular organisms through to the vertebrates.

Detailed studies of arginine biosynthesis have also been performed in molluscs. In the land snails *Otala lactea* and *Helix aspersa*, both free and protein arginine became labeled following the administration of ^{14}C-bicarbonate, and the different steps of the arginine pathway were confirmed by *in vivo* incorporation of radioactive precursors[4]. The synthesis of arginine has been reported from gastropods molluscs and seems to be restricted to only this or certain classes of molluscs, as it appears to be restricted to some vertebrate classes. A particularity of the arginine pathway enzymes in the snails is their relatively uniform distribution in the various tissues, to the difference of vertebrates in which they are mainly localized in the liver.

But the synthesis of arginine is not found in all groups of invertebrates. It has not been observed in the crustaceans, all enzymes of the urea cycle except arginase being either absent or present at levels below the limit of detection[6]. Insects, like crustaceans, appear to be devoid of all enzyme activities associated with the urea cycle except arginase and arginine seems to be an "essential" amino acid[7]. In spite of the fact that many insects contain ornithine, citrulline has only been reported in one species. Although arginine is the most important amino acid in the insects, these have lost the ability to synthesize arginine upon evolving from ureotelic annelid ancestors, analogous to the loss in arginine synthesis observed in the reptiles and birds as they evolved from primitive ureotelic amphibians.

The reason why arginase has been retained in those species deprived of the arginine biosynthetic pathway is not clear since it serves to degradate an amino acid which is essential for them. The role of arginase in these animals could be to regulate the level of arginine in the tissues in the event of excess arginine in the diet.

Urea Cycle and Nitrogen Excretion

The two main functions of the urea cycle are the synthesis of arginine and the detoxication of waste nitrogen. The last step of the cycle (hydrolysis of arginine by arginase yielding urea) has developed under selective pressions concerned in a large part with limited water supply and osmoregulatory problems. Aquatic animals, which can diffuse ammonia in the external environment, are mainly ammoniotelic. In contrast, terrestrial animals, which require a less toxic excretion product, are mostly ureotelic or uricotelic. It has therefore been suggested that, soon in evolution, animals have exploited the arginine pathway for ureotelism during their invasion of the land habitat[11]. This has proved to be true in most vertebrate classes and some examples are also found in invertebrates. A typical one is the report by Campbell[10] of a functional urea cycle in the land planarian, which provides evidence for the development of ureotelism associated with terrestrial invasion by a primitive animal. Other worms among annelids[5,8] and terrestrial molluscs[4] also possess the ability to synthesize urea via the urea cycle.

However, urea is a minor excretory product for most invertebrates. Even those species which possess the enzymatic equipment for urea synthesis are primarily ammoniotelic or uricotelic[3]. Many studies effected on the earthworms have shown that urea excretion, normally low beside that of ammonia, could be modulated by factors such as water supply and dietary level[5,8]. In *Lumbricus terrestris* the urea/NH_3 output is increased when the amount of water available is limited. In the same manner, during fasting, ammonia excretion by the worm decreases and that of urea increases greatly ; a parallel increase of the level of the urea cycle enzymes is observed. This increase is possibly regulated by the level of free amino acids in the tissues, enhanced during fasting as a result of accelerated protein catabolism.

Terrestrial gastropods molluscs, which possess the enzymes of the arginine pathway[4] and very high arginase activities[12], are mainly uricotelic. Speeg and Campbell[4] have demonstrated that urea produced in the land snails *Helix pomatia* and *Otala lactea* was converted by urease to CO_2 and NH_3 rather than accumulating as urea for excretion as it does in ureotelic species. They have suggested that the reaction may be directed toward maintaining the necessary alkaline conditions for the precipitation of calcium carbonate for shell formation.

Insects, in spite of the presence of arginase in their tissues[7] are mainly uricotelic[3,7], while marine echinoderms, crustaceans and molluscs are mostly ammoniotelic[3]. In these animals devoid of the arginine synthesizing system, some urea found in the excreta could result from a hydrolysis of exogenous arginine by arginase.

Since activities of the urea cycle enzymes, except arginase, have not been reported in arthropods, it could be presumed that crustaceans are unable of responding to hyperosmotic stress by synthesizing urea. However, it has been reported that the crayfish can shift within two days from ammoniotelism to ureotelism, and conversely[6]. Arginine may be the source of increased urea.

GUANIDINO COMPOUNDS FORMED FROM ARGININE

Beside arginine, invertebrates contain a great variety of guanidine derivatives. Most of them are monosubstituted guanidines, which show great differences in their structure, properties and distribution[13-15]. Their common feature is to be directly or indirectly biosynthesized from arginine by a number of reactions concerning (i) an alteration of the ornithine chain, (ii) the transfer of the amidine group to a convenient amine, or (iii) the phosphorylation of the guanidine radical (Fig. 1).

Alteration of the Ornithine Chain

Agmatine. Decarboxylation of arginine yielding agmatine has been observed in some marine crustaceans and molluscs and can explain the presence of agmatine in these animals[16]. The same process could be responsible for the formation of agmatine found in the shellfish *Cristaria plicata*[17] and in the sea anemones *Actinia equina* and *Actinia fragacea*[18], but the reaction has not been established.

α-keto-δ-guanidinovalerate and γ-guanidinobutyrate. The oxidative deamination of arginine by a L-amino acid oxidase present in many marine molluscs, crustaceans and echinoderms[16], in the fresh-water mollusc *Lymnaea stagnalis*[19], and in insects[20], yields α-keto-δ-guanidinovalerate (α-keto-δ-GVA). In the absence of catalase, the keto derivative is non-enzymatically oxidized by H_2O_2 produced in the reaction, with the formation of γ-guanidinobutyrate. This process has never been found in worms[14]. The sequence of reactions is:

L-arginine + H_2O + O_2 ⟶ α-keto-δ-GVA + NH_3 + H_2O_2

α-keto-δ-GVA + H_2O_2 ⟶ γ-guanidinobutyrate + CO_2

γ-guanidinobutyrate can be hydrolyzed by a γ-guanidinobutyrate ureohydrolase characterized in various classes of molluscs[21,22], producing γ-aminobutyrate and urea. In these animals, the reaction can be an alternative source of urea.

γ-guanidinobutyramide. The product of oxidative decarboxylation of arginine has been observed, beside the products of the oxidative deamination of the amino acid, in tissue extracts of *Lymnaea stagnalis* incubated in the presence of arginine and of oxygen[23]. The reaction can be the same as that established in *Streptomyces*[24], catalyzed by arginine oxygenase:

L-arginine + O_2 ⟶ γ-guanidinobutyramide + CO_2

Octopine. Octopine, present in most molluscs, results from the reductive condensation of arginine and pyruvate, catalyzed by octopine dehydrogenase according to the reversible reaction[25,26]:

L-arginine + pyruvate + $NADH_2$ ⇌ octopine + NAD

Octopine displays a function in the reduction of pyruvate and in the reoxidation of $NADH_2$ in the muscles of a number of marine invertebrates, principally in molluscs, most of which are devoid of

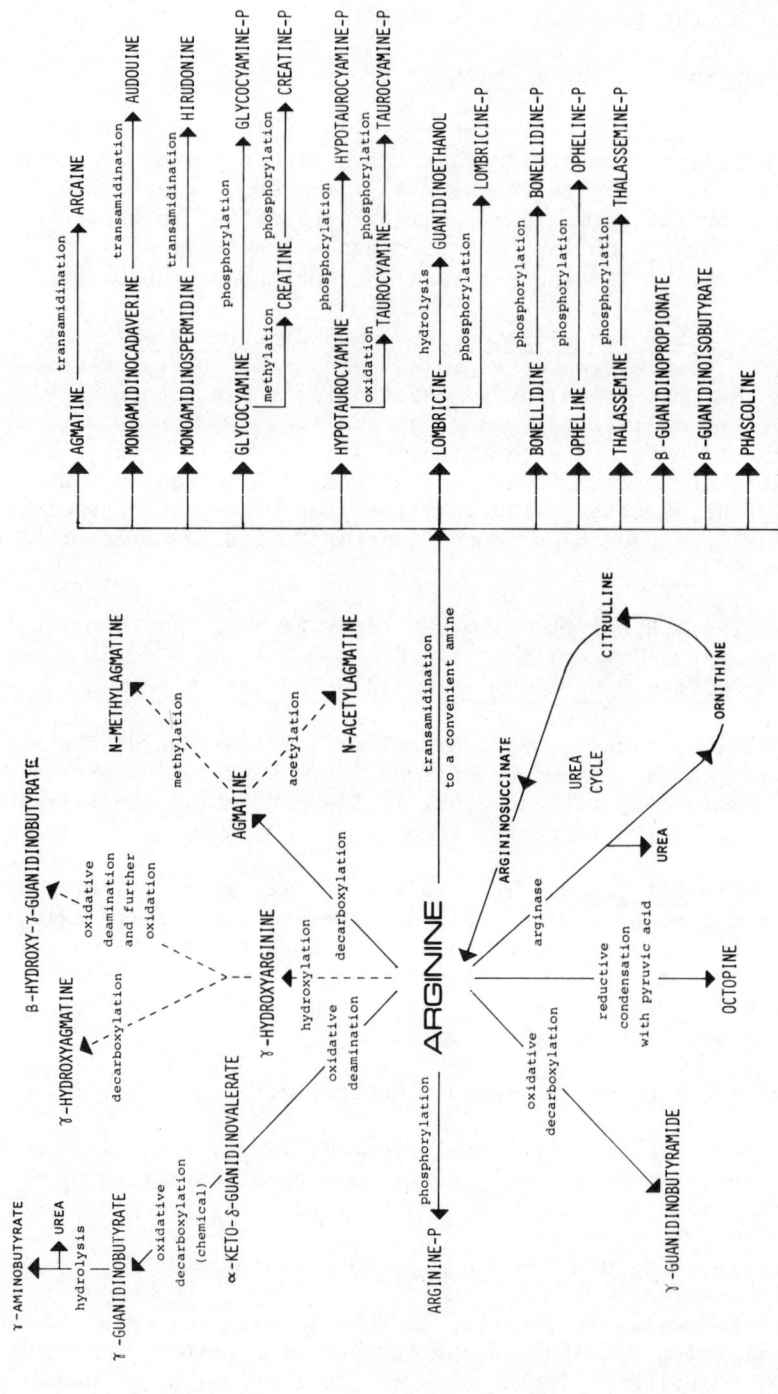

Fig. 1. Guanidino compounds from invertebrates. Relation to arginine metabolism (--- Hypothetic pathway).

lactic dehydrogenase. Its major role in muscle as a factor of regulation between phosphagen metabolism and anaerobic glycolysis, suggested by Thoai and Robin[25,26], has been confirmed by large studies concerning the function of the octopine dehydrogenase reaction in marine molluscs (see review by Gäde[27]).

Hydroxyguanidines. It is likely that γ-hydroxyagmatine and β-hydroxy-γ-guanidinobutyric acid characterized beside γ-hydroxyarginine in the sea anemone *Anthopleura japonica*[28] are, respectively, the products of the decarboxylation and of the oxidative deamination of the latter. But nothing is known upon "how" and "when" the hydroxyl group is introduced in the ornithine chain.

Transfer of the Amidine Group to a Convenient Amine

$$\text{L-arginine} + NH_2\text{-R} \longrightarrow \text{L-ornithine} + HN=C(NH_2)\text{-}NH_2\text{-R}$$

The transfer of the amidine group of arginine to a convenient amine is the biosynthetic pathway for a large number of monosubstituted guanidines. Most of them have been isolated and characterized from worms, which lack the oxidative catabolic pathway of arginine. The compounds thus synthesized can be further modified without alteration of their guanidine radical, by enzymatic or non-enzymatic processes such as methylation and oxidation. Several of them are the basis of new phosphagens. For reviews concerning these compounds, a large number of which have been discovered in our laboratory of the Collège de France, we refer the reader to Thoai and Robin[14] and to Needham[5].

Glycocyamine and creatine. Invertebrates, like vertebrates, synthesize glycocyamine by amidination of glycine from arginine. The subsequent methylation of glycocyamine into creatine has been established in most of the main phyla of invertebrates[5,14]. Methionine is supposed to be the methyl donor, via the formation of S-adenosylmethionine. Glycocyamine and creatine are phosphagen precursors in some invertebrates.

Sulfur-containing guanidines. Hypotaurocyamine (guanidinoethylsulfinic acid) is biosynthesized in worms by transfer of the amidine group of arginine to the corresponding amino precursor. Further oxidation of the sulfinic group of hypotaurocyamine yields taurocyamine (guanidinoethyl sulfonic acid)[5,14]. Thus, hypotaurocyamine is the precursor of taurocyamine in worms. The reaction has been demonstrated in annelids and in sipunculids. Both compounds are the basis of new phosphagens in worms.

Nothing is known upon the biogenesis of asterubine, the N'-dimethyl derivative of taurocyamine, the presence of which has been reported in sea-stars[29].

Guanidinoethylphosphate derivatives. Opheline, lombricine, bonellidine and thalassemine are, respectively, the methyl-, seryl-, aspartylseryl- and N,N-dimethyl-O-phosphoryl esters of guanidinoethyl phosphate[14,30]. The four compounds have been isolated from worms. It is likely that the amidine group of arginine is the precursor of the amidine group of these guanidino phosphodiesters, but the transamidination has, to date, been only demonstrated for the synthesis of lombricine from its direct amino precursor, β-aminoethyl phosphoserine[31]. Methionine was shown to be the donor of the methyl groups of opheline and thalassemine[14,30]. Lombricine can be hydrolyzed by a particulate phosphodiesterase present in the earthworm, with the formation of guanidinoethanol[32]. Guanidinoethylphosphate derivatives are phosphagen precursors in worms.

New products of pyrimidine catabolism. β-aminopropionic and β-aminoisobutyric acids are well known products of pyrimidine catabolism. Their guanidino analogues have recently proved to be the guanidine moieties of two new long-chain guanidino amides, phascoline and phascolosomine, respectively, present in large amounts in the viscera of some marine sipunculids[33]. It is now established that the carbon chain of β-guanidinopropionic acid, β-aminoisobutyric acid , phascoline and phascolosomine present in the tissues of these worms result from the catabolism of uracil and thymine, respectively, and that their amidine group is derived from arginine by transamidination (Robin and Guillou, unpublished results). Their biological role is unknown.

Diguanidino compounds. Arcaine (diamidinoputrescine), audouine (diamidinocadaverine) and hirudonine (diamidinospermidine), characterized in a number of worms, are biosynthesized by the transfer of two amidine groups of arginine to the corresponding diamine[5,14]. The reaction was shown to proceed in two steps, with the intermediary formation of a monoamidinated compound. In the case of arcaine, the intermediary compound is agmatine. Thus, agmatine is synthesized in worms by a process different from that (decarboxylation of arginine) observed in other groups of invertebrates. The diguanidino compounds do not participate in the formation of muscular phosphagens and their biological role is not clear.

Phosphagens

The substitution of the free amino group of the amidine group of arginine by the terminal phosphate group of ATP yields phosphoarginine, the phosphagen for most invertebrates. This compound accumulates in muscle during the resting periods and is utilized as a reserve of energetic phosphate for the rephosphorylation of ADP to ATP during muscular contraction. Thus, it plays a major part in arginine metabolism in invertebrates. The reversible reaction is :

$$\text{L-arginine} + \text{Mg-ATP} \rightleftharpoons \text{phosphoarginine} + \text{Mg-ADP}$$

Beside phosphoarginine, seven muscular phosphagens have been characterized in invertebrates. They result from the phosphorylation of creatine, glycocyamine, taurocyamine, hypotaurocyamine, lombricine, opheline and thalassemine (see reviews by Robin[34,35]), catalyzed by specific phosphagen kinases. The N'-phosphoryl derivative of bonellidine could not be characterized in *Bonellia*[14], but the simultaneous presence of bonellidine and of a phosphagen kinase active on this compound in the trunk muscle of the worm (Robin, unpublished data) makes it likely that bonellidine is the phosphagen precursor in *Bonellia*. Except for phosphocreatine, the distribution of which is somewhat larger, these phosphagens are new compounds found only in worms. Since the guanidine basis participating in the structure of these phosphagens are likely to be formed from arginine, as reported above, arginine can be considered as metabolically so important in their biosynthesis as in that of phosphoarginine.

CONCLUSION

From the above survey, we can see (i) that the metabolism of arginine in invertebrates is far from being so directly implicated in the urea cycle pathway as it is in most invertebrates and (ii) that arginine is a primordial link between nitrogen metabolism and the biosynthesis of guanidine derivatives in invertebrates. It is the only guanidino compound which can synthesize its amidine group *de novo*, and it participates, directly or indirectly, in the formation of most, if not all, other guanidine derivatives present in these animals.

Invertebrates are a priviledged field for the study of guanidine derivatives, due to the number and structural diversity of their guanidine constituents. These are implicated in a variety of metabolism, the study of which can lead to a better knowledge of guanidine derivatives in vertebrates. Many compounds first identified in invertebrates (e.g. taurocyamine, α-keto-δ-guanidinovaleric acid, γ-guanidinobutyric acid, hydroxy-guanidines) have been later characterized in vertebrate tissues and biological fluids.

REFERENCES

1. H.A. Krebs and K. Henseleit, Untersuchungen über die Harnstoffbildung im Tierkörper, Z. Physiol. Chem. 210 : 33 (1932).
2. O. Meyerhof, Uber die Verbreitung der Argininphosphorsäure in der Muskulatur der Wirbellosen, Arch. Sci. Biol. Italy, 12 : 536 (1928).
3. E. Baldwin, "Dynamic Aspects of Biochemistry", 2nd ed., University Press, Cambridge (1953).
4. J.W. Campbell and S.H. Bishop, Nitrogen Metabolism in Molluscs, in : "Comparative Biochemistry of Nitrogen Metabolism", vol. 1, J.W. Campbell ed., Academic Press, London (1970).

5. A.E. Needham, Nitrogen Metabolism in Annelida, in : "Comparative Biochemistry of Nitrogen Metabolism", vol. 1, J.W. Campbell ed., Academic Press, London (1970).
6. R. Hartenstein, Nitrogen Metabolism in Non-Insect Arthropods, in : Comparative Biochemistry of Nitrogen Metabolism", vol. 1, J.W. Campbell ed., Academic Press, London (1970).
7. J.J. Corrigan, Nitrogen Metabolism in Insects, in : "Comparative Biochemistry of Nitrogen Metabolism", vol. 1, J.W. Campbell ed., Academic Press, London (1970).
8. M. Florkin, Nitrogen Metabolism, in : "Chemical Zoology", vol. 4, M. Florkin and B.T. Scheer, eds., Academic Press, New York (1969).
9. P.A. Janssens and C. Bryant, The Ornithine-Urea Cycle in Some Parasitic Helminths, Comp. Biochem. Physiol. 30 : 261 (1969).
10. J.W. Campbell, Arginine and Urea Biosynthesis in the Land Planarian : its Significance in Biochemical Evolution, Nature 208 : 1299 (1965).
11. P.P. Cohen and G.W. Brown Jr., Ammonia Metabolism and Urea Biosynthesis, in : "Comparative Biochemistry", vol. II, M. Florkin and H.S. Mason, eds., Academic Press, New York (1960).
12. S. Gaston and J.W. Campbell, Distribution of arginase activity in molluscs, Comp. Biochem. Physiol. 17 : 259 (1966).
13. N.V. Thoai, Nitrogenous bases, in : "Comprehensive Biochemistry", vol. 6, M. Florkin and E.H. Stotz, eds., Elsevier Publishing Co., Amsterdam (1965).
14. N.V. Thoai and Y. Robin, Guanidine Compounds and Phosphagens, in : "Chemical Zoology", vol. 4 : Annelida, Echiura and Sipuncula, M. Florkin and B.T. Scheer, eds., Academic Press, New York and London (1969).
15. L. Chevolot, Guanidine Derivatives, in : "Marine Natural Products" vol. 4, P.J. Scheuer ed., Academic Press, London (1981).
16. N.V. Thoai, J. Roche and Y. Robin, Métabolisme des Dérivés Guanidylés. I. Dégradation de l'Arginine chez les Invertébrés Marins, Biochim. Biophys. Acta, 11 : 403 (1953).
17. T. Suzuki and S. Muraoka, New Guanidyl Derivatives and Amino Acids in the Extract of Shell-Fish *Cristaria plicata* Leach, J. Pharm. Soc. Japan, 74 : 171 (1954).
18 Y. Guillou and Y. Robin, Présence de α-N-acétylagmatine chez des Cnidaires, *Actinia equina* et *Actinia fragacea*, Compt. Rend. Soc. Biol. 173 : 576 (1979).
19. Y. Robin and N.V. Thoai, Métabolisme Oxydatif de la L-Arginine chez la Limnée, *Limnaea stagnalis*. I. Oxydation par la L-Aminoacideoxydase, Compt. Rend. Soc. Biol. 151 : 2093 (1957).
20. I. Garcia , J. Roche and M. Tixier, Sur le Métabolisme de la L-Arginine chez les Insectes. I., Bull. Soc. Chim. Biol. 38 : 1423 (1956).
21. R. Baret, M. Mourgue, A. Broc and J. Charmot, Etude Comparative de la Désamidination de l'Acide γ-Guanidobutyrique et de l'Arginine par l'Hépatopancréas ou le Foie de Divers Invertébrés, Compt. Rend. Soc. Biol. 159 : 2446 (1965).

22. Z. Porembska, I. Gasiorowska and I. Mochnacka, Isolation of Arginase and Guanidinobutyrate Ureohydrolase from the Hepatopancreas of *Helix pomatia*, Acad. Biochim. Pol. 15 : 171 (1968).
23. N.V. Thoai, Y. Robin and L.A. Pradel, Métabolisme Oxydatif de la L-Arginine chez la Limnée, *Limnaea stagnalis* L. II. Oxydation en Guanidobutyramide, Compt. Rend. Soc. Biol. 151 : 2097 (1957).
24. N.V. Thoai and T.T. An, Sur une Nouvelle Amidase Spécifique : la Guanidobutyramidase, Compt. Rend. Soc. Biol. 150 : 1722 (1956).
25. N.V. Thoai and Y. Robin, Métabolisme des Dérivés Guanidylés. VIII. Biosynthèse de l'Octopine et Répartition de l'Enzyme l'Opérant chez les Invertébrés. Biochim. Biophys. Acta, 35 : 446 (1959).
26. Y. Robin and N.V. Thoai, Métabolisme des Dérivés Guanidylés. X. Métabolisme de l'Octopine : son Rôle Biologique. Biochim. Biophys. Acta, 52 : 233 (1961).
27. G.Gäde, Biological Role of Octopine Formation in Marine Molluscs, Marine Biology Letters, 1 : 121 (1980).
28. S. Makisumi, Guanidino Compounds from a Sea-Anemone, *Anthopleura japonica* Verril, J. Biochem. 49 : 284 (1961).
29. D. Ackermann, Asterubin, eine Schwefelhaltige Guanidinverbindung der Belebten Natur, Z. Physiol. Chem. 232 : 206 (1935).
30. N.V. Thoai, Y. Robin and Y. Guillou, A New Phosphagen, N'-Phosphorylguanidinoethylphospho-O-(α-N,N-Dimethyl)Serine (Phosphothalassemine), Biochemistry, 11 : 3890 (1972).
31. R.J. Rossiter, T. Gaffney, H. Rosenberg and A.H. Ennor, Biosynthesis of Lombricine, Nature, 185 : 383 (1960).
32. Y. Robin, Répartition et Métabolisme des Guanidines Monosubstituées d'Origine Animale, Thèse de Doctorat ès Sciences Naturelles, Paris (1954).
33. Y. Guillou and Y. Robin, Phascolin (N-(3-Guanidinopropionyl)-2-Hydroxy-n-Heptylamine) and Phascolosomine (N-(3-Guanidinoisobutyryl)-2-Methoxy-n-Heptylamine), Two New Guanidino Compounds from Sipunculid Worms. Isolation and Structure, J. Biol. Chem. 248 : 5668 (1973).
34. Y. Robin, Phosphagens and Molecular Evolution in Worms, BioSystems, 6 : 49 (1974).
35. Y. Robin, Les Phosphagènes des Animaux Marins, in : "Actualités de Biochimie Marine", vol. 2, Y. Le Gal, ed., Centre National de la Recherche Scientifique, Paris (1980).

α-GUANIDINOGLUTARIC ACID AND EPILEPSY

Akitane Mori, Yoko Watanabe, Shoichiro Shindo,
Masayuki Akagi and Midori Hiramatsu

Department of Neurochemistry, Institute for Neurobiology,
Okayama University Medical School, Okayama, 700 Japan

The application of metallic cobalt to the cerebral cortex results in an experimental model of chronic focal epilepsy[1,2] Many biochemical studies on this focus tissue have been performed to explore seizure mechanism[3-6], though the induction phenomenon itself is not well understood.

We[7] first analysed guanidino compounds from an epileptogenic cerebral cortex 24 hours after cobalt application, and an unknown high peak identical to the peak of authentic α-guanidinoglutaric acid peak appeared in the chromatogram. We then isolated the unknown substance by paper chromatography, extracted it with water, dried it in vacuo, converted it into its dimethylpyrimidyl derivative by reaction with acetylacetone, and analysed it by our adaptation[8] of the GC/MS technique. We found that the mass spectrum of the substance was identical to the dimethylpyrimidyl derivative of α-guanidinoglutaric acid butylester ($M^+ = 365$). Fig. 1 shows the chemical structure of α-guanidinoglutaric acid identified by us.

We analysed α-guanidinoglutaric acid in the cobalt treated area, as well as in the contralateral area. This substance increased markedly in focus tissue, concomitant with the appearance of paroxysmal discharges 3 or 4 hours after cobalt treatment, reaching a maximum 24 hours after. In addition we also detected very small amounts of α-guanidinoglutaric acid (less than 0.02 μmol/g) in cortex of untreated animals. At this stage, an increase in the substance was observed in the contralateral area. α-Guanidinoglutaric acid was not detected at 37 and 42 days after cobalt treatment, after which time no seizure was observed visually or recorded electroencephalographically (Fig. 2)

In another experiment, we observed that α-guanidinoglutaric

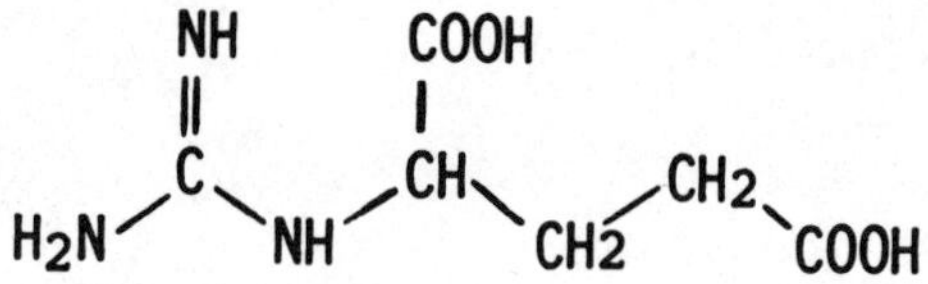

Fig. 1. Structure of α-guanidinoglutaric acid.

acid could induce paroxysmal discharges in EEG. Rabbits (N=5) were anesthetized with ether, and fixed in a stereotaxic apparatus. We applied a piece of filter paper (4 × 4 mm) soaked in 0.2 M α-guanidinoglutaric acid-saline solution on the sensory motor cortex. An EEG was recorded from cotton electrodes put on the surface of the cerebral cortex around the focus, and it was found that high voltage slow waves and atypical spike and slow wave complex appeared in the electroencephalogram about 10 minutes after application, spike and slow waves about 30-40 minutes after, and an atypical spike and high voltage slow wave burst about one hour after (Fig. 3). Thereafter, these spike and high voltage slow waves were observed continuously for about 3 hours.

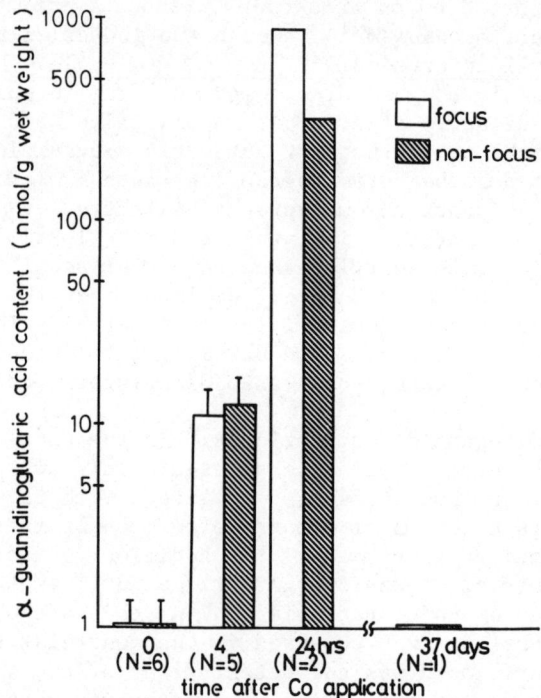

Fig. 2. Effect of cobalt on α-guanidinoglutaric acid levels of cat cerebral cortex.

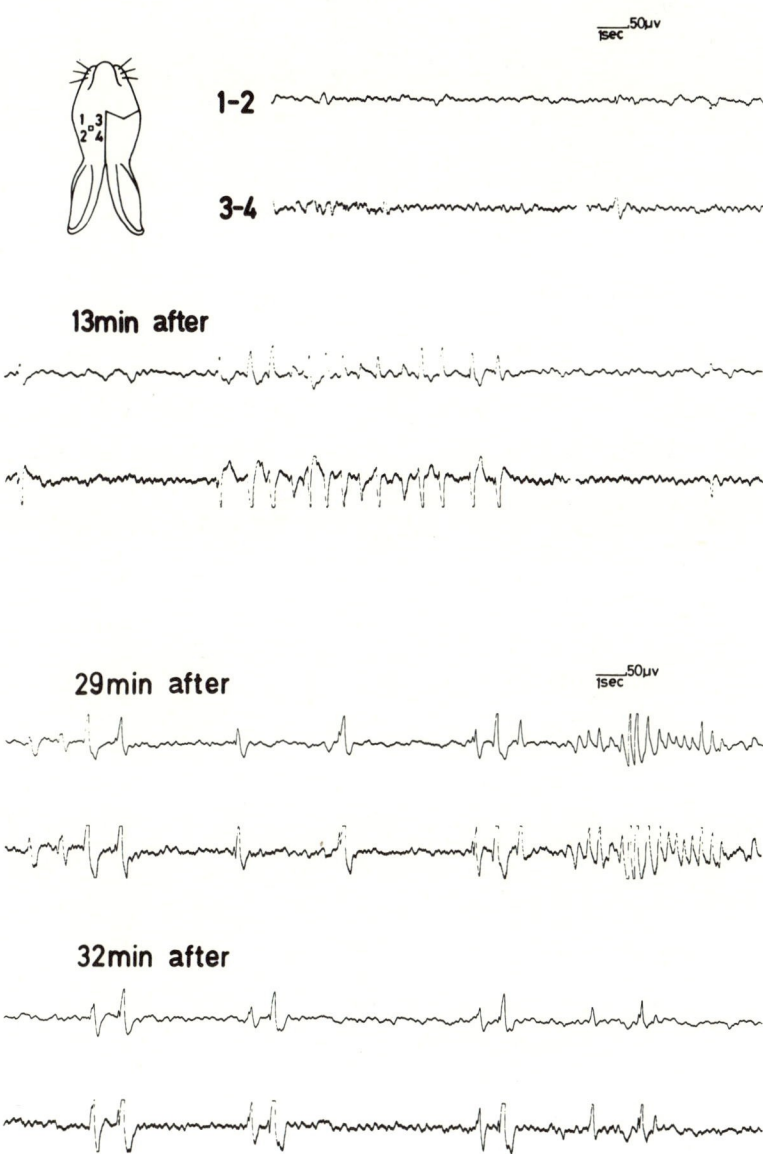

Fig. 3. EEG recording of rabbit after topical application of α-guanidinoglutaric acid.

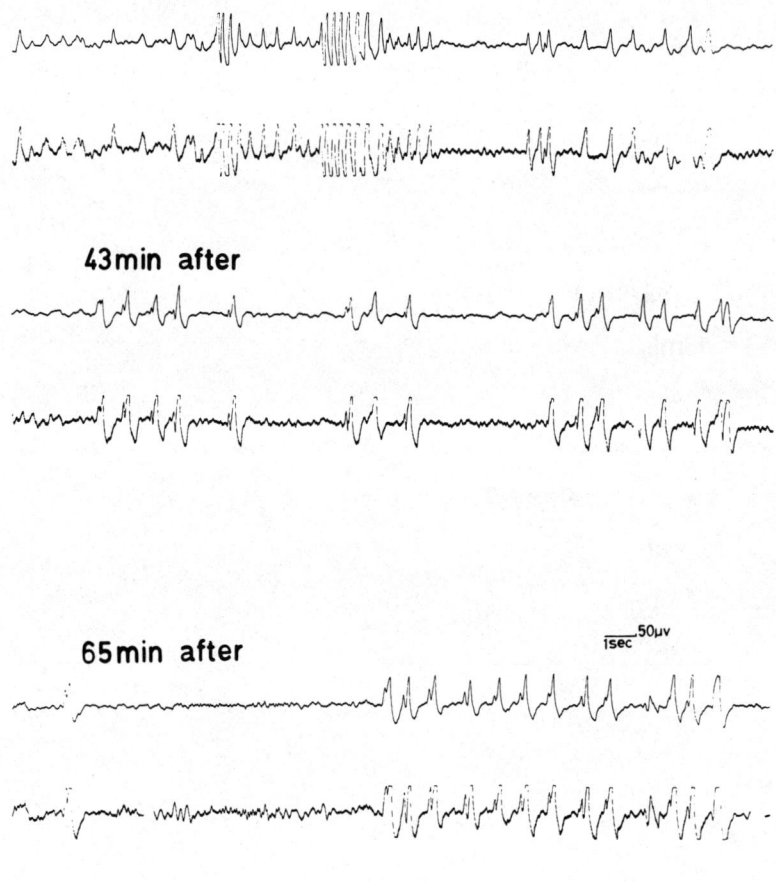

Fig. 3. EEG recording of rabbit after topical application of α-guanidinoglutaric acid (continued).

Koyama[3] analysed amino acids in cobalt-induced epileptogenic cat cortex and found a decrease in glutamic acid. Other investigators[9-12] have reported significant decreases in glutamic acid content in the epileptogenic brain of experimental animals, as well as human patients. These findings suggest that α-guanidinoglutaric acid may be produced from glutamic acid by the transamidination reaction in the epileptogenic cerebral cortex. Therefore we studied, in addition, α-guanidinoglutaric acid formation in the brain.

In this experiment, we incubated 0.025 M glutamic acid (or α-ketoglutaric acid) and 0.1 M arginine (or other possible amidine donors, i.e. creatine, creatine phosphate, creatinine, guanidine, glycocyamine, argininic acid, canavanine, urea) with brain homogenate in 0.1 M Tris-HCl buffer (pH 8.0) at 37°C for 24 hours. The reaction was stopped by adding an isovolume of 10% TCA. The supernatant after centrifugation (10,000 r.p.m. for 10 minutes) was passed through an Amberlite CG-120 (H^+ form) column. The fraction, containing amino acids, was collected, dried in vacuo, adjusted to pH 7.0 ~ 7.5, and passed through a column of Dowex 1×8 (CH_3COO^- form). The column was then eluted with 0.5 N acetic acid and the eluate was analysed by a guanidino compound analyser (JASCO, G-520). Table 1 shows that α-guanidinoglutaric acid was

Table 1. α-GGA formation from several substrates

Substrates	α-GGA formation
Glu + Arg	−
Glu + Urea	−
Glu + GAA	−
α-KG + Arg	+
α-KG + CR	−
α-KG + CR-P	−
α-KG + CRN	−
α-KG + G	−
α-KG + GAA	−
α-KG + AGA	−
α-KG + Urea	−
α-KG + CAN	−
Control	−

α-GGA:α-guanidinoglutaric acid, Glu:glutamic acid, α-KG:α-ketoglutaric acid, Arg:arginine, GAA:guanidinoacetic acid, CR:creatine, CR-P: creatine phosphate, CRN:creatinine, G:guanidine, AGA:argininic acid, CAN: canavanine.

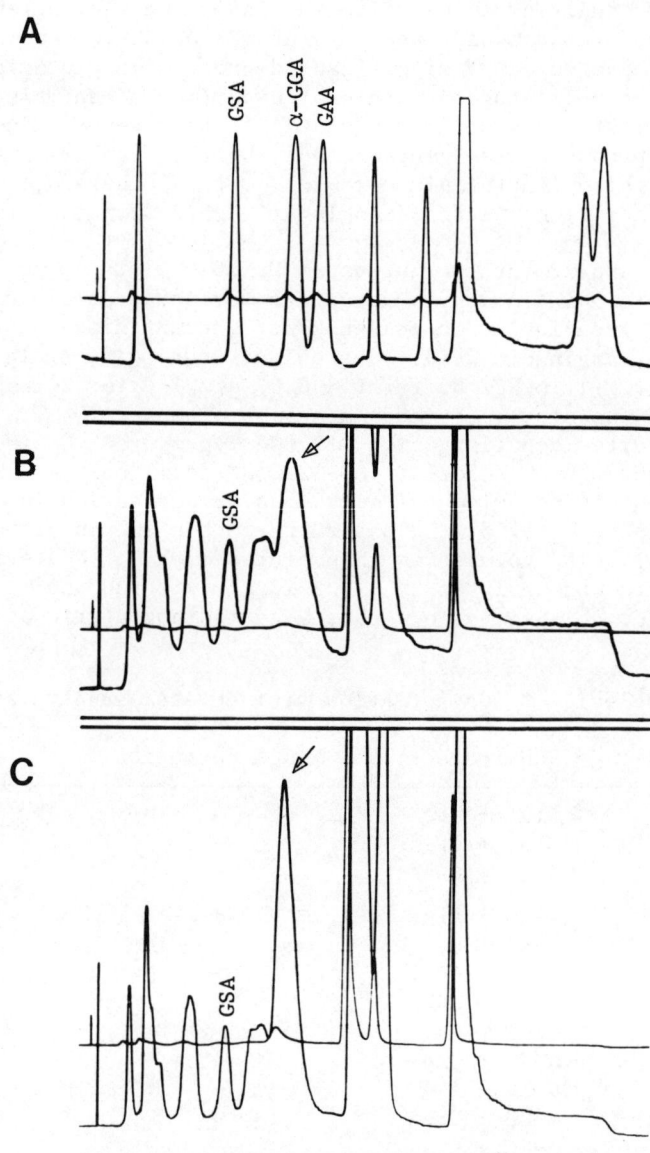

Fig. 4. Chromatogram of reaction mixture
(substrates:α-ketoglutaric acid and arginine).
  A : authentic α-guanidinoglutaric acid
  B : reaction mixture (substrates:α-keto-
      glutaric acid and arginine)
  C : B + authentic α-guanidinoglutaric acid

not produced from glutamic acid and arginine by a transamidination reaction, and that urea or guanidinoacetic acid (glycocyamine) did not play a role as an amidine donor in the transamidination reaction. However, we did observe a peak identical to authentic α-guanidinoglutaric acid in the chromatogram shown in Fig. 4. This experimental result suggests the following reaction:

$$\text{HOOC-CO-CH}_2\text{-CH}_2\text{-COOH} \quad + \quad \text{NH}_2\text{-}\overset{\text{NH}}{\underset{\|}{\text{C}}}\text{-NH-CH}_2\text{-CH}_2\text{-CH}_2\text{-}\overset{\text{NH}_2}{\underset{|}{\text{CH}}}\text{-COOH}$$

α-ketoglutaric acid   arginine

$$\downarrow$$

$$\text{H}_2\text{N-}\overset{\text{NH}}{\underset{\|}{\text{C}}}\text{-NH-CH-CH}_2\text{-CH}_2\text{-COOH} \quad + \quad [\text{-CH}_2\text{-CH}_2\text{-CH}_2\text{-}\overset{\text{NH}_2}{\underset{|}{\text{CH}}}\text{-COOH}]$$
$$\qquad\qquad\quad\; \underset{\text{COOH}}{|}$$

α-guanidinoglutaric acid

This reaction was not accerelated by the addition of 1 mM $CoCl_2$, suggesting that the cobalt ion did not relate directly to the reaction. We hope to explain the reaction mechanism by more detailed studies in the near future.

In conclusion, we isolated and identified a new guanidino compound, α-guanidinoglutaric acid, in the cobalt-induced epileptogenic focus of cat cerebral cortex. This substance increased markedly, concomitant with the appearance of paroxysmal discharges, reaching a maximum 24 hours after cobalt treatment. This substance also was detectable in small quantity in normal cortex. We applied α-guanidinoglutaric acid topically on the sensory motor cortex of cat, and observed paroxysmal discharges in EEG recordings. We found that α-guanidinoglutaric acid was produced from α-ketoglutaric acid and arginine in brain homogenate, though the reaction mechanism is still unexplained.

REFERENCES

1. L. M. Kopeloff, Experimental epilepsy in the mouse, Proc. Soc. Exp. Med. 104:500 (1960).
2. R. S. Dow, The production of cobalt experimental epilepsy in the rat, Electroencephalogr. Clin. Neurophysiol. 14:399 (1962).
3. I. Koyama, Amino acids in the cobalt-induced epileptogenic and nonepileptogenic cat's cortex, Can. J. Physiol. Pharmacol. 50:740 (1972).
4. N. M. Van Gelder, Glutamate dehydrogenase, glutamic acid decarboxylase, and GABA aminotransferase in epileptic mouse cortex, Can. J. Physiol. Pharmacol, 52:952 (1974).

5. P. R. Dodd, H. F. Bradford, A. S. Abdnlghani, D. W. G. Cox and Continho-Netto, Release of amino acids from chronic epileptic and subepileptic foci in vivo, Brain Res. 193: 505 (1980).
6. S. M. Ross and C.R. Craig, Studies on γ-aminobutyric acid transport in cobalt experimental epilepsy in the rat, J. Neurochem. 36:1006 (1981).
7. A. Mori, M. Akagi, Y. Katayama and Y. Watanabe, α-Guanidinoglutaric acid in cobalt induced epileptogenic cerebral cortex of cats, J. Neurochem. 35:603 (1980).
8. A. Mori, T. Ichimura and H, Matsumoto,Gas chromatography-mass spectrometry of guanidino compounds in brain, Anal. Biochem. 89:393 (1978).
9. E. L. Peters and D. B. Tower, Glutamic acid and glutamine metabolism in cerebral cortex after seizures induced by methionine sulfoximine, J. Neurochem. 5:80 (1959).
10. S. Berl, D. P. Purpura and H. Waelsch, Amino acid metabolism in epileptogenic and nonepileptogenic lesions of the neocortex (cat), J. Neurochem. 4:311 (1959).
11. N. Mison-Crighel, N. Luca and E. Crighel, The effect of an epileptogenic focus, induced by topical application of mescaline on glutamic acid, glutamine and GABA in the neocortex of the cat, J. Neurochem. 11:333 (1965).
12. N. M. Van Gelder, A. L. Sherwin and T. Rasmussen, Amino acid content of epileptogenic human brain : Focus versus surrounding regions, Brain Res. 40:385 (1972).

GUANIDINO COMPOUNDS IN HYPERARGININEMIA

B. Marescau, A. Lowenthal, H.G. Terheggen[*], E. Esmans[**]
and F. Alderweireldt[**]

Laboratory of Neurochemistry, Born-Bunge Foundation
Universitaire Instelling Antwerpen, Wilrijk, Belgium
[*]Children's Hospital, Neuss, Federal Republic of Germany
[**]Rijksuniversitair Centrum Antwerpen, Laboratory for
Organic Chemistry, B-2020 Antwerpen, Belgium

INTRODUCTION

The patients studied are three sisters affected with hyperargininemia. The clinical and biochemical picture has been reported in full[1-3]. Like all patients affected with urea cycle diseases, the first clinical symptoms seen in a patient with hyperargininemia are irritability, coma and epilepsy. All the patients described either by us or in the literature are still alive. The age of our patients is 10, 13 and 17 years. The evolution of the hyperargininemic patient is characterized by neurological degradation, like spasticity.

ARGININE ACCUMULATION

Patients with hyperargininemia accumulate arginine in their body fluids in consequence of the arginase deficiency. Table 1 shows clearly the increase of arginine in serum, urine and cerebrospinal fluid (CSF).

The arginine accumulation is not only seen in the body fluids but also in erythrocytes and leucocytes[4] (Table 2), as a consequence of the intracellular arginase deficiency.

Table 1. Arginine levels in body fluids: serum and CSF in μmol/100 ml, urine in μmol/g. creatinine

Cases	Serum	CSF	Urine
A.W.	64	5,3	1.420
M.W.	99,6	9,5	13.700
I.W.	157,9	11,3	17.787
Controls	9,2 (±2,2)	1,4 (±0,7)	25-34

Table 2. Arginine concentration in erythrocytes (nmol/10^9 cells) and leucocytes (μmol/10^9 cells)

Cases	Erythrocytes	Leucocytes
A.W.	12,4	2,2
M.W.	14,5	1,7
I.W.	12,4	1,8
Controls	0,47 (±0,27)	0,38 (±0,18)

IDENTIFICATION OF MONOSUBSTITUTED GUANIDINO COMPOUNDS IN URINE OF PATIENTS WITH HYPERARGININEMIA

The arginine accumulation is reflected in the high excretion values of the guanidino compounds. The guanidino compounds are considered to be the catabolites of arginine. To reach a better understanding of the physiology and the metabolism of the guanidino compounds in patients with hyperargininemia, we have controlled the identity of the urinary guanidino compounds. This identification was done by a combination of three different chromatographic techniques: liquid cation-exchange chromatography, thin-layer chromatography (TLC) on cellulose plates and coupled gas-chromatography - mass spectrometry (GC-MS).

Liquid cation exchange chromatography was performed on a single column amino acid analyzer and the elution procedure followed was according to Efron[5]. In our experiments we have analysed the free monosubstituted guanidino compounds in parallel with the amino acids. Figure 1 gives the elution patterns. The guanidino positive peaks were collected individually and desalted for identification studies with TLC and GC-MS. By applying these techniques the identity of guanidinoacetic acid, N-α-acetylarginine, argininic acid, γ-guanidinobutyric acid and arginine was confirmed[6]. The guanidino compound present in the urine, having the highest concen-

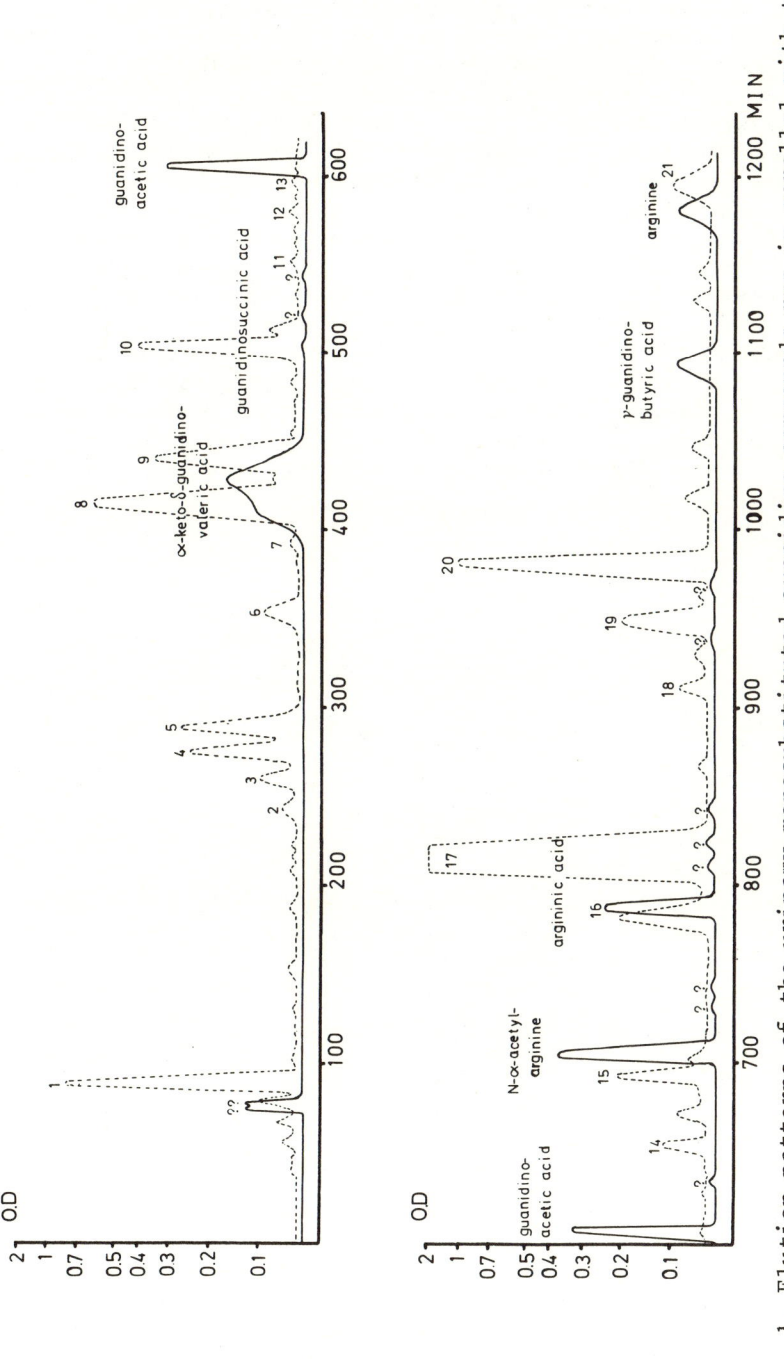

Fig. 1. Elution patterns of the urinary monosubstituted guanidino compounds run in parallel with the amino acids of patients with hyperargininemia.

? = unknown monosubstituted guanidino compounds
1.urea,2.aspartic acid,3.threonine,4.serine,5.glutamine,6.glutamic acid,7.citrulline,8.glycine,9.alanine,10.cystine,11.methionine,12.isoleucine,13.leucine,14.tyrosine,15.β-aminoisobutyric acid,16.ethanolamine,17.ammonia,18.ornithine,19.lysine,20.histidine,21.arginine.

tration has been identified in our laboratory as the keto analogue of arginine, namely α-keto-δ-guanidinovaleric acid[7]. Indeed with liquid chromatography we were able to show that enzymatically synthesized α-keto-δ-guanidinovaleric acid (α-keto-δ-GVA) had the same retention time and the same shape of elution pattern. The shouldered elution peak was caused by the fact that α-keto-δ-GVA exists in equilibrium state with its cyclic form. The identity was confirmed by TLC as seen in figure 2. The first sample applied is a standard mixture of γ-guanidinobutyric acid (γ-GBA) and arginine (Arg) taken as reference; the second is the unknown peak isolated from urine with ε-guanidinocaproic acid (ε-GCA) as internal standard; the third is standard α-keto-δ-GVA and ε-GCA; the fourth is a mixture of the unknown urine fraction and standard α-keto-δ-GVA. No internal standard was added to this mixture. Its identity was also confirmed by GC-MS studies. After acylation of the dimethylpyrimidyl derivative of the unknown peak, isolated from urine, the mass spectrum shown in figure 3 is found. So far we have identified besides α-keto-δ-GVA, all the monosubstituted guanidino compounds present in high concentration.

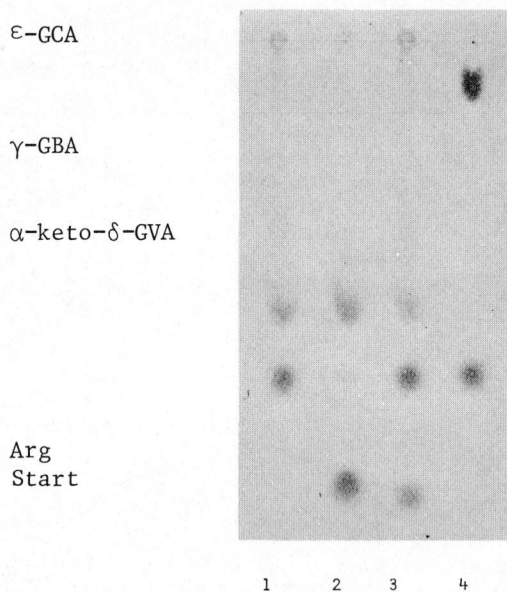

Fig. 2. Thin layer chromatography of :

    1. standard of arginine and γ-guanidinobutyric acid
    2. unknown urinary peak and ε-GCA as internal standard
    3. standard α-keto-δ-GVA and ε-GCA as internal standard
    4. mixture of unknown peak and standard α-keto-δ-GVA
    Solvent: n-propanol/1M acetic acid (150 ml + 50 ml)

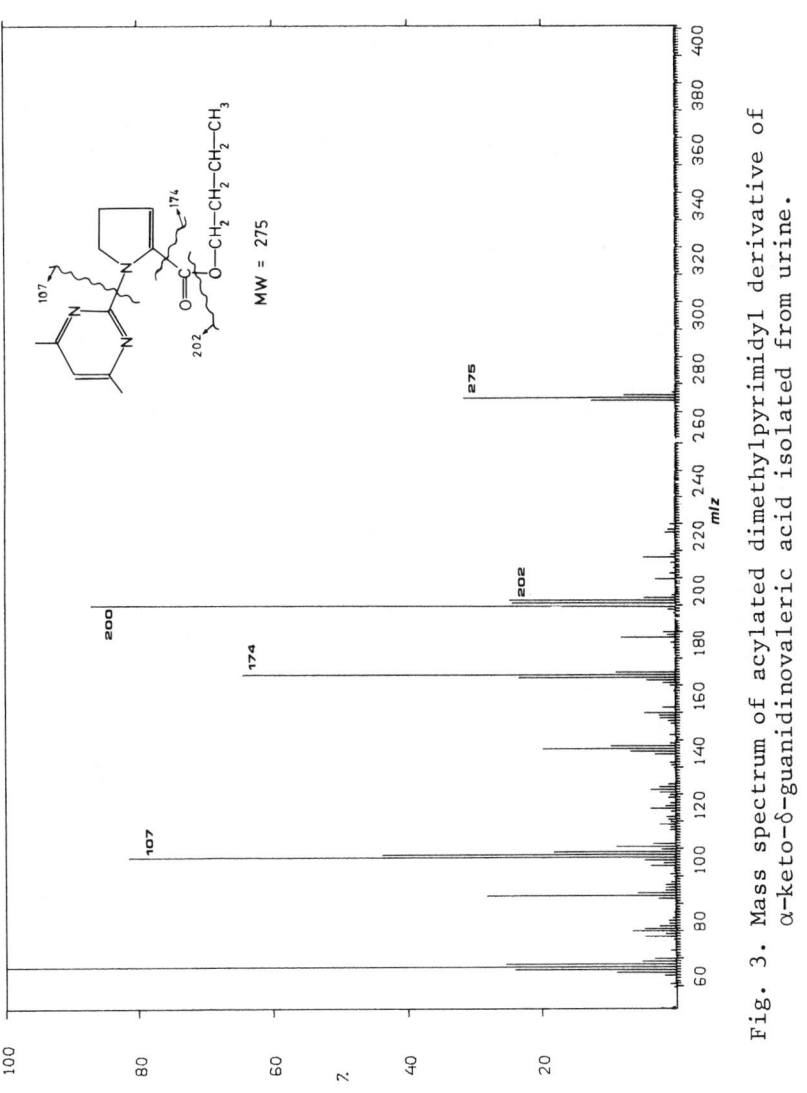

Fig. 3. Mass spectrum of acylated dimethylpyrimidyl derivative of α-keto-δ-guanidinovaleric acid isolated from urine.

Table 3. Excretion values of some monosubstituted guanidino compounds in urine of patients with hyperargininemia (μmol/g. creatinine), after low protein diet.

	controls n=8	A.W.	M.W.	I.W.
α-keto-δ-GVA[a]	traces	2228	6410	904
GAA[b]	182-1213	745	2764	1182
N-α-AA[c]	19-50	734	4271	649
Arg A[d]	3-36	355	1069	61
γ-GBA[e]	traces-31	123	233	19
GSA[f]	14-97	traces	traces	traces
Arg[g]	10-73	88	127	497

a = α-keto-δ-guanidinovaleric acid, b = guanidinoacetic acid, c = N-α-acetylarginine, d = argininic acid, e = γ-guanidinobutyric acid, f = guanidinosuccinic acid, g = arginine.

Table 4. Excretion values of guanidinosuccinic acid in urine of patients with a urea cycle disease or hyperammonemia (μmol/g. creatinine).

cases	guanidinosuccinic acid
ornithine carbamyl transferase deficiency (n=2)	-
citrullinemia (n=4)	-
argininosuccinic aciduria (n=3)	-
argininemia (n=4)	-
HHH syndrome[a] (n=1)	-
transient hyperammonemia (n=1)	-
controls	traces-100

a = patient with hyperammonemia, hyperornithinemia and homocitrullinuria.

URINARY EXCRETION VALUES OF MONOSUBSTITUTED GUANIDINO COMPOUNDS

The excretion values of some monosubstituted guanidino compounds are given in Table 3. The excretion values for the patients with hyperargininemia are 10 to 100 times higher than in controls, except for guanidinosuccinic acid. The values for guanidinosuccinic acid in patients with hyperargininemia are even lower than in controls. The same phenomenon was observed in patients with other urea cycle disorders: low concentrations of guanidinosuccinic acid are excreted compared to controls (Table 4). In this study the detection limit is a critical parameter.

CONCLUSION

To conclude we wish to discuss the physiological rôle played by the guanidino compounds in patients with hyperargininemia. Patients with hyperargininemia are characterized by an interruption of the urea cycle in its last step. Normally nitrogen is excreted as urea: the two nitrogen atoms of urea are derived from arginine. Instead of an arginine hydrolysis, in patients with hyperargininemia, arginine is catabolized along other pathways and apparently nitrogen is partially excreted as guanidino compounds. Arginine can be catabolized by a transamination reaction to form α-keto-δ-GVA. By the oxidative decarboxylation of α-keto-δ-GVA, γ-guanidinobutyric acid can be formed. Reduction of α-keto-δ-GVA, e.g. by a dehydrogenase, can give argininic acid. As an alternative, transamidination of arginine can give guanidinoacetic acid[8] and γ-guanidinobutyric acid[9]. Acetylation of arginine can also give N-α-acetylarginine. The biosynthesis of guanidinosuccinic acid is still unclear. There are however two hypotheses. Some authors say that guanidinosuccinic acid is a catabolite of arginine[10] and that it can be formed by transamidination[11]. Others however believe that guanidinosuccinic acid is synthesized by an alternative pathway[12].

Before we can ascertain that the monosubstituted guanidino compounds are physiological nitrogen waste products in patients with hyperargininemia more knowledge must be acquired about the toxicity of the excreted guanidino compounds. In fact, other amino acid disorders have taught us that the accumulated products may be related to the neurological symptoms. Perhaps the monosubstituted guanidino compounds are also related to the neurological symptoms observed in patients affected with hyperargininemia.

ACKNOWLEDGEMENTS

The authors express their thanks to Dr. Ch. Charpentier (Paris), Dr. J. P. Farriaux (Lille), Dr. T. Grisar (Liège), Dr. A. L. Beaudet (Houston), Dr. M. Yoshino (Kurume) and Dr. B. A.

Gordon (London-Ontario) for kindly providing us with urine samples of their patients.

REFERENCES

1. H. G. Terheggen, A. Schwenk, A. Lowenthal, M. Van Sande and J. P. Colombo. Hyperargininämie mit Arginasedefekt. Eine neue familiäre Stoffwechselstörung. I. Klinische Befunde, Z. Kinderheilk. 107:298 (1970).
2. H. G. Terheggen, A. Schwenk, A. Lowenthal, M. Van Sande and J. P. Colombo, Hyperargininämie mit Arginasedefekt. Eine neue familiäre Stoffwechselstörung. II. Biochemische Untersuchungen, Z. Kinderheilk. 107:313 (1970).
3. H. G. Terheggen, A. Lowenthal, F. Lavinha and J. P. Colombo, Familial hyperargininemia, Arch. Dis. Child. 50:57 (1975).
4. B. Marescau, J. Pintens, A. Lowenthal, H. G. Terheggen and K. Adriaenssens, Arginase and free amino acids in hyperargininemia, J. Clin. Chem. Clin. Biochem. 17:211 (1979).
5. M. L. Efron, Automation in analytical chemistry, in: "Technicon Symposia", p.637, Mediad, New York (1965).
6. B. Marescau, A. Lowenthal, E. Esmans, Y. Luyten, F. Alderweireldt, and H. G. Terheggen, Isolation and identification of some guanidino compounds in urine of patients with hyperargininemia by liquid and thin-layer chromatography and gas chromatography - mass spectrometry, J. of Chromat. 224:184 (1981).
7. B. Marescau, J. Pintens, A. Lowenthal, E. Esmans, Y. Luyten, G. Lemière, R. Domisse, F. Alderweireldt and H. G. Terheggen, Isolation and identification of 2-oxo-5-guanidinovaleric acid in urine of patients with hyperargininemia by chromatography and gas chromatography - mass spectrometry, J. Clin. Chem. Clin. Biochem. 19:61 (1981).
8. H. Dubnoff and H. Borsook, A micromethod for the determination of glycocyamine in biological fluids and tissue extracts, J. Biol. Chem. 138:381 (1941).
9. J. Pisano, D. Abraham and S. Udenfriend, Biosynthesis and disposition of γ-guanidinobutyric acid in mammalian tissues, Arch. Biochem. Biophys. 100:323 (1963).
10. G. Perez, A. Rey and E. Schiff, The biosynthesis of guanidinosuccinic acid by perfused rat liver, J. Clin. Invest. 57:807 (1976).
11. B. D. Cohen, Guanidinosuccinic acid in uremia, Arch. Int. Med. 126:846 (1970).
12. S. Natelson and J. E. Sherwin, Proposed mechanism for urea nitrogen re-utilization: Relationship between urea and proposed guanidine cycles, Clin. Chem. 25:1343 (1979).

GUANIDINOSUCCINIC ACID AND THE ALTERNATE UREA CYCLE

B.D. Cohen and H. Patel

Bronx-Lebanon Hospital
1276 Fulton Avenue - Bronx, N.Y. 10456, U.S.A.

"What seems to account for the general derangement and suffering (in uremia) is the fact that urea ... or the elements of which it is formed are abundant in the blood". Richard Bright, 1833.

Disturbance in nitrogen balance is the hallmark of renal insufficiency. Unlike sodium, phosphate and the myriad other minerals which the kidney regulates and balances, nitrogen cannot be reutilized by animal organisms and is restored to the environment via the urine without mechanisms for reabsorption (except for the small quantity of amino acids which are filtered and can be reutilized). Reduced filtration leads to azotemia and sets in motion a variety of hormonally regulated adaptions.

Several years ago Bricker (1) proposed a hormonal hypothesis as justification for some of the bizarre compensatory phenomena observed in uremia. Phosphate is the prototype, the retention of which leads to increases in circulating parathormone which results, then, in reduced renal tubular reabsorption and restoration of balance. The hormone compensates for reduced filtration by encouraging tubular rejection. There is no tubular reabsorption of nitrogen, however, so balance, if it is to be achieved, must be accomplished via increased storage.

Some twenty-five years ago it was shown, by incubating rat liver slices in plasma from uremic animals, that urea formation increases in renal insufficiency (2). Concurrently, several investigators reported increases in hepatic glucose production in uremia (3) leading to the condition known as azotemic pseudodiabetes.

More recently these observations were confirmed and vindicated by the finding of increased serum glucagon in uremia (4).

Increasing urea generation in uremia can scarcely be considered an adaption to nitrogen retention, most of which occurs in the form of urea. Except for a small quantity which is degraded by bacterial ureases, a process shown to be unaltered in uremia, the bulk of urea is presumably synthesized solely for the elimination of excess nitrogen and enters into no further biochemical activities.

Recently, Natelson and Sherwin (5) have challenged this hypothesis with the introduction of the canaline or guanidine cycle, a pathway for the conversion of urea nitrogen to the more stable storage form, creatinine. The pathway begins via the enzymatic oxidation of urea to hydroxyurea, a reaction which is known to occur in reverse (6). It proceeds by a series of steps analogous to the urea cycle to produce creatine from canavanine with guanidinosuccinic acid (GSA) as an inert, overflow byproduct.

If this is the case then GSA can be used as a marker measuring the activity of the guanidine cycle and should increase in the body fluids whenever there is increased urea generation, such as uremia, and decrease in states where urea production is inhibited. Several years ago we showed increased GSA excretion in uremia, prerenal azotemia and high protein feeding and disappearance of urinary GSA in the presence of inborn errors of urea production (7).

As for the reactivity of GSA, we have lately reported data indicating that it circulates unbound to plasma protein, concentrates intrahepatically, does not diffuse intracellularly or react with cytosol protein or enzymes and is excreted wholly via filtration (8). Injected intravenously it disappears from the circulation of normal rats in 2 hours. Instilled intraperitoneally it clears within 24 hours requiring 4-5 days to be wholly recovered from the urine. Following bilateral nephrectomy the serum concentration of an injected load levels off at a concentration consistent with an extracellular distribution without degradation or metabolism.

In the data which follows the 24-hour GSA excretion for each animal is calculated per mg of creatinine. Each figure recorded represents the mean ratio of the excretion by injected animals compared to simultaneously collected controls, i.e. a ratio of 2.0 means that the treated rats averaged twice the GSA excretion of untreated controls.

The measurements shown in Table I confirm urea as a source of GSA and, therefore, support the Natelson proposal. Urea increases the production of GSA in proportion to the amount injected

Table I. Excess GSA excretion following intraperitoneal instilllation of the listed substances in the quantities indicated in mM.

Hydroxyurea 5	(24)	1.1±0.3
Urea 10	(36)	2.2±0.6
Canavanine 1.25	(8)	2.2±0.5
Hydroxyurea 10	(20)	2.4±0.7
Urea 10 hydroxyurea 5	(32)	3.1±0.9
Urea 10 hydroxyurea 10	(12)	4.4±1.5
Urea 20	(40)	4.3±1.4
Urea 10 aspartate 1.25	(26)	2.5±0.7
Urea 10 homoserine 1.25	(32)	2.5±1.4
Urea 10 methionine 1.25	(12)	0.8±0.1
Urea 20 methionine 1.25	(9)	2.5±1.6

Figures in parentheses indicate the number of rats.

and hydroxyurea appears to be an intermediate. Small quantities of hydroxyurea do not increase GSA synthesis while larger amounts affect GSA production to the same extent as urea. This suggests that the equilibrium between hydroxyurea and urea favors urea formation as previously reported (6).

On the other hand, when urea levels are concurrently increased, such as when both are injected (or in uremia), the oxidation reaction proceeds increasing the incorporation of exogenous hydroxyurea into GSA. The oxidation reaction, therefore, appears to have slower kinetics and accelerates when urea concentration increases in the body fluids such as in uremia.

Aspartate and homoserine, while participants in the guanidine cycle, do not increase GSA output and, therefore, do not enter into rate-limiting steps in the biosynthetic pathway yielding GSA as a

Table II. Ratio of GSA and urea output of hormone-treated and control rats.

	GSA	Urea
Prolactin (28)	1.02±0.27	0.95±0.13
Glucagon (52)	0.99±0.25	1.13±0.52
Glucagon + insulin (60)	1.19±0.39	1.23±0.28
Glucagon + cortisone (32)	1.40±0.50	1.25±0.21
Urea (84)	1.90±0.45	2.42±0.72

byproduct. Canavanine, the immediate precursor of GSA, has an effect comparable to that of urea. Methionine, which is not an intermediate in the guanidine cycle and does not directly react with GSA (8), reduces GSA excretion. One possible explanation relates to the pH optimum of the canavaninosuccinic acid reductase reaction which is 8.7 compared to that of 6.5 for the CSA lyase. A reduced pH resulting from the excess of sulfur-bearing amino acid would favor, therefore, conversion to canavanine rather than GSA.

The quantity of urea injected produces a blood urea nitrogen concentration in the rats which is comparable to that seen in uremia (mean of 9 injections = 160 mg/dl) returning to preinjection levels at 24 hours. Calculations of the average daily urea output in 32 injected animals compared to an equal number of controls indicates that, of 290 mg urea nitrogen injected, 190 mg is recovered leaving 100 mg unaccounted for and possibly shunted into the guanidine cycle (the amount of this detoured into GSA is negligibly small being on the average less than 0.1 mg).

Table II shows the urinary GSA and urea excess output in response to the injection of a variety of hormones which both alter nitrogen metabolism and increase in uremia. The insulin, prolactin and urea were injected intraperitoneally. Glucagon was given subcutaneously to half the animals and intraperitoneally to the other half following four days of subcutaneous injections of cortisone to induce a catabolic state after the method of

THE ALTERNATE UREA CYCLE

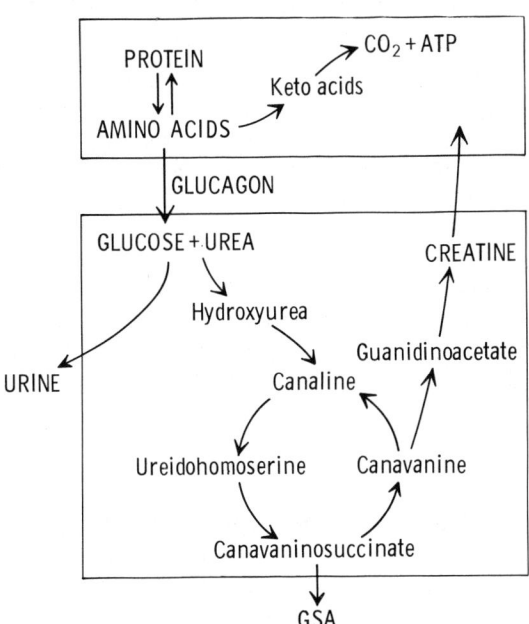

Figure 1. The upper box signifies muscle, the lower box is liver.

McClean and Gurney (9). The significant increase following glucagon presumably results from induction of the enzymes of the guanidine cycle comparable to that induced by glucagon on the urea cycle. Blood studies show a prompt and reversible rise in blood sugar and urea following intraperitoneal glucagon.

Figure 1 summarizes the proposed cycle. Where urine output of urea is blocked, a larger share is shunted into creatine and muscle storage. Figure 2 compares the two hormonally induced adaptive mechanisms for regulating nitrogen and phosphate in states of reduced renal glomerular filtration.

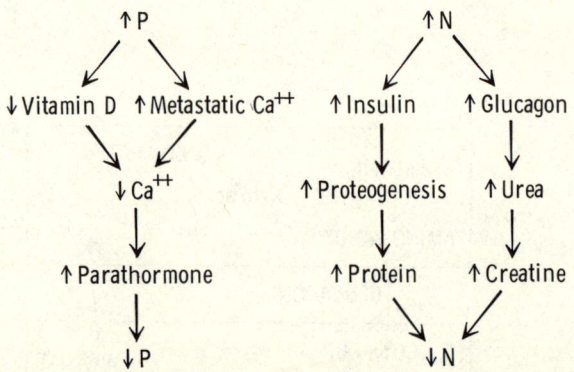

Figure 2. A schematic diagram indicating the role of hormones in achieving phosphate and nitrogen balance in uremia.

Apart from the startling proposition that there exist circumstances in which urea is not biochemically inert, the guanidine cycle can serve an useful practical function in the management of uremia. Much is made of the possibility of nitrogen assuming toxic forms in uremia and considerable interest is generated in an effort to reduce intake without compromise to nutrition. The guanidine cycle may represent a physiologic adaption to this need which is only partially successful since its prime function is the conservation of nitrogen in situations of dietary deficit. If this is so and if we can manipulate urea oxidation via hormones such as glucagon, excess labile nitrogen in uremia could be stored as muscle protein or creatine.

REFERENCES

1. N. S. Bricker, and L. G. Fine, The Trade-off hypothesis : current status. Kidney Int. 13:55 (1978).
2. A. L. Sellers, J. Katz, and J. Marmorston : Effect of bilateral nephrectomy on urea formation in rat liver slices. Am. J. Physiol. 191:345 (1957).
3. B. D. Cohen, Abnormal carbohydrate metabolism in renal disease. Ann. Intern. Med. 57:204 (1962).

4. G. L. Bilbrey, G. R. Faloona, M. G. White and J. P. Kochel, Hyperglucagonemia of renal failure. J. Clin. Invest. 53:841 (1974).
5. S. Natelson, J.E. Sherwin, Proposed mechanism for urea nitrogen re-utilization : relationship between urea and proposed guanidine cycles. Clin. Chem. 25:1343 (1979).
6. M. Colvin, and F. H. Bono, Jr., The enzymatic reduction of hydroxyurea to urea by mouse liver. Cancer Res. 30:1516 (1970).
7. I. M. Stein, B. D. Cohen, and R. S. Kornhauser, Guanidinosuccinic acid in renal failure, experimental azotemia and inborn errors of the urea cycle, N. Engl. J. Med. 280:926 (1926).
8. H. Patel, and B. D. Cohen, The guanidine cycle, (In press).
9. P. McClean, and M. W. Gurney, Effect of adrenalectomy and of growth hormone on enzymes concerned with urea synthesis in rat liver. Biochem. J. 87:96 (1963).

GUANIDINOSUCCINIC ACID EXCRETION IN ARGININOSUCCINIC ACIDURIA

H. Böhles , B.D. Cohen[*] , D. Michalk

Universitätskinderklinik, Erlangen, F.R. Germany, and
Bronx Lebanon Hospital[*], New York, U.S.A.

Guanidinosuccinic acid (GSA) is a normally occurring metabolite in serum and urine of human beings. An intact pathway of urea formation seems to be necessary for its formation because from different inborn defects of urea synthesis the excretion of only minute amounts of GSA is reported (Stein et al. 1969). States of nitrogen retention, especially renal failure, show increased concentrations of GSA in serum and urine (Stein et al. 1969). We want to present our data on the urinary GSA excretion in a patient with argininosuccinic aciduria during the first two weeks after the introduction of ketoanalogues of essential amino acids into the dietary regime.

The patient was a 22 year old female. She was reported as being entirely well until 6 years of age, when she had "liver swelling and yellow jaundice" of several weeks duration. According to the mother the patient never totally recovered. She avoided meat, milk and dairy products, complaining that these foods produced nausea and put herself on a vegetarian diet.
At the age of 12 years she began to have psychomotor seizures, sometimes occurring several times a day. Her IQ at the time of admission was 56. Despite of a daily dose of 900 mg 2-propylvalerianic acid and 350 mg diphenylhydantoine her seizures occurred almost daily and sometimes several times a day. In view of the striking history of protein intolerance in this patient an inborn error of the urea cycle was suspected. This possibility was confirmed by thin layer and column chromatography of plasma and urine, which revealed metabolites with chromatographically identical characteristics to pure argininosuccinic acid and its anhydrides. The diagnosis of argininosuccinic acid lyase deficiency was proved

by enzymatic assay on the patients erythrocytes, which showed an activity of about 10 % of that found in 5 healthy females of similar age.

Protein handling was evaluated in a tolerance test consisting of the administration of 1 g/kg protein (total 50 g). The protein was supplemented with 2 g L-arginine. The plasma ammonia level rose to about 600 µg/dl after 5 hours. The GSA excretion before loading was low (1.93 mg/24hrs.) and did not differ significantly from that after the load (1.43 mg/24 hrs.). This is different from Dr. Cohen's observation in 7 normal subjects on a 125 g protein diet for 3 days. GSA excretion rose from 10.3 ± 5.1 mg/24 hrs. to 58.7 ± 37.9 mg/24 hrs.

After a 6 day baseline period on a low protein diet (0.5 g/kg/day) about 14 g of analogues of 5 essential amino acids were administered along with the other constituents listed in table 1. The supplement was given in 6 equal portions every 2 hours during daytime. Plasma ammonia levels of initially 90 µg/dl fell during the supplementation period to about 40 µg/dl (Fig. 1).

As major abnormalities in plasma amino acids in argininosuccinic aciduria are described elevations of argininosuccinic acid, glutamine, alanine and glycine, while the branched chain amino acids (leucine, isoleucine and valine) are low. As a result of the supplementation with ketoanalogues of essential amino acids changes of various plasma amino acids occurred. Except for argininosuccinic acid, the concentrations of glutamine, alanine and glycine fell, while the branched chain amino acids rose (Figs. 2, 3). In contrast to the findings in plasma, the argininosuccinic acid excretion fell from about 18 000 µmoles/24 hours to 6000 µmoles/24 hours. Before the therapeutic approach with ketoanalogues of essential amino acids the GSA excretion was low of about 2 mg/24 hrs. Immediately after introduction of the supplement there was a transient increase of GSA excretion up to normal values (8 mg/24 hrs.) which lasted only about 4 days (Fig. 4). Possible errors in urine collection were excluded. This transient increase of GSA excretion was also present when related to creatinine (130 µmoles/g creatinine).

Our data show that the mean GSA excretion is low throughout and comparable to the values obtained in other subjects with inborn errors of the urea cycle (Stein et al. 1969). There appears to be no effect of protein loading unlike in normal controls, who increase promptly the production of GSA on protein feeding. Also a supplementation with arginine had no effect on GSA excretion. This is in accordance to the data from Colombo et al. (1976), who showed that in a child with argininemia no measurable amount of GSA was detected in spite of a high arginine concentration and even after a load with arginine. Hence arginine does not seem to be primarily

Table 1. Composition of the ketoanalogue supplement.

COMPOSITION

Administered per 24 hours:

Amino acid analogues: CO-Valine-Ca CO-Leucine-Ca CO-Isoleucine-Ca CO-Phenylalanine-Ca OH-Methionine-Ca	13.98 gm
Amino acids: L-Threonine L-Thryptophane L-Lysine L-Histidine	2.16 gm
Carbohydrates: Xylitol Sorbitol Sucrose Starch	38.4 gm
Others, including water:	3.06 gm

Energy supply from the supplement:

Analogues	49.2 Cal
Amino acids	10.2 Cal
Carbohydrates	153.6 Cal
	213.0 Cal

Nitrogen intake from amino acids:	388.2 mg

related to GSA formation. We cannot confirm the idea presented by Menyhart et al. (1975) that an elevated argininosuccinic acid concentration might activate a new enzyme converting argininosuccinic acid into ornithine and GSA, as argininosuccinic aciduria should be the ideal test model.

The identification of GSA in the urine from normal humans and animals indicates that it is a product from normal metabolic

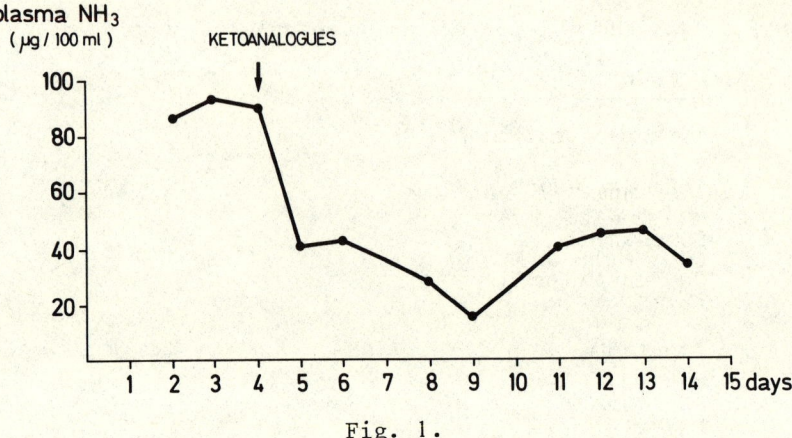

Fig. 1.

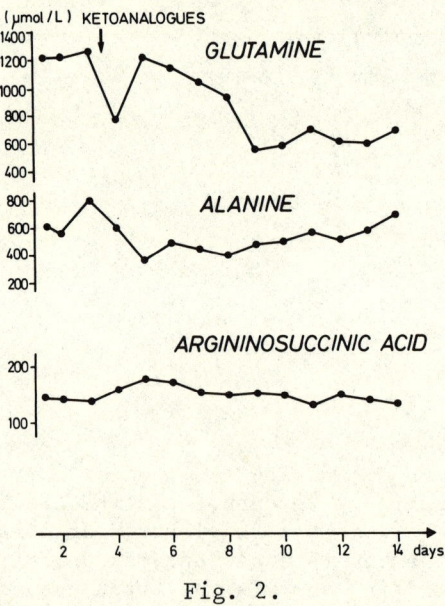

Fig. 2.

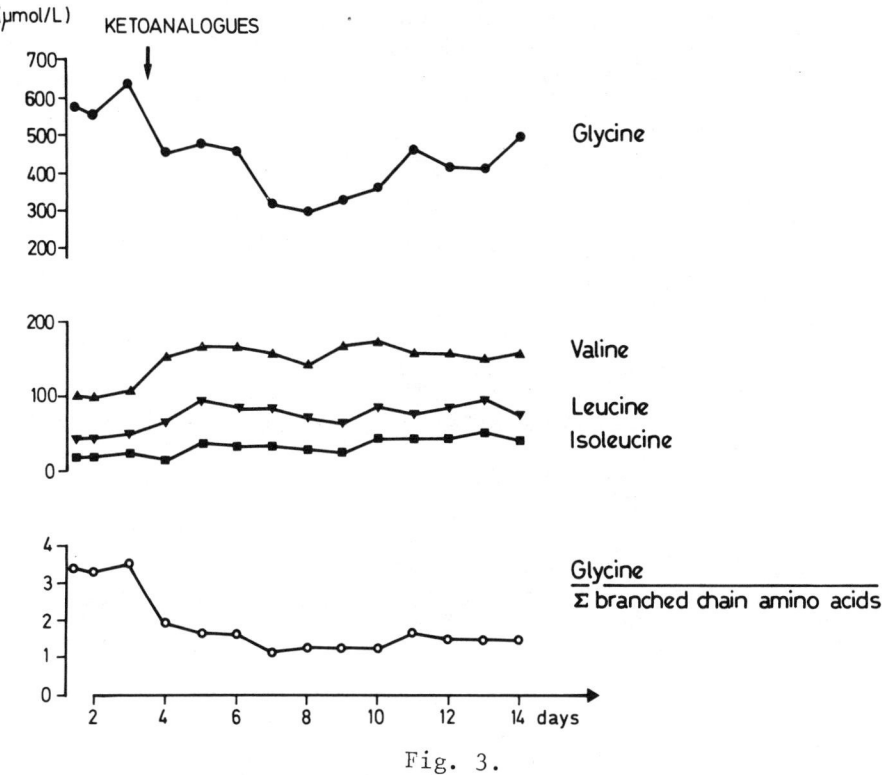

Fig. 3.

activity. Elevated levels in both, urine and serum in states of N-retention as demonstrated by a protein load on one hand or by renal insufficiency on the other (Kopple 1971) suggest an increased rate of GSA synthesis. Dobbelstein et al. (1971) could demonstrate that de novo synthesis occurs and increased levels are not derived from the urea molecule. The increased GSA excretion after the introduction of the ketoanalogue supplement thus might reflect increased nitrogen retention and anabolism. As ketoanalogues of essential amino acids may not only be transaminated to their corresponding amino acid, but also undergo decarboxylation, this described effect may be a consequence of rapidly adapting metabolic processes. Threshold concentrations of metabolites to trigger GSA formation may have been reached only transiently.

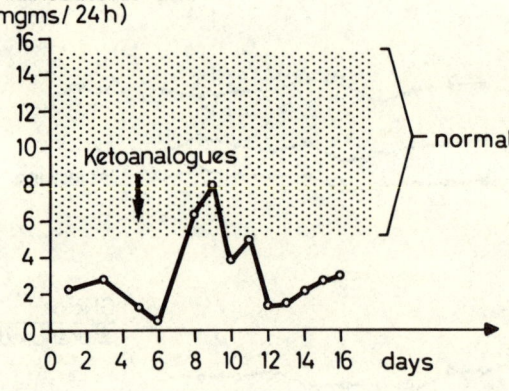

Fig. 4.

REFERENCES

Colombo, J.P., Bachman, C., Terheggen, H.G., Lavinha, F., Lowenthal, A., 1976, Argininemia, in: "The Urea Cycle", S. Grisolia, R. Baguena, F. Mayor, eds., John Wiley & Sons, New York.

Dobbelstein, H., Edel, H.H., Schmidt, M., Schubert, G., Weinzierl, M., 1971, Guanidinobernsteinsäure und Urämie, Klin. W'schr. 49:348.

Kopple, J.D., Gordon, S., Wang, M., 1971, Effect of chronic uremia protein intake and hemodialysis on guanidinosuccinic acid levels, Abstracts of Am. Soc. Nephrol. 5:41.

Menyhart, J., Grof, J., Somogyi, J. Abstracts, 9th Ann. Meeting Soc. Clin. Invest. Rotterdam 1975.

Stein, J.M., Cohen, B.D., Kornhauser, R.S. 1969, Guanidinosuccinic acid in renal failure, experimental azotemia and inborn errors of the urea cycle, New Engl. J. Med. 280:926.

METABOLIC PATHWAY OF GUANIDINO COMPOUNDS

IN CHRONIC RENAL FAILURE

Hiroshi Mikami, Yoshimasa Orita, Akio Ando, Masamitsu
Fujii, Takeo Kikuchi, Kazuo Yoshihara, Akira Okada, and
Hiroshi Abe

The first Department of Medicine, Osaka University
Medical School
1-1-50 Fukushima, Fukushima-ku, Osaka, 553 Japan

INTRODUCTION

Guanidino compounds are considered to be one of the metabolic products of protein catabolism. Serum concentrations of some guanidino compounds, especially methylguanidine (MG)[1] and guanidinosuccinic acid (GSA)[2] are increased in the uremic state. Recently, MG and GSA were found to have uremic toxicity in vitro[3]. Giovanetti[4], et al., for example, brought out many uremic symptoms in normal dogs by repeatedly injecting MG until the MG concentration reached the uremic state. Cohen[5] proposed the metabolic pathway of MG. He suggested that the metabolic origin of MG is mainly creatine and partially of arginine (Arg). The present authors have also suggested that there might be two metabolic origins of MG[6]. One is possibly Arg itself (or the metabolites of arginine); the other is creatinine (Cr). The present study aims to clarify the metabolic precursor of guanidino compounds, especially MG (Fig.1).

MATERIALS AND METHODS

Male Sprague-Dawley rats weighing 200-250g were introduced to the experimental uremic state using the procedure of Platt[7]. After the operation, the rats were fed with ordinary laboratory chow until their serum Cr level reached 2 mg/dl (mean 28 days). The control rats were chosen to match in weight to the experimental rats. The normal and uremic rats were kept in a continuous intravenous hyperalimentation system without oral intake. The basal solution of intravenous nutrients contains 2.13g of amino acids mixture, vitamins

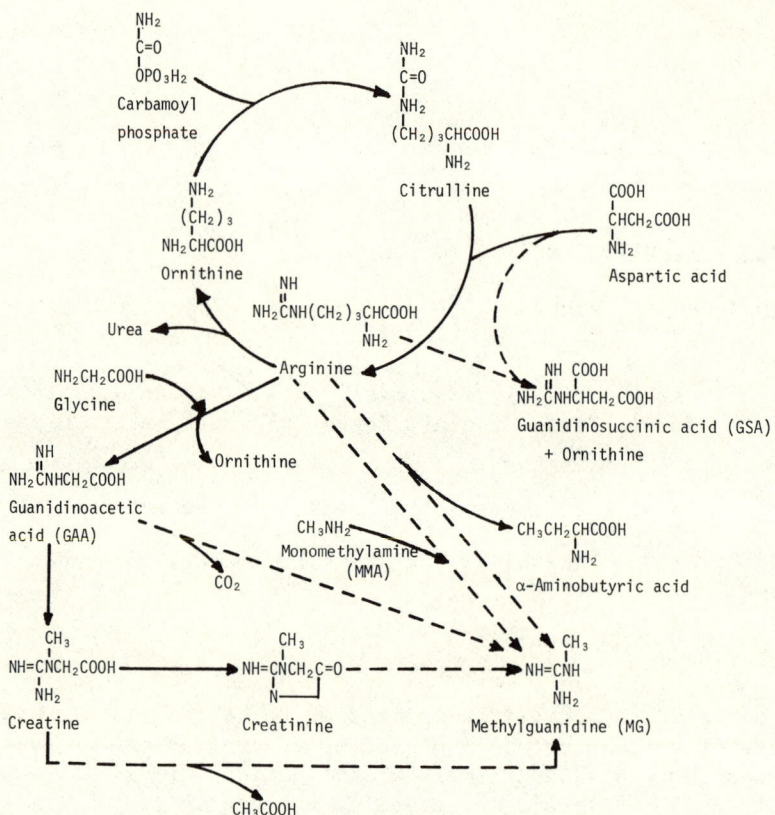

Fig.1. Our speculated pathways for the production of guanidino compounds and the relationship to the urea cycle.

and trace elements with 50 ml volume per day (50 Cal). Amino acid composition was adjusted to maintain the constant body weight of all the rats. The rats were kept in individual metabolic cages and given the basal nutrient solution for three days (Period I). However, the first day of the period I, the rats were given only half volume of the solution to avoid overhydration. After the period I, the rats were supplied with the basal nutrient solution mixed with one of the six different solutions that contain one of the possible metabolic affecting substances of guanidino compounds (Arg, Arg + monomethylamine, guanidinoacetic acid, creatine, Cr and urea) for four days (Period II). A dose of Cr or urea was given, so that the serum concentrations of the normal rats would reach the uremic state (Cr, 5 mmole/day; urea, 25 mmole/day). The molar dose of Arg or creatine or Arg + MMA was made equal to that of the Cr. However, the dose of guanidinoacetic acid (GAA) was lower because of its low water solubility (2.0 mmole/day). Urine was collected at the end of period I and at the end of period II. At the end of period II, blood samples were withdrawn from the abdominal aorta, and we removed the liver and the femoral muscle; all for the determination of guanidino compounds. Guanidino compounds were determined with Automated Guanidine Analyzer G-520 (Japan Spectroscopic Co., Tokyo, Japan)[8,9].

RESULT

Table 1 showed the serum concentration of urea nitrogen and Cr in normal and uremic rats infused with six different solutions which contain possible metabolic affecting substances. In the urea infused normal rats, BUN was increased to the uremic level (52.8 ± 1.7 mg/dl). In the MMA + Arg infused normal rats, the BUN level became higher (78.1 ± 22.8 mg/dl). Serum Cr and BUN remained unchanged when each of the four kinds of solutions (Arg, GAA, Arg + MMA, creatine) were infused in normal rats respectively. In the creatine infused uremic rats, the serum Cr concentration increased to 10.6 mg/dl. Table 2 shows the serum concentrations of guanidino compounds after the infusion of the six different kinds of solutions

Table 1. Effects of intravenous infusion of several substances on body weight and blood chemistry in normal and uremic rats

Rat used	Substance Infused	Rat No.	Body Weight Increase (g)	BUN (mg/dl)	s-Cr (mg/dl)
Normal	Control	5	-7.0 ± 8.6	25.0 ± 3.4	0.73 ± 0.05
	Arginine	6	-3.3 ± 2.8	34.5 ± 5.6	1.03 ± 0.09
	GAA	5	-6.7 ± 1.7	30.0 ± 4.9	0.97 ± 0.13
	Creatine	4	-11.5 ± 1.3	20.1 ± 5.6	1.11 ± 0.40
	Creatinine	5	-5.6 ± 2.3	22.4 ± 3.2	9.98 ± 1.21
	MMA+Arg	3	-6.7 ± 4.4	78.1 ± 22.8	0.76 ± 0.20
	Urea	4	-3.0 ± 1.8	52.8 ± 1.7	0.87 ± 0.25
Uremic	Control	5	-8.3 ± 2.9	75 ± 25	2.0 ± 0.8
	Arginine	4	-33.6 ± 12.1	139 ± 16	2.0 ± 0.3
	GAA	4	-10.8 ± 8.7	119 ± 62	3.0 ± 2.0
	Creatine	3	-8.3 ± 2.9	82 ± 52	10.6 ± 1.6
	Creatinine	3	-8.3 ± 2.9	112 ± 31	28.7 ± 13.2
	MMA+Arg	3	-23.3 ± 8.8	117 ± 34	2.9 ± 1.1
	Urea	4	-28.8 ± 15.5	243 ± 33	2.0 ± 0.5

Mean ± S.E.

Table 2. Effects of intravenous infusion of several substances on serum concentration of guanidino compounds in normal and uremic rats.

Rat used	Substance Infused	Rat No.	GSA	GAA	MG
Normal	Control	5	3.6 ± 2.2	51.4 ± 14.4	n.d.
	Arginine	6	12.5 ± 6.7	221.8 ± 27.8	0.03 ± 0.03
	GAA	6	n.d.	/	n.d.
	Creatine	4	n.d.	42.9 ± 11.1	n.d.
	Creatinine	5	n.d.	68.9 ± 10.2	2.40 ± 0.68
	MMA+Arg	3	41.0 ± 16.6	81.0 ± 8.5	n.d.
	Urea	4	31.8 ± 15.3	124.2 ± 18.6	n.d.
Uremic	Control	5	45.0 ± 33.9	51.6 ± 26.4	2.5 ± 2.4
	Arginine	4	167.6 ± 88.4	114.5 ± 32.7	2.7 ± 0.5
	GAA	4	107.9 ± 168.0	/	8.2 ± 11.8
	Creatine	3	245.1 ± 378.9	49.4 ± 34.4	16.6 ± 13.6
	Creatinine	3	123.4 ± 129.1	64.4 ± 61.5	31.6 ± 18.2
	MMA+Arg	3	289.3 ± 155.5	50.4 ± 20.3	6.0 ± 3.2
	Urea	4	438.1 ± 285.0	43.4 ± 37.6	1.7 ± 1.0

n.d.=not detectable ; / = not measured ; Mean ± S.E., (µg/dl)

in normal and uremic rats. The serum GSA levels rose in urea infused (normal rats, 31.8 ± 15.3 µg/dl; uremic rats, 438.1 ± 285.0 µg/dl) and Arg + MMA infused rats (normal rats, 41.0 ± 16.6 µg/dl; uremic rats, 289.3 ± 155.5 µg/dl). The serum GAA levels rose in the Arg infused rats (normal rats, 221.8 ± 27.8 µg/dl; uremic rats, 114.5 ± 32.8 µg/dl). Serum MG concentrations increased remarkably in the Cr infused normal (2.40 ± 0.68 µg/dl) and uremic rats (31.6 ± 18.2 µg/dl). On the other hand, creatine infusion resulted in an elevation of serum MG levels in the uremic rats only. The Cr or creatine infusion increased urinary excretions of MG in the normal and uremic rats (Fig.2). In the normal rats, urinary MG excretions increased 100-fold with the Cr infusion and 8-fold with the creatine infusion. Cr infusion caused the most remarkable increase in urinary MG excretions in both normal and uremic rats. Creatine has also the same effect, but less than Cr. Urinary excretions of GSA in uremic rats were higher than in normal rats, and the results were comparable with MG (Fig.3). A remarkable increase of urinary GSA excretions was observed in both normal and uremic rats with the use of urea infusion. In the Cr infused normal and uremic rats, urinary excretion of GSA also tended to increase. Urinary excretions of GAA in normal rats were higher than in the uremic rats. In the Arg infused uremic rats, GAA excretions tended to increase slightly.

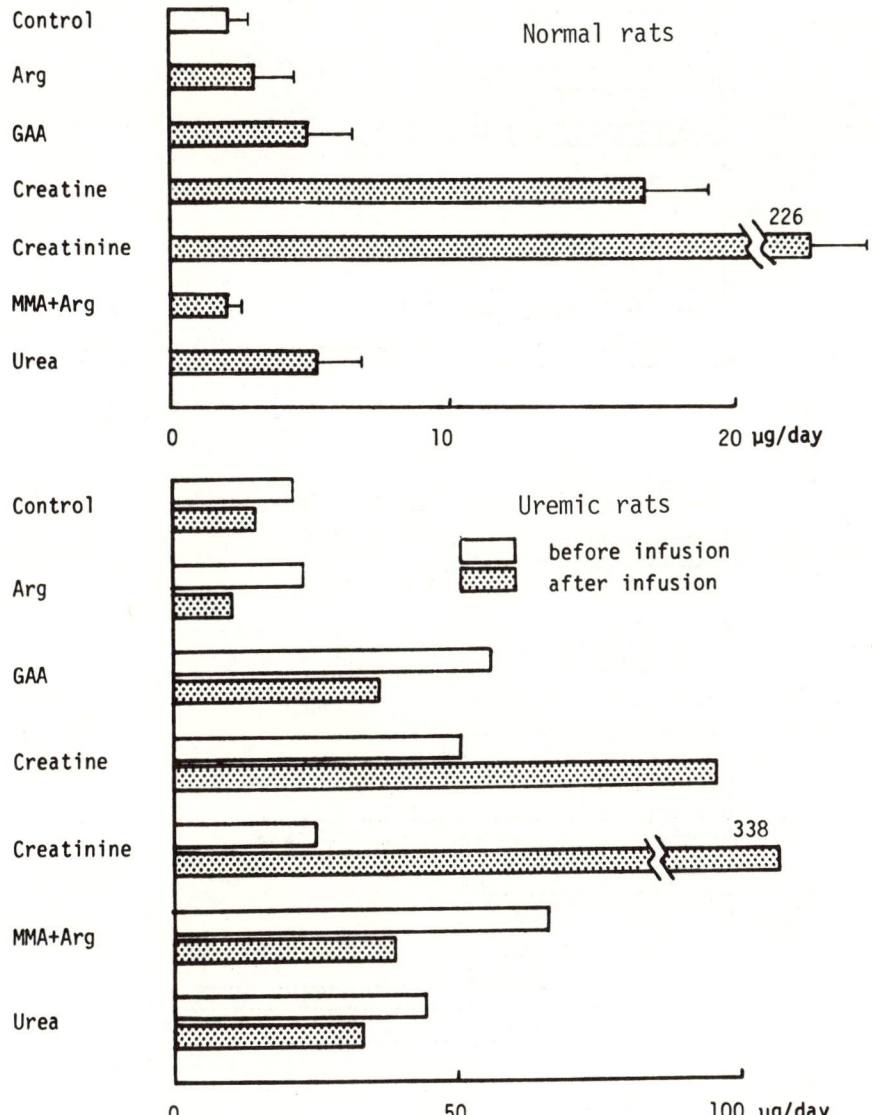

Fig.2 Effects of intravenous infusion of several substances on urinary excretion of methylguanidine.

There were less urinary excretions of guanidine in the uremic rats than in the normal rats. The urinary excretion of guanidine increased remarkably only in the GAA infused normal and uremic rats (Fig.4). Table 3 shows tissue concentration of guanidino compounds. Generally, GSA concentration in the liver was higher than in the muscle in both normal and uremic rats, but MG concentration in the muscle tended to be higher than in the liver. Tissue concentration

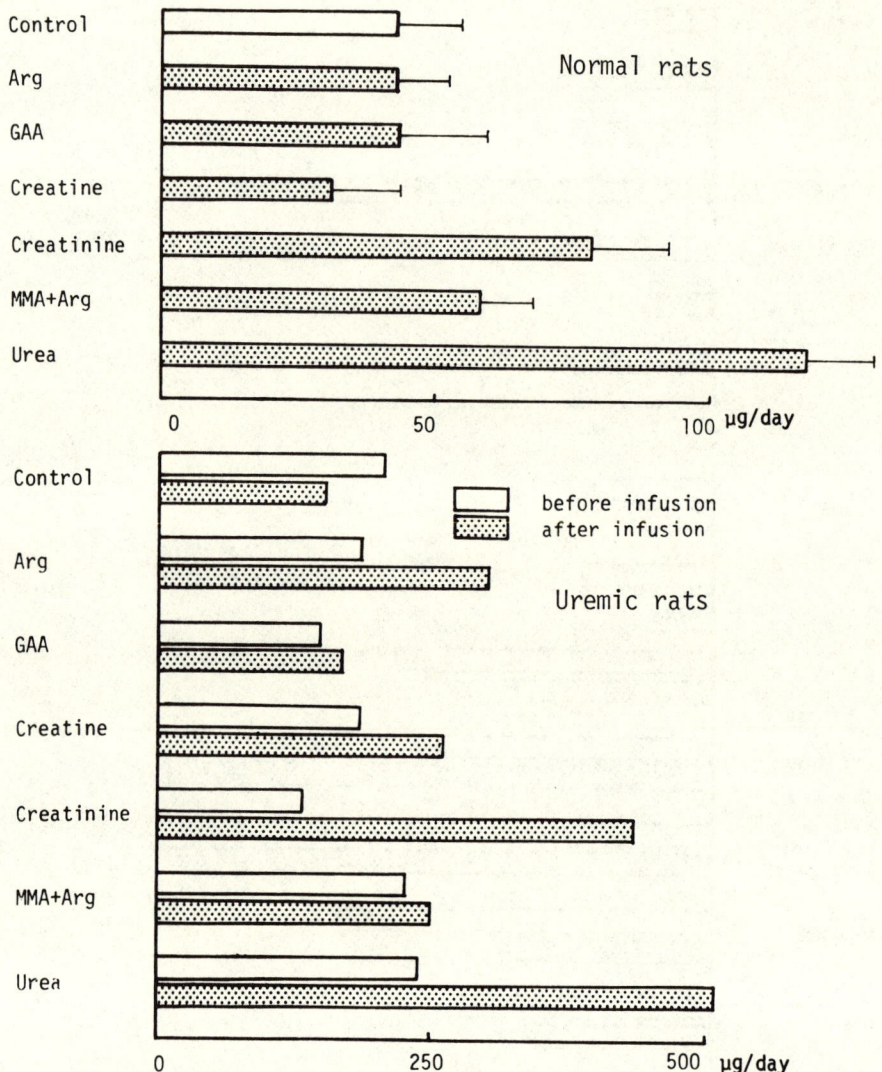

Fig.3 Effects of intravenous infusion of several substances on urinary excretion of guanidinosuccinic acid.

of GAA and guanidine were only negligibly different in the liver and in the muscle. A remarkable increase in the liver GSA concentration was observed only in the urea infused uremic rats. In the Cr infused uremic rats, MG concentration in the liver was exactly the same as in the muscle.

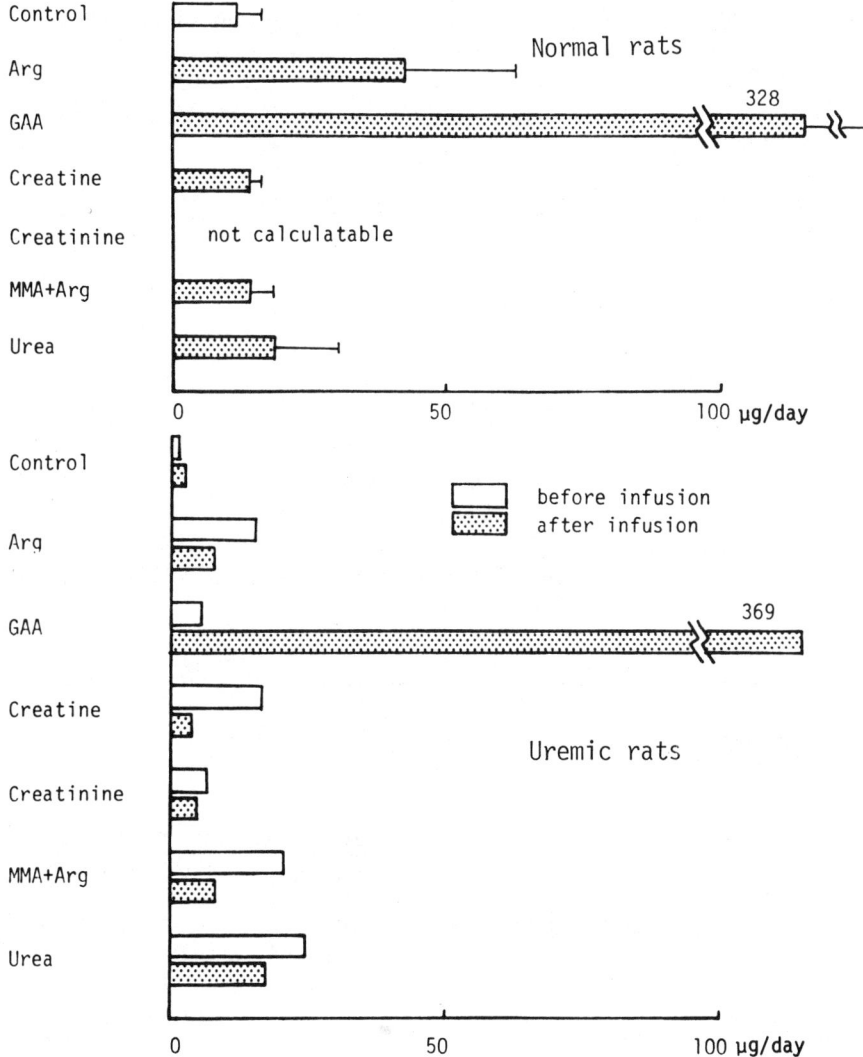

Fig.4 Effects of intravenous infusion of several substances on urinary excretion of guanidine.

DISCUSSION

The study of the metabolic pathway of guanidino compounds, especially MG and GSA, appears to be very important in the control of patients with renal failure. This is largely because MG and GSA are considered to be important uremic toxins. Urinary excretion of

Table 3 Effects of intravenous infusion of several substances on tissue concentration of guanidino compounds in normal and uremic rats.

Rat used	Substance Infused	MG Liver	MG Muscle	GSA Liver	GSA Muscle	GAA Liver	GAA Muscle
Normal	Control	n.d.	0.07	0.09	0.17	4.87	0.51
	Arginine	n.d.	0.10	0.55	n.d.	2.68	1.68
	GAA	n.d.	0.01	0.10	0.01	/	/
	Creatine	/	/	/	/	/	/
	Creatinine	0.21	/	0.47	/	14.45	/
	MMA+Arg	n.d.	0.22	0.15	n.d.	1.70	1.35
	Urea	n.d.	0.12	1.59	n.d.	0.73	0.93
Uremic	Control	0.02	0.15	2.08	0.05	0.86	1.41
	Arginine	0.09	0.14	14.0	1.79	0.64	1.20
	GAA	0.26	0.33	6.93	0.22	/	/
	Creatine	0.16	0.16	4.65	0.21	0.73	0.88
	Creatinine	1.34	1.34	10.79	0.74	0.56	0.99
	MMA+Arg	0.14	0.34	6.72	0.79	0.44	4.55
	Urea	0.23	0.17	113.3	4.60	1.09	1.70

n.d. = not detectable; / = not measured; Mean ± S.E., (μg/g wet weight).

MG and GSA increases in uremia, but urinary GAA excretion decreases in uremia. GAA is produced from Arg and glycine by transamidination mainly in the kidney. GAA and creatinine have been shown to inhibit the enzyme Arg-glycine amidinotransferase[10]; therefore the decrease of GAA production in uremia is caused by impairment of the enzyme. Cohen[11] reported that GSA was produced by transamidination from Arg to aspartate in the liver. The increase of GSA production in uremia may be caused by the diversion of Arg no longer used in the production of GAA. However, in the present study, the increase of GSA concentration in urine, serum and liver was observed only in the urea infused normal and uremic rats. This finding suggests that the increase of GSA production in uremia is closely related to the retention of urea rather than the retention of Arg by repression of Arg-glycine amidinotransferase. Cohen and the present authors have proposed that part of the MG produced from Arg was not through Cr. In the present study, the production of MG was increased only by Cr and creatine infusion and the increase of MG production was larger in the Cr infusion than in that of creatine. Additionally, serum Cr levels rose in the creatine infused uremic rats. These results suggests that MG may be produced from Cr. We could not demonstrate another metabolic pathway of MG other than Cr,

because the amount of MG production through the pathway may be
too small to detect. On the other hand, we found that guanidine
production was increased by the GAA infusion. This result shows
that GAA is easily changed to guanidine. If guanidine could be
metabolized to MG by methylation, then the aforementioned hypothesis
may be possible. None the less, most of MG is produced through Cr
because, in the GAA infused rats, (though guanidine was much
produced), MG production did not increase.

SUMMARY

The metabolic pathways of guanidino compounds have not been
clarified. This study aimed to clarify the metabolic pathways of
guanidino compounds, especially MG and GSA which are accepted as
uremic toxins. Arg, MMA, GAA, creatine, Cr and urea, which is a
possible precursor of MG, were intravenously administered to the
normal and experimental uremic rats treated with total parenteral
nutrition. Guanidino compounds in urine, serum, liver and muscle
were measured in all rats. MG production was increased by Cr and
creatine infusion, and the increase of MG production was larger in
the case of Cr. GSA production was increased only in the urea
infusion. GAA infusion caused the increase of guanidine, but did
not cause the increase of MG. These results suggest that most MG
is produced through Cr, and that GSA production may be closely
related to urea.

REFERENCES

1. S.Giovannetti, P.L.Balestri, and G.Barsotti, Methylguanidine in uremia, Arch. Intern. Med., 131:709 (1973).
2. I.M.Stein, B.D.Cohen, and R.S.Kornhauser, Guanidinosuccinic
   acid in renal failure, experimental azotemia and inborn
   errors of the urea cycle, New Engl.J.Med., 280:926 (1969).
3. S.Giovannetti, L.Cioni, P.L.Balestri, and M.Biagini,
   Evidence that guanidines and some related compounds cause
   hemolysis in chronic uremia, Clin.Sci., 34:141 (1968).
4. S.Giovannetti, M.Biagini, P.L.Balestri, R.Navelesi,
   P.Giognoni, A.de Matteis, P.Ferro-Milone, and C.Perfetti,
   Uremia-like syndrome in dogs chronically intoxicated with
   methylguanidine and creatinine, Clin.Sci., 36:445 (1969).
5. B.D.Cohen, Uremic toxins, Bull.N.Y.Acad.Med., 51:1228
   (1975).
6. Y.Orita, Y.Tsubakihara, A.Ando, K.Nakata, Y.Takamitsu,
   Y.Fukuhara, and H.Abe, Effect of arginine or creatinine
   administration on the urinary excretion of methylguanidine,
   Nephron, 22:328 (1978).
7. R.Platt, M.H.Roscoe, and F.W.Smith, Experimental renal
   failure, Clin.Sci., 11:217 (1952).

8. Y.Yamamoto, A.Saito, T.Manji, H.Nishi, K.Ito, K.Maeda, K.Ohta, and K.Kobayashi, A new automated analytical method for guanidino compounds and their cerebrospinal fluid levels in uremia, Trans.Am.Soc.Artif.Intern.Organs, 24:61 (1978).
9. A.Ando, H.Mikami, T.Kikuchi, Y.Orita, and H.Abe, An automated analytical method for guanidino compounds; Determination of optimal deproteinization method, Jpn.J.Clin.Chem., 9:191 (1980).
10. J.B.Walker, End product repression in the creatine pathway of the developing chick embryo, in:"Advances in Enzyme Regulation", Pergamon, New York (1963).
11. B.D.Cohen, Guanidinosuccinic acid in uremia, Arch.Intern.Med., 126:846 (1970).

GUANIDINO COMPOUNDS AND HEMODIALYSIS

Yohji Ochiai, Shinya Abe, Teruo Yamada,
Keiichi Tada, Futami Kosaka

Department of Anesthesiology
Okayama University Hospital and Intensive Care Unit
Okayama 700, Japan

INTRODUCTION

Despite a large number of reports that have appeared recently in regard to guanidino compounds and their role in causing uremia and aggravating renal failure, many points are still unclear. There are very few reports on guanidino compounds in acute renal failure, and those that are available have deficiencies related to the clinical assessment or etiological research. Furthermore, many of the patients admitted to the Intensive Care Unit have renal failure in a high risk situation of multiple organ failure, thus contributing to the difficulty of treatment by dialysis. The clinical picture varies, and can include those with infective foci, decreased immunological capacity, and unstable circulatory and respiratory system, abnormality of the central nervous system, and hepatic failure or disseminated intravascular coagulation. The pathophysiology of acute renal failure was investigated to elucidate the miscellaneous complicating factors by analyzing guanidino compounds in patients requiring hemodialysis.

METHOD

The series consisted of patients admitted to the Intensive Care Unit of Okayama University Medical School since 1980 for acute renal failure. Of these, data before and after hemodialysis were taken from 22 patients who had a total number of 183 dialysis treatments.

The measurement was performed using JASCO G-520 guanidino compounds autoanalyzer in 12 patients (G-520 group), and Shimadzu LC-3A type HPLC in 10 patients (LC-3A group). In G-520 group,

pretreatment consisted of withdrawing 1 ml of whole blood, diluting
it to 10 times with 0.01 N-HCL, centrifuging with an Amicon centri-
flo-25, and taking the supernatant for analysis. In LC-3A group,
whole blood was deproteinized by adjusting the final concentration
to approximately 10 % with 50 % TCA (Trichloro acetate). The
supernatant after proper centrifugation was then subjected to
analysis. This HPLC used a step-gradient method. The reaction
fluid was 0.5 % NaOH and 0.6 % ninhydrin. In G-520 group, we
measured area of each peak curve on recorder and calculated the
area ratio between standard fluid and specimen fluid. In LC-3A
group, each value was obtained automatically by Shimadzu chroma-
topac C-RIA.

The dialysis machine used for hemodialysis was a Nipro NF-01
model. The blood flow (Q_B) was 80-120 ml/min, and the dialysis
fluid flow (Q_D) was 500 ml/min, at a negative pressure of
150-200 mmHg. In some cases, direct hemoperfusion was also used.

RESULTS

1) Eleven types of guanidino compounds were detected :
guanidino succinic acid (GSA), guanidino acetic acid (GAA),
non-acetyl arginine (NAA), creatinine (G), creatine (Cr), taurocyamine (TAU), guanidino propionic acid (GPA), and guanidino
butyric acid (GBA). In some cases, some of the compounds were
not detected. Moreover, differences in the sensitivity of the
two types of machines used in measurement led to some substances
not being detected. We, therefore, studied each value which has
no significant difference of sensitivity between these two
measurement machines.

2) We studied correlations between various guanidino compounds
in terms of the clinical course in each case. GSA has a direct
correlation with both BUN and Crea. For example, in 5 cases out
of 7 in which GSA and BUN were detected, they showed a good direct
correlation. In 8 cases out of 11, GSA and Crea also showed good
direct correlation. MG showed good correlation with creatinine
in 3 cases out of 6, but not with BUN (Table 1)

3) In the induction phase of hemodialysis, we investigated
the correlation between BUN and GSA, between BUN and MG, between
Crea and MG (Table 2).

4) There existed no consistent tendency in serum level
of various guanidino compounds in the induction phase of hemodialysis. Some showed quite a high value, some showed a low value
and some have not even appeared in this phase. But two typical
types did exist; one group has gradually increased as the clinical
gravity progressed. Another group has rapidly disappeared as the
renal failure has been ameliorated. GSA has been detected in an

Table 1. Correlation with the Guanidino Compounds.

( ) : Number of Patients

		GSA		MG	
BUN	P	(5)	P	(2)
	N	(1)	N	(2)
	O	(1)	O	(2)
Crea	P	(8)	P	(3)
	N	(1)	N	(0)
	O	(2)	O	(3)

* P : Positive correlation
  N : Negative correlation
  O : No correlation.

Table 2. Correlation with the Guanidino Compounds

	GSA	MG
BUN	r = 0.691 (n = 13)	r = 0.110 (n = 7)
Crea	r = 0.805 (n = 10)	r = 0.886 (n = 10)

r : correlation coefficient

Table 3. Elimination Rate (%)

BUN	36.4 ± 6.2 (n = 28)
Crea	34.2 ± 8.1 (n = 30)
GSA	46.9 ± 12.8 (n = 39)
GAA	51.6 ± 7.7 (n = 9)
MG	31.3 ± 10.1 (n = 19)

Table 4. Elimination Rate (%/hr.)

	HD	HD + DHP	
BUN	7.28 ± 1.23 (n = 28)	8.15 ± 0.93 (n = 6)	$P < 0.1$
Crea	6.84 ± 1.62 (n = 30)	8.63 ± 1.53 (n = 14)	$P < 0.01$
GSA	9.38 ± 2.55 (n = 39)	13.99 ± 3.98 (n = 12)	$P < 0.001$
MG	6.26 ± 2.02 (n = 19)	9.58 ± 1.52 (n = 5)	$P < 0.01$

HD : hemodialysis
HD + DHP : hemodialysis with direct hemoperfusion

early stage in almost all cases except several cases of drug induced hepatic failure. In 7 cases out of 22, MG has not been detected at all throughout course. There is a general tendency that the appearance of MG was a little bit later than that of GSA, and disappearance of MG tended to be a bit quicker in convalescent stage. The MG value was 1.75 ± 0.29 n mol/ml (n = 6).

5) Of the guanidino compounds, GSA and MG in particular play an important role as uremic toxins. It is possible to eliminate them by dialysis. The elimination rate for guanidino compounds by 5 hours dialysis was GSA 46.9 %, GAA 51.6 %, and MG 31.3 % (Table 3).

6) In the group of patients with hepatic failure and drug poisoning, direct hemoperfusion gave significantly better elimination of Crea, GSA, and MG in comparison to simple hemodialysis. However, there was no significant difference in the elimination rate of BUN (Table 4).

7) Since each case of acute renal failure has various underlying disorders, in cases where rebound in the levels of GSA, MG, BUN and Crea occurred between each dialysis, there was no definite trend in the rate of increase in 48 hours after each dialysis, however, some cases did show a fixed rebound rate.

8) In a case of hepato-renal failure due to drug poisoning, GSA and MG were not detected in the blood. GAA was present in high concentrations at 10 times the level compared to patients without hepato-renal failure. GAA showed a marked decrease with time that paralleled the improvement in the patient's condition.

9) In 4 cases of eclampsia treated with hemodialysis, full cure without residual after effects was achieved. In two of these cases, dialysis for guanidino compounds was performed but neither GSA nor MG was detected.

10) In contrast to the fact that GAA, GPA and GBA were detected in approximately 70 % pf all cases, TAU and G were detected in only a small number of cases.

DISCUSSION

The causative factors of acute renal failure are different so there are differences in the guanidino compounds detected. Furthermore, during the unstable induction period of dialysis, the nutritional and metabolic state are important factors and the progress of catabolism results in a high rate of increase of guanidino compounds. The varying toxicities of guanidino compounds do not always have an intimate relationship with the clinical signs and clinically, factors other than the guanidino compounds need to be considered. It is possible that their toxicity as a uremic toxin is easily amplified during the unstable period of instituting dialysis for acute renal failure. This is important in stabilizing the respiratory system and in understanding metabolism and the effects on the nervous system. These points are though to have great differences from the pathological picture as seen in chronic renal failure.

In regard to the elimination rates obtained with dialysis, MG was difficult to remove. The problems of binding to protein as shown by Giovannetti (1974) and by Ando (1976), and the pre-treatment protocol for the measurement of the guanidino compounds need further analysis.

In cases of hepatic coma and widespread hepatic damage, there was no production of GSA or MG, nor was GAA degraded. This is a high level obstruction of the urea cycle. It is possible that GAA is related to disturbances in consciousness.

In the cases of eclampsia treated by dialysis it was expected that clinical signs, such as fitting, which are due to substances that could be removed by dialysis, would occur. Guanidino compounds could not be detected in the blood of two patients with eclampsia as shown by Bartholomew (1957). However, there still remain problems of relating the timing of the blood sampling at the convulsion phase, and the fact that liquor has not yet been analysed.

The timing of dialysis is different with the various kinds of acute renal failure. In order to choose the most appropriate time, we consider that the elimination and rebound rates of the guanidino compounds are useful. BUN, Crea, Creatinine clearance and the BUN/Crea ratio are, of course, excellent indices for deciding whether or not to dialyse, but analysis of guanidino compounds also needs to be done. Furthermore, it is essential to understand the complicated pathological status of renal failure that occurs during multiple organ failure of various etiologies and to institute more specific patient care. For example, analysis of guanidino compounds is necessary in cases such as hepatic failure. Early detection of uremic toxin can be taken as an index of the physiological function of the kidney in determining the extent of reversibility of the renal damage. In this regard, the fluctuations in blood levels of the guanidino compounds are important in observation of the course and prognosis of patients with acute renal failure.

In conclusion, the pathology of acute renal failure is clearly different from that of chronic renal failure and further research should be continued to clarify.

REFERENCES

Ando, A., Nakata, K., Tsubakihara, Y. Tanaka, T. and Abe, Y., 1976, Guanidino Compounds as Uremic Toxin, Saishin-Igaku, 31:1695.
Bartholomew, R. A., Calvin, E. D., Grimes, W. H., Jr., and Fish, J. S., 1957, Facts pertinent to the Etiology of Eclamptogenic Toxemia. A Summation of Previous Observations, Amer. J. Obstet. Gynec., 74:64.
Giovannetti, S., and Barsotti, G., 1974, Dialysis of Methylguanidine, Kidney International, 6:177.

EFFECT OF GUANIDINO COMPOUNDS ON HEN EGG DEVELOPMENT

Sonoko Seki, Noriyuki Yuyama and Midori Hiramatsu[*]

Department of Physiology Kanagawa Dental College, Yokosuka, Japan and Department[*] of Neurochemistry, Institute for Neurobiology, Okayama University Medical School, Okayama, Japan

INTRODUCTION

It has been reported that levels of guanidinosuccinic acid,[1] methylguanidine[2], β-guanidinopropionic acid[3] and taurocyamine in serum of uremic patients are significantly elevated compared to normal control subjects. Also, brain levels of taurocyamine and methylguanidine were found to be increased in experimental uremia produced by bilateral urethral ligation[4]. Because of these results, methylguanidine, β-guanidinopropionic acid and guanidinosuccinic acid are becoming of interest as potential "uremic toxins".

Furthermore, taurocyamine, guanidinoacetic acid, γ-guanidinobutyric acid, N-acetylarginine and methylguanidine induce violent convulsions on intracisternal injection into rabbit, dog or cat. In addition taurocyamine was found increased in CSF of epileptic patients. Such data suggest that these guanidino compounds are toxic substances which may cause seizures when they accumulate under certain circumstances in nervous tissue. There is a further possibility that metabolic anomalies of guanidino metabolism may have an adverse effect on fetal development.

In the present paper, we present studies on the effects of 24 naturally occurring guanidino compounds on the development of embryonic chicks.

EXPERIMENTAL METHODS

Animals

Fertilized eggs, weighing 50-70 g, were obtained from a commercial firm and were incubated in a special incubator designed by us[6]. Embryonic development was ascertained daily with the aid of a special egg checker which allowed examination of the embryo when the egg is held against the light.

Guanidino Compounds

The following compounds were dissolved in distilled water, and 90 ng of each of the below mentioned guanidino compounds was injected into 5 eggs : N-acetylarginine, α-N-acetyl-γ-hydroxyarginine, agmatine sulfate, L-α-amino-γ-guanidinobutyric acid hydrochloride, L-α-amino-β-guanidinopropionic acid hydrochloride, L-arginine, L-arginic acid, canavanine, creatine, creatinine, guanidinoacetic acid, guanidine hydrochloride, guanidinoadipic acid, γ-guanidinobutyramide, γ-guanidinobutyric acid, α-guanidinoglutaric acid, γ-guanidino-β-hydroxybutyric acid, β-guanidinopropionic acid, guanidinosuccinic acid, ε-guanidinocaproic acid, L-homoarginine hydrochloride, γ-hydroxyarginine, 1-methylguanidine hydrochloride, and taurocyamine.

Injection Method

The eggs were cleaned with distilled water and were wiped with sterilized gauze. The injection area was sterilized with 70 % ethylalcohol, and then a small hole was drilled in the mid portion of the egg. One hundred µl of guanidino compound was slowly injected manually to a depth of approximately 0.1 mm. The hole was then closed with a small square of paraffin film and adhesive tape.

Observation of Development

Three stages of embryonic development (which is completed in 21 days) were distinguished as follows ;
(1) Early stage, comprising the first 7 days, in which the arterial system begins to differentiate and the vascular system becomes extended.
(2) Middle stage, the following 8-14 days, during which ossification occurs.
(3) Last embryonic stage is the period 15-21 days (hatching), when allantoic respiration changes to pulmonary respiration. These three stages of development were verified daily by means of a light transmission detector. When a deformity was suspected, the

Table 1. Effect of Guanidino Compounds on the Developmental Stage of Chick (1)

Guanidino compound	N	Dead eggs in developmental stage 0-7	8-14	15-21 (days)	Hatched eggs	Deformity
Arginine	5	2	0	0	3	0
Creatinine	5	1	1	0	3	0
Guanidinoglutaric acid	5	2	0	0	3	0
Methylguanidine	5	1	1	0	3	0

Table 2. Effect of Guanidino Compounds on the Developmental Stage of Chick (2)

Guanidino compounds	N	Dead eggs in developmental stage 0-7	8-14	15-21 (days)	Hatched eggs	Deformity
N-Acetylarginine	5	1	0	0	4	0
N-Acetyl-γ-hydroxyarginine	5	1	0	0	4	0
Agmatine	5	1	0	0	4	0
α-Amino-β-guanidinopropionic acid	5	0	1	0	4	0
Arginic acid	5	1	0	0	4	0
Canavanine	5	1	0	0	4	0
Creatine	5	0	1	0	4	0
Guanidinoadipic acid	5	0	1	0	4	0
γ-Guanidinobutyric acid	5	1	0	0	4	0
ε-Guanidinocaproic acid	5	1	0	0	4	0
γ-Hydroxyarginine	5	0	1	0	4	0
Taurocyamine	5	1	0	0	4	0

eggs were broken to examine for any embryonic abnormality. All chicks after hatching were also examined for possible deformities.

Table 3. Effect of Guanidino Compounds on the Developmental Stage of Chick (3)

Guanidino compound	N	Dead eggs in developmental stage			Hatched eggs	Deformity
		0-7	8-14	15-21 (days)		
α-Amino-β-guanidinopropionic acid	5	0	0	0	5	0
Guanidinoacetic acid	5	0	0	0	5	0
Guanidine-HCl	5	0	0	0	5	0
γ-Guanidinobutyramide	5	0	0	0	5	0
γ-Guanidino-β-hydroxybutyric acid	5	0	0	0	5	0
β-Guanidinopropionic acid	5	0	0	0	5	0
Guanidinosuccinic acid	5	0	0	0	5	0
Homoarginine	5	0	0	0	5	0

RESULTS

Mortality during Egg Stage

Some embryos died in eggs 6-11 days after injection of some of the 24 guanidino compounds. Two out of five eggs exhibited mortality at the early stage or at the beginning of the middle stage in the groups which received respectively one of the following substances : L-arginine, creatinine, guanidinoglutaric acid and 1-methylguanidine hydrochloride (Table 1). One out of 5 embryos died at this same stage in group administered N-acetylarginine, α-N-acetyl-γ-hydroxyarginine, agmatine sulfate, L-α-amino-β-guanidinopropionic acid hydrochloride, L-arginine, creatine, guanidinoadipic acid, γ-guanidinobutyric acid, ε-guanidinocaproic acid, γ-hydroxyarginine and taurocyamine (Table 2). The remaining guanidino compounds did not induce lethality (Table 3).

Deformity after Hatch

All chicks which survived the embryonic period appeared healthy on hatching.

DISCUSSION

During the course of development during the 21 days, the mesoderm on days 1-2 begins to give rise to the vascular system and the heart, thus during this period the arterial system begins to differentiate. On days 3-4, the vascular system is established in general, the digestive system and the respiratory system just begins to differentiate, and also the development of the genital organs occurs around this time. On day 6, the sexual differentiation of the genital gland begins, and at about day 6, the ectoderm begins to differentiate ; from the 7th to the 10th day, the vascular net work begins to invade the yolk and the first signs of feathers appear. During days 11-12, the bones start to ossify. The embryo then enlarges rapidly and on days 15-16, the beak approaches the air cell. On days 17-18, the respiratory and cirulatory systems are complete. By day 19, the air cell is considerably enlarged, but circulation of the blood does not yet occur. On day 20, the beak is in the air cell, and allantoic respiration changes to pulmonary respiration. This is the normal course of development of a chick embryo.

In our experiment it was found that one or two embryos out of five which died after administration of the guanidino compounds, almost all died between 6-8 days, when the ectoderm begins to differentiate. Arginine, creatinine, α-guanidinoglutaric acid and methylguanidine were found to be more toxic than any of the other guanidino compounds.

α-Guanidinoglutaric acid is known to increase in the cobalt focus of cat and to induce convulsions after application on motor cortex of rabbit[7]. This substance is therefore clearly toxic when it accumulates in nervous tissue. Methylguanidine is also thought to be a uremic toxin. On the other hand both arginine and creatinine are naturally present in all mammalian tissues. Because all four compounds seem to cause embryonic death at around the same stage, we would like to suggest that they cause the lethality by the same mechanism. It would seem the only feature these compounds have in common, is that they are guanidino substances. However, this by itself cannot explain their toxicity since many of the other compounds we tested were also of this type but did not cause embryonic death. At this time therefore the toxicity of these compounds at a specific and critical stage (6-8 days) of embryonic development is difficult to explain. One very tentative conclusion may be that an excess of arginine itself, may interfere with ectoderm differentiation and that increases of guanidinoglutaric acid, creatinine or methylguanidine induce such a build up. Especially this last compound may be imagined to interfere with the transfer of the urea group from arginine to recipient substances.

SUMMARY

Guanidino compounds were administered to groups of 5 hen eggs in a dose of 90 ng/egg, and their effect on the development of eggs was followed daily.

The results of these experiments were as follows ; two out of five eggs died during the early stage or at the beginning of the middle stage in the groups administered arginine, creatinine, guanidinoglutaric acid, or methylguanidine. One of 5 eggs died at the same stage in the groups administered N-acetyl-arginine, N-acetyl-γ-hydroxyarginine, agmatine, α-amino-γ-guanidino-butyric acid, arginic acid, canavanine, creatine, guanidino-adipic acid, γ-guanidino butyric acid, ε-guanidino caproic acid, γ-hydroxy-arginine or taurocyamine. No embryos were found affected in the groups administered any of the other guanidino compounds. If chicks survived the embryonic stage, they appeared normal in all respects on hatching.

REFERENCES

1. P. P. Kamoun, J. M. Pleau and N. K. Man, Semiautomated method for measurement of guanidinosuccinic acid in serum, Clin. Chem., 18:355 (1972).
2. L. R. I. Baker and R. D. Marshall, A Reinvestigation of methylguanidine concentrations in sera from uraemic subjects, Clin. Sci., 41:563 (1971).
3. I. M. Stein, G. Perez, R. Johnson and N. B. Cummings, Serum levels and urinary excretion of methylguanidine in chronic renal failure, J. Lab. Clin. Med., 77:1020 (1971).
4. M. Matsumoto, H. and A. Mori, Guanidino compounds in the sera of uremic patients and in the sera and brain of experimental uremic rabbits, Biochem. Med. 16:1 (1976).
5. A. Mori, in press.
6. S. Seki, K. Ohe, T. Hirata, et al., Inhibitory effect of methionine sulfoximine on chick development, Kanagawa Shigaku, 1:16 (1967).
7. A. Mori, Y. Watanabe, S. Shindo and M. Hiramatsu, α-Guanidino-glutaric acid and epilepsy, Symposium on Urea Cycle Diseases, Okayama, Japan (1981).

IV. FREE COMMUNICATIONS

α-KETOGLUTARATE INDUCED TRANSAMINATION DURING ISCHEMIC EXERCISE

C. Cerri[*], F. Fici[**] and G. Scarlato[*]

[*] University of Milano, Medical School Neurology Department and [**] Istituto Lusofarmaco d'Italia
via Francesco Sforza 35 Milano, Italy

INTRODUCTION

Administration of α-ketoglutarate (KGT) has been reported to reduce lactacidemia both in pathological conditions (Scarlato et al. 1981) and after exercise in normal subjects (Cerretelli et al. 1981). KGT could reduce lactacidemia by increasing the catabolism of pyruvate via either the Krebs cycle or by transamination. To discriminate between those two possibilities we investigated the effect of IV administration of KGT in subjects performing a standard ischemic work. As reported by Munsat (1970) ligation of the arm while performing a squeezing exercise with the hand blocks the utilization of pyruvate by the Krebs cycle. If KGT reduces lactacidemia favoring pyruvate utilization by the Krebs cycle, then administration of KGT should not change the blood level of lactate after exercise. Our results suggest that this is not the case.

MATERIALS AND METHODS

Ten healthy volunteers (age between 22 to 40) were studied. KGT (8mg/kg) was administred intravenously immediately before exercise in a first series of experiments and immediately following exercise in a second series. Isotonic saline solution was used as placebo. Each subject performed four ischemic work tests; between them he (or she) rested for 45 minutes. The order of administration of KGT and placebo and the time of administration (before or after exercise) were randomized to avoid interferences by preceeding tests, although four tests performed without any intravenous drug administration at 45 minutes intervals by two subjects gave superimposable results. In the same subject basal values before each test were equivalent. Venous blood samples were drawn from an indwelling venous catheter (Angiocath, Deseret Co, Sandy Utah) without arm ligation, and kept in ice until assayed (which was done within 20 minutes from collection). Samples were drawn before exercise and after 1,3,5,10,15 minutes. The ischemic exercise was performed according to Munsat (1970). Ammoniemia was determined according to Reichelt et al. (1964), pyruvicacidemia to Czok and Lamprecht (1974), and lactacidemia to Noll (1974).

RESULTS

Intravenous administration of KGT in healthy subjects at rest results in moderate reduction of lactacidemia with contemporary increase in pyruvicacidemia and ammoniemia. The effect is evident after one minute and reaches its peak in five, vanishing after ten minutes (table 1)

No changes were detected administrating saline placebo solution before or after ischemic exercise (table 2).

Ischemic work followed by intravenous KGT administration diminished the increase in lactic acid slightly increasing pyruvic acid and increasing the ammonia production (table 3).

Performance of the test immediately after KGT administration abolished the increase in lactic acid, did not change the increase in pyruvic acid while it dramatically increased the ammonia production (table 4).

Table 1. Effects of Intravenous Administration of α-Ketoglutarate at Rest

Time	Lactacidemia*	Pyruvicacidemia*	Ammoniemia*
0	14.0 ± 2.2	0.52 ± 0.03	40 ± 2
3	6.8 ± 1.8	0.58 ± 0.04	45 ± 5
5	8.0 ± 1.9	0.60 ± 0.03	48 ± 4
7	9.6 ± 2.0	0.55 ± 0.05	44 ± 3
10	13.3 ± 1.6	0.52 ± 0.04	41 ± 2

Time in minutes after administration
Values (mean ± SD) as *mg/dl of plasma and *µg/dl of plasma

Table 2. Ischemic Work Test in Control Subjects

Time	Lactacidemia*	Pyruvicacidemia*	Ammoniemia*
0	12.4 ± 2.1	0.45 ± 0.08	40 ± 5
1	39.6 ± 2.0	0.66 ± 0.09	75 ± 6
3	42.2 ± 2.9	0.91 ± 0.10	80 ± 9
5	30.6 ± 3.5	0.94 ± 0.12	70 ± 8
10	22.2 ± 3.0	0.80 ± 0.09	60 ± 9
15	14.4 ± 2.2	0.69 ± 0.06	51 ± 6

Time in minutes after exercise.
Values (mean ± SD) as *mg/dl or *µg/dl of plasma

Table 3. α-Ketoglutarate Administration after Ischemic Exercise

Time	Lactacidemia*	Pyruvicacidemia*	Ammoniemia*
0	12.8 ± 1.9	0.50 ± 0.05	41 ± 2
1	34.7 ± 2.0	1.00 ± 0.08	86 ± 5
3	41.2 ± 2.1	1.10 ± 0.07	90 ± 6
5	30.6 ± 1.5	1.19 ± 0.09	88 ± 5
10	18.1 ± 2.1	0.65 ± 0.03	60 ± 3
15	13.6 ± 1.5	0.55 ± 0.06	45 ± 4

Time in minutes after exercise
Values (mean ± SD) as *mg/dl or *µg/dl plasma

Table 4. Ischemic Work Test after α-Ketoglutarate Administration

Time	Lactacidemia*	Pyruvicacidemia*	Ammoniemia*
0	13.0 ± 1.7	0.51 ± 0.04	42 ± 8
1	14.0 ± 1.2	0.90 ± 0.06	96 ± 9
3	18.0 ± 2.0	0.95 ± 0.19	120 ± 9
5	14.0 ± 1.2	0.88 ± 0.07	98 ± 7
10	13.1 ± 2.9	0.70 ± 0.09	85 ± 8
15	12.9 ± 1.9	0.60 ± 0.10	70 ± 9

Time in minutes after exercise
Values (mean ± SD) as ✱ mg/dl or * μg/dl plasma

DISCUSSION

KGT has been used to reduce lactacidemia in patients with ophthalmoplegia plus (Scarlato et al. 1981). The molecular defect underlying hyperlactacidemia in this syndrome has not yet been clarified, although Nemni et al. (1981) reported decreased cytochrome oxidase activity and alterations of cytochrome aa_3 spectral scan in patients with chronic external ophthalmoplegia and muscular mitochondria morphologic abnormalities. Understanding how KGT reduces lactacidemia not only offers information about the role of this metabolite in regulating cellular metabolism but could offer further insight in the pathogenesis of the ophthalmoplegia plus syndrome. KGT is one of the substrates of the Krebs cycle, is necessary for the malate-aspartate shuttle in mitochondria and plays a crucial role in many transamination reactions allowing use of aminoacids as energy source. Our experimental procedure blocked the Krebs cycle locally and allowed us to evaluate mainly the effect of KGT on transamination of pyruvate and ornithine. As pointed out by Stumpf and Parks (1980), KGT could divert ornithine from urea cycle into a new cycle through glutamic acid semialdehyde, glutamate and KGT again, producing two NADH and one NH_3 mole for each ornithine mole utilized. Two glutamate moles are produced for each ornithine transaminated to glutamic acid semialdehyde, one directly by ornithine-α-ketoglutarate transaminase, the second as result of the following reduction of glutamic acid semialdehyde by glutamic acid semialdehyde dehydrogenase. As result glutamate is available to transaminate pyruvic acid to alanine producing an "extra" KCT. Pyruvate transamination thus

results in formation of two NADH, although at expenses of increased
ammonia production, whereas the dehydrogenation of pyruvate to
lactate results in only one. This pathway has been demonstrated
so far only in isolated mitochondria, a preparation lacking of the
cytosoluble urea cycle enzymes (Stumpf and Parks 1980), therefore
without possibility of ornithine utilization in the urea cycle.
A sudden KGT load in the muscle cell could shift the utilization
of ornithine toward the cycling through KGT particularly if no
energy from the Krebs cycle is available to the cell. Furthermore
KGT enhancing the malate-aspartate shuttle (Cerretelli et al. 1981)
displaces the redox state of the cytosol towards oxidation; as a
consequence the equilibrium of the pyruvate dehydrogenase reaction
is shifted to the left. Our data suggest that this series of events
take place and activate transamination of pyruvate which enters a
"non lactacid" anaerobic energy pathway (Robin and Hance 1980) thus
decreasing lactate production.

In conclusion KGT diverts ornithine from urea cycle offering
an anaerobic way for energy production. As consequence urea
production decreases and ammonia production increases. The decrease
of lactic acid could be due both to displacement towards oxidation
of the cytosol redox state and to increase pyruvate transamination.
In patients with increased lactacidemia due to impaired aerobic
energy production KGT could reduce lactacidemia through a similar
mechanism. Preliminary data show that ammoniemia increases after
KGT administration at a rate comparable to the decrease of lactacidemia in some patient with ophthalmoplegia plus syndrome.

REFERENCES

Cerretelli, S., Marconi, C., and Sassi, G., 1981, The effects of
  pyridoxine alphaketoglutarate on human maximal aerobic and
  anaerobic performance. Europ. J. Physiol., in press.
Czok, R., Lamprecht, W., 1974, in "Methoden der Enzymatischen
  Analyse", Verlag Chemie Weinheim, p 1491.
Munsat, T.L., 1970, A standardized forearm ischemic exercise test,
  Neurology, 20 : 1178.
Nemni, R., Mitsumoto, H., Bradley, W.G., 1981, Morphological and
  biochemical analysis in chronic progressive external ophthalmoplegia, J. Neuropathol. Exptl. Neurol., 40 : 350.
Noll, F., 1974, in "Methoden der Enzymatischen Analyse" Verlag
  Chemie, Weinheim, p 1521.
Reichelt, K.L., Kvamme, E., Tveit, B., 1964, The enzymatic determination of ammonia in blood and tissue, Scan. J. Clin. Inv.,
  16 : 433.
Robin, E.D., Hance, A.J., 1980, Skeletal muscle as a facultative
  anaerobic system. in "Exercise Bioenergetic and Gas Exchange"
  P. Cerretelli and B.J. Whipp eds. Elsevier Amsterdam p 101.

Scarlato, G., Pellegrini, G., Moggio, M., 1981, Treatment of
ophthalmoplegia plus with α-Ketoglutarate, J. Neurol., in press.

Stumpf, D.A., Parks, J.K., 1980, Urea cycle regulation : Coupling
of ornithine methabolism to mitochondrial oxidative phosphor-
ilation, Neurology, 30 : 178.

EFFECT OF PYRIDOXINE-2-OXOGLUTARATE ADMINISTRATION IN PATIENTS
WITH ADVANCED CIRRHOSIS : CONTROL OF AMMONIA PYRUVATE AND LACTATE
HIGH PLASMA CONCENTRATIONS

F. Salerno, M.C. Lorenzini, M. Conti, R. Abbiati
and F. Fici

Third Medical Clinic, University of Milan
via Pace 15, 20122 Milan - Italy

INTRODUCTION

The treatment of cirrhotic patients with impending coma is often limited, being the pathogenesis of hepatic encephalopathy barely known. The currently most employed therapeutic measures are directed towards reducing hyperammonemia, although a causative relation between elevated ammonia concentrations in plasma or CSF and the encephalopathy has not been firmly established[1]. Cirrhotic liver has a lowered ability to clear ammonia, therefore it is necessary to inhibit ammonia production or to enhance its extrahepatic utilization. Oxoglutaric acid, an intermediate of the trycarboxilic acid cycle (TAC), has been used in order to reduce high ammonia concentrations because it can be aminated to form glutamic acid in extrahepatic tissues[2,3]. Administration of oxoglutaric acid in pharmacological doses could affect other metabolites as well as ammonia, however there are no studies evaluating such effects in cirrhotic patients. Since changes in the amino acid pattern or in the concentrations of some products of glucose oxidization could be involved in the pathogenesis of the hepatic encephalopathy, we studied the effect of both acute and chronic administration of oxoglutaric acid on ammonia, amino acids, glucose, lactic and pyruvic acids in cirrhotic patients with hyperammonemia.

MATERIALS AND METHODS

Twenty-two patients with liver cirrhosis and basal hyperammonemia were studied. None of them had received any sort of pharmacological treatment in the previous week.

Twelve patients were infused with a solution of pyridoxine-2-oxoglutarate (P-2-O) (Glutarase - Istituto Lusofarmaco, Milano), 4.6 g/250 ml of saline, at the rate of 2 ml/min, starting at 8.30 a.m. after an overnight fasting. At 0,60 and 120 minutes, blood samples were taken in heparinized plastic tubes. After 3 days, 8 out of these 12 patients underwent a control study by infusion of a not distinguishable saline solution : the procedures were the same as those outlined above. Another group of 10 patients entered a randomised double-blind cross-over trial of one month treatment with P-2-O 600 mg orally three times daily and pyridoxine alone 340 mg t.i.d. Blood samples were obtained before and at the end of both kinds of treatment. Ammonia was determined on fresh plasma rapidly separated by centrifugation at 4° C by the method of Miller[4]. Amino acids were determined on plasma deproteinized with sulfosalicylic acid employing a Carlo Erba 3A28M amino acid analyzer[5]. Lactic, pyruvic and oxoglutaric acids were determined on plasma deproteinized with trichloroacetic acid by enzymatic spectrophotometric method . Statistical significances were evaluated with Student's t test for both paired and unpaired analysis.

RESULTS

All patients showed resting hyperammonemia ranging from 44.1 to 220.6 µmol/l (normal values in our laboratory : 9.8 - 27.8). Hyperpyruvicemia was found in 16 patients and hyperlactacidemia in 17 patients. Table 1 shows the effect of P-2-O or saline infusion on the biochemical parameters investigated.

Plasma oxoglutaric acid increased significantly during the infusion (from 17.8 ± 2.7 to 161.9 ± 16.1 µmol/l, $P < .001$), while

Table 1. Plasma levels of AMMONIA, PYRUVATE, LACTATE and GLUCOSE in patients infused with P-2-O (a) or saline (b)

(time)		0	+ 60	+ 125
AMMONIA (µmol/l)	a	92 ± 11	83 ± 10	75 ± 10
	b	124 ± 20	124 ± 22	123 ± 21
PYRUVATE (µmol/dl)	a	17 ± 2.7	6.6 ± 1.8	2.6 ± 1
	b	21 ± 3	20 ± 3.1	21 ± 3
LACTATE (mmol/l)	a	2.1 ± 0.1	2.1 ± 0.1	1.7 ± 0.2
	b	2.5 ± 0.3	2.3 ± 0.2	2.3 ± 0.3
GLUCOSE (mmol/l)	a	5.2 ± 0.4	5.5 ± 0.5	5.2 ± 0.2
	b	5.1 ± 0.4	5.0 ± 0.4	5.3 ± 0.3

Values are reported as mean ± S.E.M.

Table 2. Plasma amino acid levels (μmol/l) found in 8 cirrhotic patients before and after P-2-O infusion for 120'

	Controls	patients 0'	120'
TAU	44 + 5	61 + 8	46 + 5 *
ASP	16 + 5	22 + 5	14 + 4 **
THR	121 + 18	132 + 21	126 + 20
SER	85 + 13	122 + 19	119 + 19
$AspNH_2$	53 + 6	68 + 8	111 + 10 *
GLU	58 + 9	221 + 52	196 + 44
$GluNH_2$	395 + 35	520 + 61	437 + 38 *
PRO	230 + 28	295 + 40	204 + 26
GLY	212 + 25	224 + 27	223 + 31
ALA	338 + 37	391 + 52	304 + 48
VAL	273 + 19	177 + 18	176 + 22
MET	15 + 2	43 + 5	56 + 5
ILE	63 + 6	56 + 8	47 + 6
LEU	173 + 27	119 + 9	114 + 11
TYR	47 + 4	145 + 11	128 + 7 *
PHE	55 + 4	93 + 12	87 + 11
ORN	87 + 9	77 + 8	75 + 8
LYS	126 + 13	110 + 11	107 + 11
HYS	96 + 12	82 + 9	77 + 9
ARG	67 + 8	120 + 18	125 + 16

Values are reported as mean + S.E.M.

* = P <.05   ** = P <.01 (time 120' vs time 0').

it did not change during saline administration (from 14.6 + 2.3 to 14.1 + 2.2) (data not shown in the table). Ammonia decreased in 8 patients and remained unchanged in the other 4, so its final level was the 82 % of the initial one (P< .05, pre-treatment vs post-treatment values). Lactic acid slightly decreased, but the final mean level was not significantly different from the initial one. In contrast, pyruvic acid decreased markedly (from 17 + 2.7 to 2.6 + 1 μmol/dl, P <.001).

Table 2 shows the amino acid concentrations found at 0, 120 minutes in 8 patients investigated. A significant decrease was well evident for taurine, aspartic acid, glutamine and tyrosine, whereas asparagine significantly increased. Saline infusion did not affect amino acid pattern (data not shown) as well as ammonia, lactate and pyruvate concentrations (see table 1).

Table 3. Changes in ammonia lactate pyruvate and glycemia after oral administration of P-2-O or Pyridoxine (P) to 10 cirrhotic patients for 30 days

	before	after P-2-O	P
AMMONIA (µmol/l)	86 ± 16	68 ± 9	79 ± 12
PYRUVATE (µmol/dl)	7.7 ± 0.7	4.7 ± 0.9*	6.3 ± 0.9
LACTATE (mmol/l)	1.6 ± 0.2	1.1 ± 0.1	1.4 ± 0.2
Glycemia	6.2 ± 0.6	5.6 ± 0.4	5.9 ± 0.4

Values are reported as mean ± S.E.M.

* = $P < .05$ (vs "before" values)

In patients treated for 30 days with oral administration of P-2-O, we observed a decrease of ammonia, pyruvic acid, lactic acid and glycemia (table 3). Administration of pyridoxine alone in the same patients failed to modify these parameters. As far as the amino acid pattern is concerned, P-2-O treatment significantly increased glutamic acid (231 ± 32 vs 127 ± 31 nmol/l, $P < .05$) and alanine (401 ± 43 vs 330 ± 14 nmol/l, $P < .05$), and decreased glutamine (180 ± 32 vs 401 ± 70 nmol/l, $P < .01$). Neutral and basic amino acids did not change.

DISCUSSION

High ammonia levels, frequently found in patients with severe liver disease, are not always easy to lower, because the cirrhotic liver is failing in order to metabolize ammonia by urea cycle[7].

The present study shows that oxoglutaric acid, infused in high concentrations as pyridoxine-2-oxoglutarate, lowered ammonia plasma levels in the 66 % of our hyperammonemic subjects. This result is in agreement with those previously obtained by other authors with P-2-O[8] or with other products of salification of oxoglutaric acid[2,3].

The hypoammonemic effect was related neither to the basal levels of ammonia, nor to the clinical conditions, the age or the sex of the patients, so it remains unclear why some patients were unresponsive. The mechanism whereby acute administration of oxoglutaric acid reduced ammonia levels might lie in the potential amination of the compound to form glutamate and glutamine in the extrahepatic tissues, as experimentally shown by Schlienger[9] in abdominally eviscerated rats ; although we found no changes in plasma glutamate levels. However, plasma levels of glutamate may not necessarily reflect its tissue levels in a study of acute administration of the precursor, being plasma half-life of glutamate less than one minute[10]. Accordingly, in the group of patients given P-2-O orally for one month we observed a significant rise of both glutamate and alanine plasma concentrations. This result seems to reflect better the intramitochondrial consumption of oxoglutaric acid to form glutamate, which in turn activates glutamate-pyruvate transaminase allowing alanine synthesis and accumulation.

Other remarkable effects of P-2-O were the striking decrease of pyruvate concentrations, obtained with both kinds of administration, and the significant decrease of lactate concentrations obtained only with the chronic administration. High levels of pyruvic and lactic acid are frequently found in patients with advanced cirrhosis[11]. Hyperpyruvicemia seems to be the consequence of a state of intramitochrondrial high NADH/NAD ratio with consequent inhibition of the pyruvate-dehydrogenase[12]. Hyperlactacidemia is the result of at least two causes : increased peripheral production due to hyperpyruvicemia and, sometimes, to the presence of hyperventilatory alkalosis[13], and defective disposal by the cirrhotic liver[14].

The mechanism whereby oxoglutaric acid administration lowers pyruvate concentrations could be through both an enhanced synthesis of alanine for the increased availability of glutamate and a restoring effect on the pyruvate-dehydrogenase activity. Pyruvate oxidization to acetyl CoA, mediated by pyruvate-dehydrogenase, could be reactivated in our cases by the following chain of events : the consistent dosage of oxoglutaric acid administered may enhance the activity of the tricarboxylic acid cycle, by replenishing the intermediates ; acetyl CoA is better utilized to synthetise citric acid by the increased availability of oxalacetic acid : then pyruvate-dehydrogenase equilibrium is shifted to the synthesis of acetyl CoA. Improvement of the function of the TAC through the administration of 2-oxoglutarate to rats with experimentally induced ketosis has been already envisaged by other authors[15]. Why the rapid pyruvic acid decrease obtained during two hours of P-2-O infusion was not coupled with a similar lactic acid decrease is attributable to the delayed clearance of the cirrhotic liver, since a significant fall of lactic acid high levels was observed during P-2-O infusion in diabetic patients after muscular exercise (Alpi O. et al. in preparation) and in normal dogs with portacaval anastomosis (personal unpublished observation). This hypothesis is also supported by the finding that plasma lactate significantly decreased, when cirrhotic patients were given P-2-O orally for 30 days. This result may be the simple consequence of the reduced pyruvic acid pool, even though an improved hepatic clearance cannot be ruled out.

Whatever the mechanism of action might be, the reduction of plasma lactic acid levels is a further favourable effect of the administration of oxoglutaric acid in pharmacological doses. Hyperlactacidemia becomes evident in cirrhotic patients when the liver function progressively decays ; it can lead to a situation of lactic acidosis[11,16] with the potential role in the pathogenesis of hepatic encephalopathy[17]. Treatment of lactic acidosis is frequently ineffective in such patients[18], thus a protective effect of pyridoxine-2-oxoglutarate given chronically could be relevant.

REFERENCES

1. G. O. Walker and S. Schenker, Pathogenesis of hepatic encephalopathy with special reference to the role of blood ammonia, Am. J. Clin. Nutr. 23:619 (1970).
2. B. Peter, M. Imler, J. Tongio, J. L. Schlienger, and J. Stahl, Influence de l'alpha-cétoglutarate de l'ornithine sur l'hyperammoniémie provoquée des cirrotiques, Ann. Gastroent. Hepatol. 16:179 (1974).
3. M. Michel, P. Oge, and L. Bertrand, Action de l'alphacétoglutarate d'ornithine sur l'hyperammoniémie du cirrhotique, Presse Med. 79:867 (1971).
4. G. E. Miller and J. D. Rice jr, Determination of the concentration of blood ammonia by use of cation exchange method, J. Lab. Clin. Med. 60:170 (1962).
5. D. M. Spackman, W. H. Stein, and S. Moore, Automatic recording apparatus for use in the chromatography of amino acids, Anal. Chem. 30:1190 (1958).
6. H. U. Bergemeyer, "Methods of enzymic analysis", Academic Press Inc., New York (1974).
7. A. H. Lockwood, J. M. Mc Donald, R. E. Reiman, A. S. Gelbard, J. S. Laughlin, T. E. Duffy, and F. Plum, The dynamics of ammonia metabolism in man : effects of liver disease and hyperammonemia, J. Clin. Invest. 63:449 (1979).
8. F. Ghinelli, G. Magnani, G. Pedretti, G. Pelosi, P. Perinotto, and D. Sacchini, Effetto dell'alfachetoglutarato di piridossina sull'iperammoniemia dei cirrotici, Epatologia 26: 261 (1980).
9. J. L. Schlienger, A. Frick, and M. Imler, Effect of ornithine and alphaketoglutarate on hyperammonemia in hepatectomized and abdominal eviscerated rats, in:"III Ammoniak symposium", G. Wewalka and B. Dragosics eds., Verlag-Stuttgart (1978).
10. D. H. Elwyn, W. J. Launder, H. C. Parikh and E. M. Wise jr,

Roles of plasma and erythrocytes in interorgan transport of amino acids in dogs, Am. J. Physiol. 222:1333 (1972).
11. R. E. Heinig, E. F. Clarke, and C. Waterhouse, Lactic acidosis and liver disease, Arch. Int. Med. 139:1229 (1979).
12. J. Bremer, Pyruvate dehydrogenase, substrate specificity and product inhibition, Eur. J. Biochem. 8:535 (1969).
13. M. N. Berry and J. Scheuer, Splanchnic lactic acid metabolism in hyperventilatory metabolic alkalosis and schock, Metabolism 16: 537 (1967).
14. P. J. Woll and C. O. Record, Lactate elimination in man : effects of lactate concentrations and hepatic dysfunction, Eur. J. Clin. Invest. 9:395 (1979).
15. L. Garbin, M. Plebani, and P. M. Terribile, Effect of ACP (pyridoxine-2-oxoglutarate) on CCl intoxication and in streptozotocin-induced ketosis in rat, Acta Vitamin. Enzymol. 6:175 (1977).
16. H. Connor, H. F. Woods, J. D. Murray, and J. C. G. Ledingham, Lactate metabolism in liver disease, Eur. J. Clin. Invest. 7:244 (1977).
17. R. Mulhausen, A. Eichenholz, and A. Blumentals, Acid-base disturbances in patients with cirrhosis of the liver, Medicine 46:185 (1967).
18. P. B. Oliva, Lactic acidosis, Am. J. Med. 48:209 (1970).

TREATMENT OF PYRUVIC AND LACTIC ACIDAEMIA IN OPHTHALMOPLEGIA PLUS

G. Scarlato, M. Moggio, F. Fici[*] and C. Cerri

Clinica Neurologica II^a Universita' di Milano
[*] Istituto Lusofarmaco d'Italia
via Francesco Sforza 35 Milano, Italy

INTRODUCTION

Disorders of lactate and pyruvate metabolism are a relatively common feature in a number of different diseases frequently associated with mitochondrial abnormalities in muscle, brain, liver and heart. As it is well-known, pyruvate sits at the center of any ordinary chart of intermediary metabolism. The formation of pyruvate is primarily from glucose by the process of glycolyses and its utilization is by transamination, reduction, carboxylation or oxidation processes. Therefore pyruvate and lactate metabolism are under control of several enzymatic mechanisms.

Clinically, deficiencies of the above mentioned enzymatic activities lead to many different syndromes. Hereditary deficiencies in the pyruvate dehydrogenase complex (PDHC) or the ketoglutarate dehydrogenase complex (KGDC) have been described in over 25 patients with congenital lactic acidosis and severe retardation of growth and development (Kuroda et al., 1979). PDHC and KGDHC deficiencies, with activity 35 to 50 percent of normal were reported in syndromes of spinocerebellar degeneration (Kark and Buduelli, 1979) with or without hyperlactacidemia. Other alterations of pyruvate and lactate metabolism appear to exist, particularly in patients with mitochondrial disorders (Sulaiman et al., 1974 ; Schapira et al., 1975 ; Askanas et al., 1978). They have not been defined biochemically as PDHC, KDHC or other enzymatic deficiencies, even if hyperlactacidemia and hyperpyruvicacidemia are a relatively common feature in these syndromes : a raised plasma lactate either at rest or after exertion has been reported in patients with progressive proximal myopathies due to mitochondrial abnormalities, in patients with mitochondrial myopathy, epileptic seizures and various forms of chronic progressive external ophthalmoplegia (CPEO) associated with mitochondrial abnormalities (Bradley et al., 1978). The most clearly defined syndrome of CPEO with mitochondrial myopathy is that bearing the eponymous title of Kearns and Sayre (1958) or ophthalmoplegia plus (OP). From time to time a very wide range of features have been included in this syndrome including progressive external ophthalmoplegia, diffuse skeletal muscle weakness, pigmentary retinal degeneration, retarded somatic growth, sensorineural deafness, cerebellar ataxia, cardiac conduction defect and high cerebrospinal fluid protein. In some instances mitochondrial abnormalities like those observed in muscle tissue were present in Purkinje's cells, granule cells (Schneck et al., 1973), sweat glands of the skin (Karpati et al., 1973) and liver (Gonatas et al., 1967) suggesting a multicellular mitochondriopathy. An abnormal increase of pyruvic and lactic acid in OP has been reported either at rest (Sulaiman et al., 1974 ; Reske-Nielsen et al., 1976), or after exertion (Scarlato et al., 1978).

This paper records seven observations of OP with hyperlacticacidaemia and hyperpyruvicacidaemia. All seven patients were treated with pyridoxine- α-ketoglutarate (PAK). This drug reduces pyruvic and lactic acidemia in normal subjects after muscular exercise (Cerretelli et al., 1981).

MATERIAL AND METHODS

All patients underwent a careful clinical and neurological examination. Special investigations included electroretinogram, CAT scan, EEG, CSF examination, EKG, EMG and nerve conduction velocity studies. In all patients myasthenia gravis was excluded by Tensilon test, thyroid ophthalmopathy by T_3 and T_4 determinations, Refsum disease by phytanic acid determination, abetalipoproteinemia by lipoprotein electrophoresis.

Blood lactate was determined by the method of Noll (1974). Blood pyruvate was determined by the method of Czok and Lamprecht (1974).

Each patient was subjected to a brachial biceps muscle biopsy. Cryostat sections were prepared and a battery of histological and histochemical reactions were performed (haematoxylin and eosin, modified trichrome, nicotinamide adenine nucleotide dehydrogenase, succinic dehydrogenase, ATPase pH 9.4-4.6-4.3, PAS, phosphorylase, acid phosphatase and oil red O). Samples of all muscles were treated with standard techniques for electron microscopy.

Table 1.

N° patient	AGE OF ONSET	PTOSIS CPEO	DESCENDING MYOPATHY	HEART CONDUCTION DEFECTS	RETINITIS PIGMENTOSA	HYPOACUSIA	CEREBELLAR ATAXIA TREMOR	SMALL STATURE	DIABETES	MENTAL DEFECTS	E E G ALTERATIONS	PERIPHERAL NEUROPATHY
1 ♀ 22 y.	15	++	+	0	0	0	0	0	0	0	0	0
2 ♀ 31 y.	12	+++	+	+++	0	+++	0	1,48 m.	0	0	++	0
3 ♂ 19 y.	11	+++	+++	+++	+++	+++	+++	1,52 m.	+++	+	+	0
4 ♀ 31 y.	17	+++	++	++	0	+++	+	1,55 m.	0	+	+	0
5 ♀ 13 y.	10	+++	++	+++	+++	+	+++	1,51 m.	++	+	0	0
6 ♂ 22 y.	5	++	+	+	0	+	+	0	0	0	++	0
7 ♂ 31 y.	15	++	+	+	0	+	0	0	0	0	0	0

This table summarizes the most important clinical and laboratory features of the seven patients.

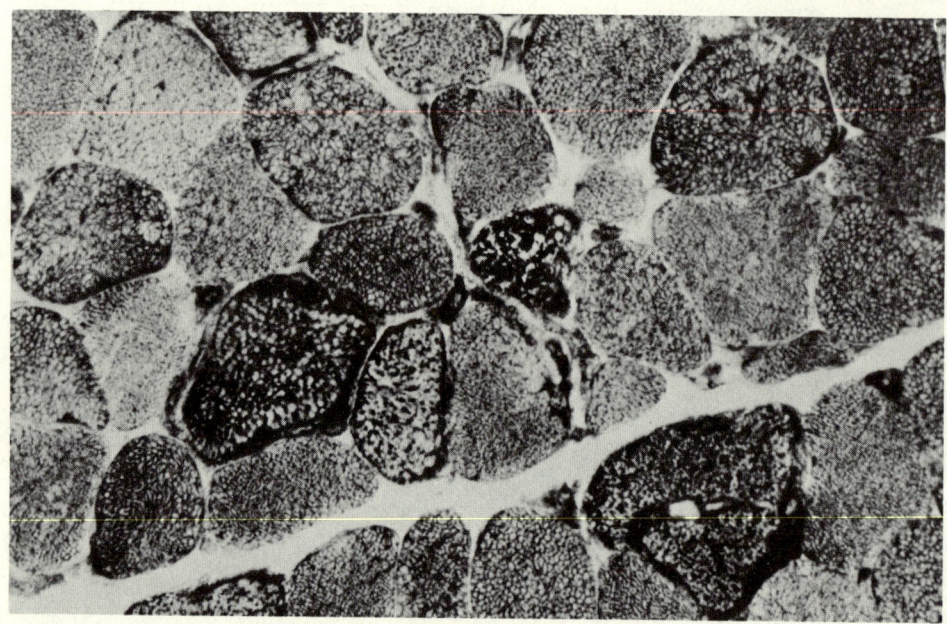

Fig. 1. Ragged red fibers : increased amount of SDH activity. Magnification 450 x

Muscle Biopsy

In all patients histological and histochemical reactions revealed the presence of a great number of "ragged red fibers", (RRF), predominantly type I. Everyone had abnormal amounts of granular red material with Gomori's trichrome stain, increased NADH and SDH activity (Fig. 1), aggregates of abnormal mitochondria at EM examination, often containing structural changes and crystal-like inclusions. The RRF had also an excess of lipid droplets and glycogen granules (Fig. 2). The myofibrils were often compressed and reduced in numer and size.

Therapeutical Trial

All patients were treated with 40 mg/Kg/die of PAK per os; blood pyruvate and lactate concentrations were determined on basal conditions and, in six cases, after two months of treatment. To further investigate pyruvate and lactate metabolism four patients (n° 1-2-6 and 7) underwent an oral glucose test load (1.75 g/Kg). Evaluation of blood pyruvate and lactate was performed before glucose load and after 30-60-90-120-150 and 180 minutes. Two

Table 2.

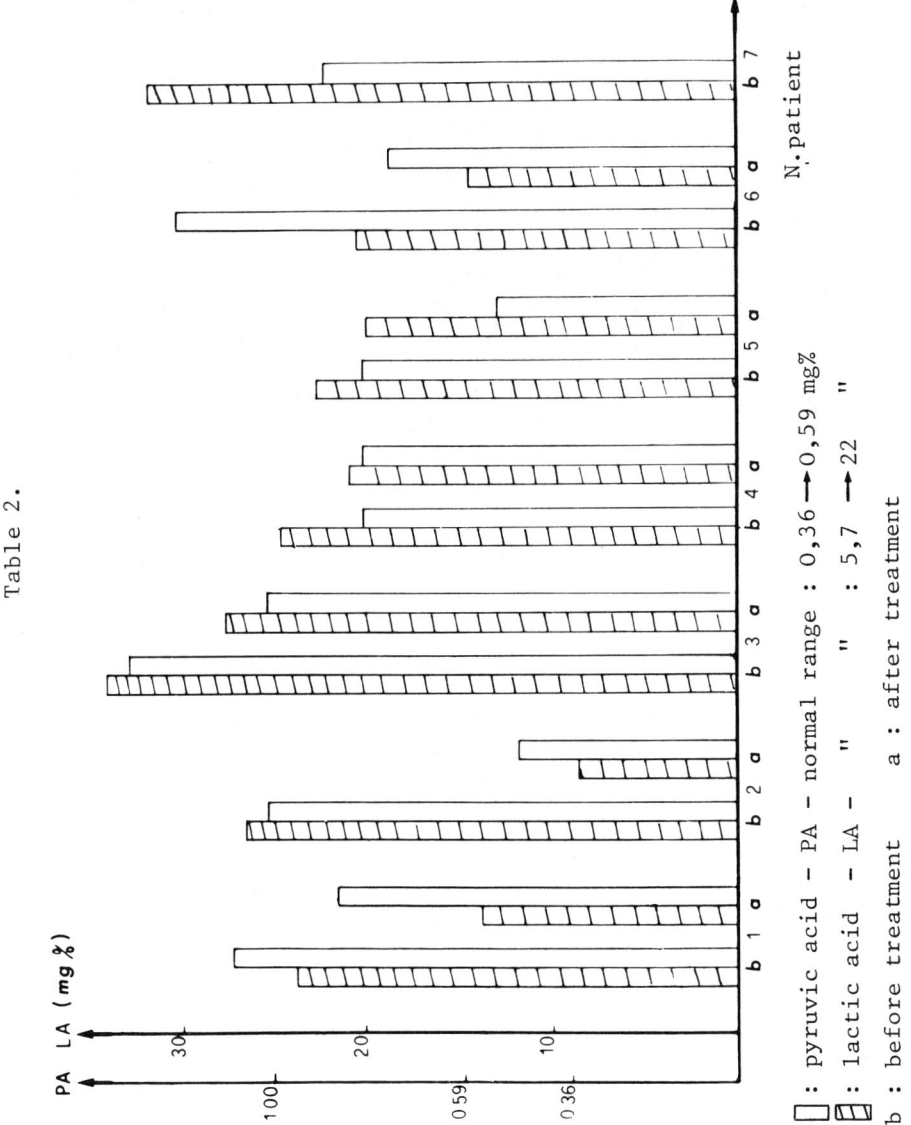

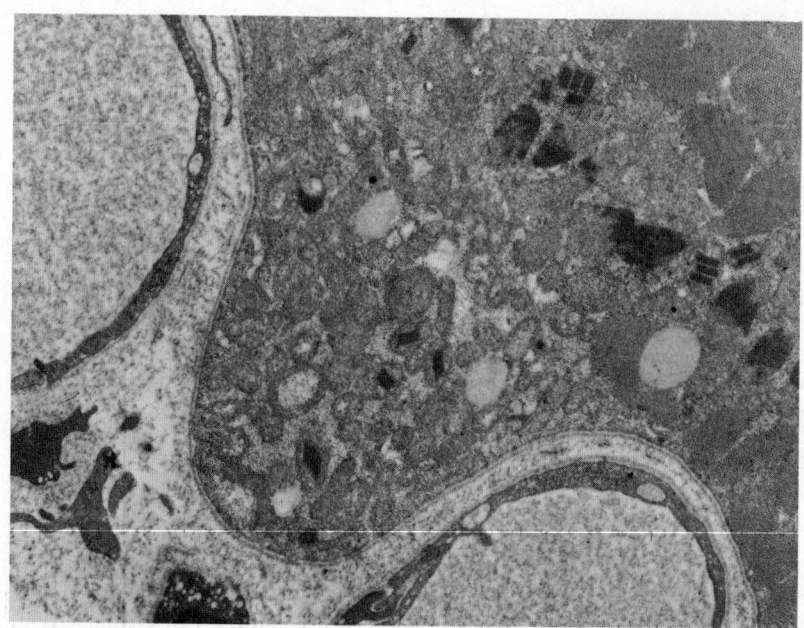

Fig. 2. Aggregates of mitochondria containing crystallike inclusions in subsarcolemmal space of a RRF. Magnification 25.000 x

of these patients (2 and 6) repeated the oral glucose test after two months of treatment. Results are reported in table 1, 2, 3, 4, 5, and 6. Statistical analysis of data is reported in Table 7.

DISCUSSION

The seven patients reported here as affected by CPEO meet the criteria for OP diagnosis recently revised by Bastiaensen et al. (1978). They showed an almost identical pattern of clinical picture, beginning at the age of 20 to 30 years with ptosis, CPEO and descending muscular weakness. The muscular weakness seemed to be myogenic in view of the EMG findings. EMG signs of neurogenic involvement were never present. Raised CSF protein in one patient and abnormalities of the EEG in five others indicated CNS involvement. Moderate to severe sensorineural deafness was found in all patients ; vestibular reflectivity was of difficult evaluation because of severe limitation of ocular movements. In two cases there was bilateral pigmentary degeneration of the retina. Endocrine abnormalities, such as diabetes, were detected

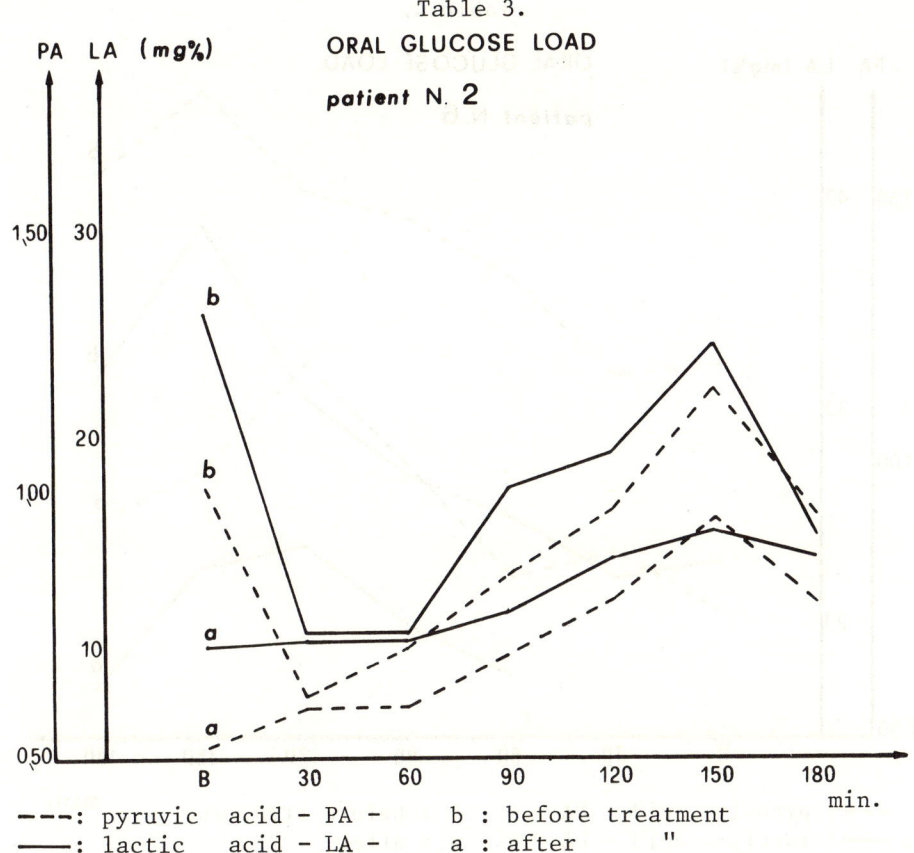

Table 3. ORAL GLUCOSE LOAD
patient N. 2

---: pyruvic acid - PA -   b : before treatment
———: lactic acid - LA -   a : after   "

in two patients ; three others had small stature. Disorders of the cardiac conduction system were present in six out of seven patients ; two patients had a cardiac pacemaker. Biochemical investigations revealed abnormal values of blood lactate and pyruvate both in basal conditions and after an oral glucose load.

The muscle biopsies confirmed mitochondrial myopathy ; all displayed a great number of RRF and prominent mitochondrial changes.

As is well known a number of human neuromuscular diseases have been described in which mitochondrial alterations of voluntary muscle are a prominent pathologic finding. The study of Ernster et al. (1959) stimulated workers to investigate mitochondrial function in human skeletal muscle diseases. Subsequent to the description of the hypermetabolic syndrome by Luft et al., (1962) other muscular disorders have been reported that fulfil the concept of "mitochondrial myopathies" (Van Wijngaarden et al , 1967).

Table 4.

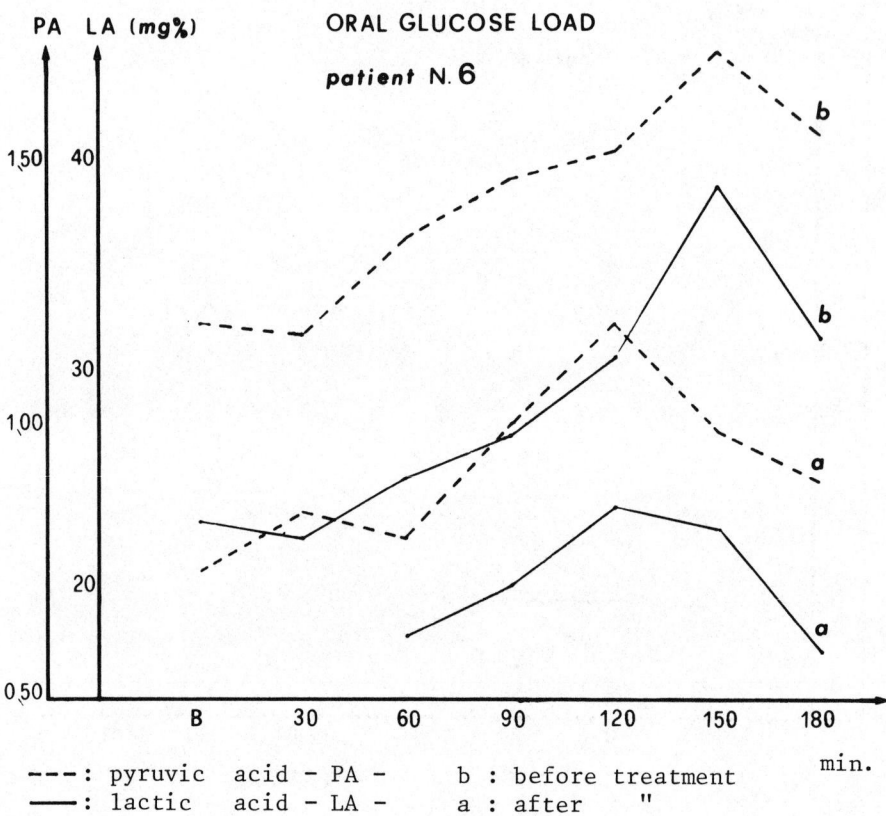

- - - : pyruvic acid - PA -   b : before treatment
——— : lactic acid - LA -   a : after   "

The designation "mitochondrial myopathy" is still valid, but only in a descriptive sense. In some cases the RRF and mitochondrial changes have been correlated with a defect in the oxidative pathways : uncoupling of the oxidative phosphorylation (Luft et al. 1962 ; Haydar et al. 1971 ; DiMauro et al. 1973a), low activity of cytochrome b (Spiro et al. 1970 ; Morgan-Hughes et al. 1977) deficient activity of cytochrome a_1/a_3 (French et al. 1972), carnitine deficiency (Engel and Angelini. 1973), poorly coupled oxidative phosporylation with alpha-glycerophosphate as substrate (DiMauro et al. 1973b) and a possible defect in NADH oxidase (Kark et al. 1972). On the other hand it is well known that experimental ischemia of skeletal muscle induces a variety of mitochondrial changes like those described in the so called mitochondrial myopathies (Hanzlikova and Schiaffino. 1977 ; Heffner et al. 1978).

With regard to the high level of lactate and pyruvate found

Table 5.

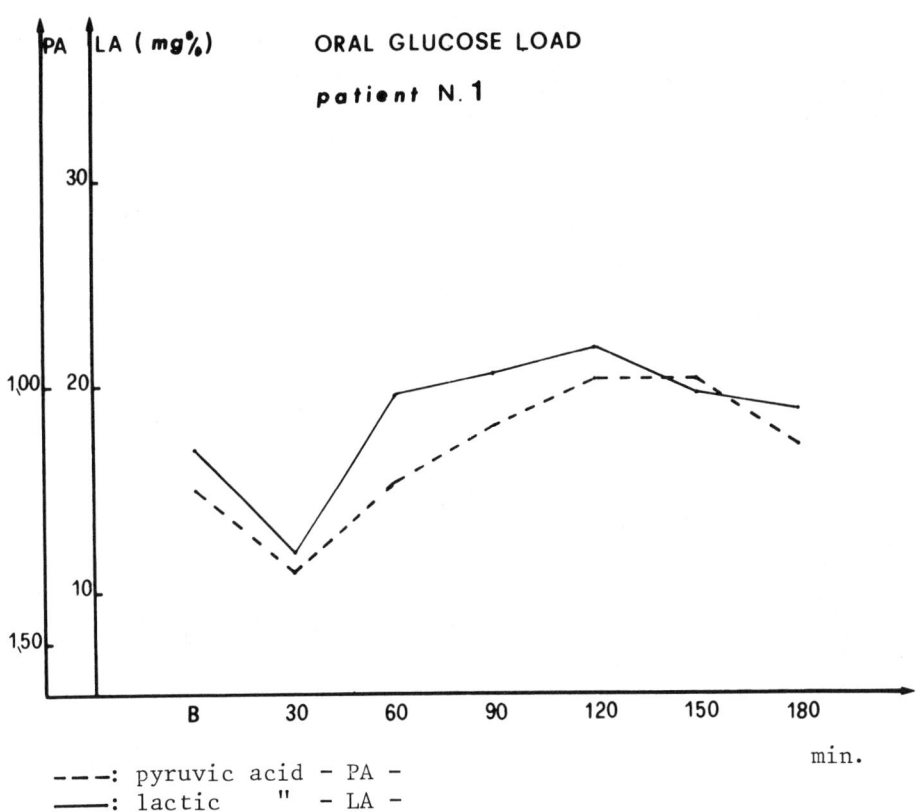

- - -: pyruvic acid - PA -
———: lactic   "   - LA -

in our patients, the lack of biochemical studies, in particular on isolated mitochondria, does not allow the conclusion of a specific enzymatic defect. Similar alterations of lactate and pyruvate have been reported by Van Wijngaarden et al. (1967) in a patient neurologically normal but whose mitochondria demonstrated loosely coupled phosphorylation. Lactic acidemia associated with muscle mitochondria disorders has been described in children Markesberry 1979) as well as in adults (Reske-Nielson et al. 1976; Scarlato et al. 1978). It is tempting to speculate that the increase of lactate and pyruvate in the blood may be the consequence of a block of the utilization of carbohydrate via the tricarboxylic acid cycle in the mitochondria. But it is clear, owing to the broad spectrum of the clinical features associated with elevation of the lactic acid level, that this association does not constitute a single entity.

In four patients with CPEO Bastiaensen et al. (1978) found signs of disturbed pyruvate and lactate metabolism ; the bioche-

Table 6.

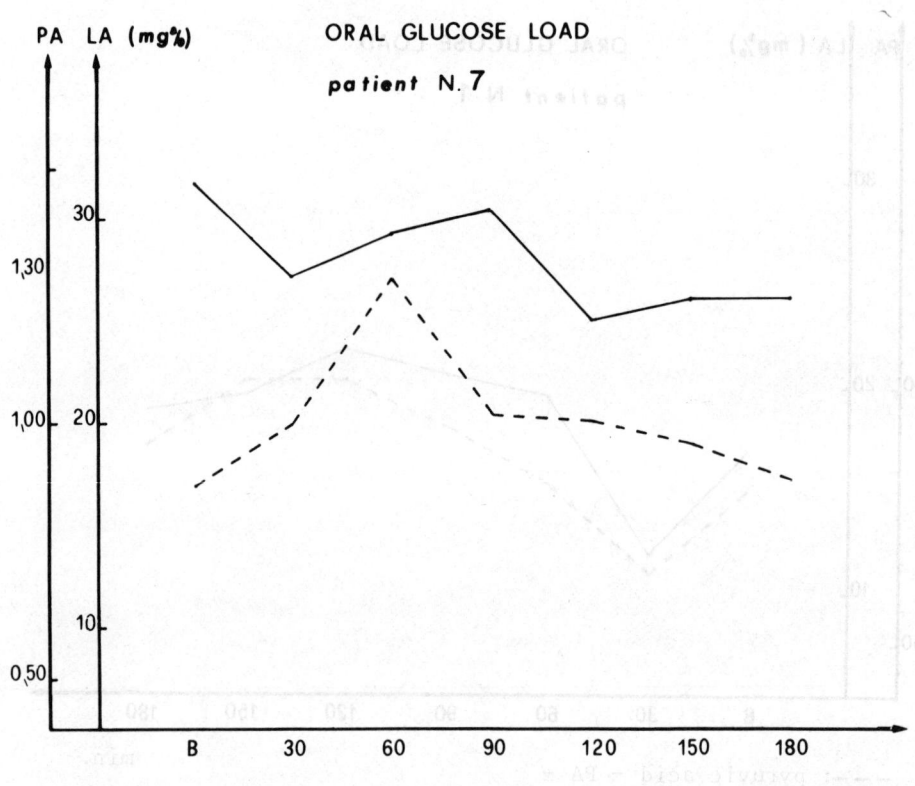

---: pyruvic acid - PA -
———: lactic acid - LA -

mical studies of the muscle homogenates showed disturbance of
pyruvate oxidation in two cases, loss of respiratory rate and
respiratory control in two others and some loose coupling of
oxidative phosphorylation as substrates in three cases. In the
author's opinion whether there is a link between a disturbed
pyruvate oxidation and the disturbed oxidative phosphorylation
is not known : both can lead to high lactate levels. Moreover the
normal fatty acid oxidation does not suggest replacement by the
fatty acid oxidation of the pyruvate oxidation in the formation
of acetyl-CoA. The statement of Bastiaensen et al. (1978) is
that in CPEO the hyperlactacidemia is not specific enough to
come to the conclusion of an inability of catabolizing pyruvate
in the citric acid cycle, without having determined the rate of
pyruvate oxidation or the process in the respiratory chain,
but it does indicate mitochondrial functional abnormality. On the
other hand in an experimental model (Sahgal et al. 1979) an
increase in muscle lactic acid does not seem to have a significant

Table 7.

STATISTICAL ANALYSIS

	lactic acid		pyruvic acid	
	b	a	b	a
m̄(mg%)	25.37	17.45	1.04	0.73
s.e.	1.85	2.71	0.08	0.09
%deviation		-31.22		-29.80
p		< 0.02		< 0.02

b : before treatment
a : after   "

role in the pathogenesis of mitochondrial pathology. The production of the mitochondrial changes, like concentric laminar bodies, fingerprint bodies and paracristalline inclusions, after intra-arterial injection of 2,4-dinitrophenol suggest that these changes represent an uncoupled oxydative phosphorylation and an intact or enhanced mitochondrial respiration (Saghal et al. 1979). Thus the increased lactate level causes tissue acidosis which further uncouples oxidative phosphorylation.

Further evidence supporting this hypothesis has been recently obtained by Nemni et al. (1981). In five patients with CPEO and morphological mitochondrial changes they found a diminished respiratory function (state 3 respiration and DNP-uncoupled respiration with NADH-linked substrates) and a decreased cytochrome oxidase (aa_3) activity of muscle homogenates.

In our CPEO patients the treatment with PAK induced a fall in lactate and pyruvate both in basal conditions and after an oral glucose load. Two mechanisms could be suggested to explain these actions : the PAK could accelerate the citric acid cycle by replenishing the intermediaries or could act by bypassing a possible defective oxidative decarboxylation present in CPEO. Moreover during mitochondrial energy deficiency states, like the CPEO, ornithine may be utilized to sustain ATP levels at the expense of decreased urea cycle flux and increased ammonia production at the glutamate dehydrogenase step. The decarboxylation of ornithine in the citric acid cycle acts through a pathway involving several enzymatic reactions. Catalytic amounts of PAK apparently initiate this sequence inducing a fall in the hyper-ammonemia observed as a consequence of failure of mitochondrial metabolism in some clinical conditions characterized as "mitochondrial encephalomyopathies" (Stumpf et al. 1979).

REFERENCES

Askanas, V., Engel, W. K., Britton, D. E., Adornato, B. T., and Eiben, R. M., 1978, Reincarnation in cultured muscle of mitochondria abnormalities, Arch. Neurol., 35:801.
Bastiaensen, L. A. K., Jaspar, H. M. J., Veerkamp, J. H., Brookelman, H., and Von Hinsbergh, V. W. M., 1978, Ophthalmoplegia plus, a real nosological entity, Acta Neurol. Scand., 58:9.
Bradley, W. G., Tomlinson, B. E., and Hardy, M., 1978, Further studies of mitochondrial and lipid storage myopathies, J. Neurol. Scie., 35:201.
Cerretelli, S., Marconi, C., and Sassi, G., 1981, The effects of pyridoxine alphaketoglutarate on human maximal aerobic and anaerobic performance, Europ. J. Physiol., in press.
Czok, R., Lamprecht, W., 1974, in "Methoden der Enzymatischen Analyse", Verlag Chemie, Weinheim p 1491.
Di Mauro, S., Schotland, D. L., Lee, C. D., Bonilla, E., and Conn, H. L., 1973a, Biochemical and ultrastructural studies of mitochondria in Luft disease : implications for "mitochondrial myopathies", Trans. Ann. Neurol. Assoc. 97:265.
Di Mauro, S., Schotland, D. L., Bonilla, E., Lee, C. P., Gambetti, P., Rowland, L. P., 1973b, Progressive ophthalmoplegia, glycogen storage and abnormal mitochondria, Arch. Neurol. 29:170.
Engel, A. G., Angelini, C., 1973, Carnitine deficiency of human skeletal muscle with associated lipid storage myopathy : a new syndrome. Science 179:899.
Ernster, I., Ikkos, D., and Luft, R., 1959, Enzymatic activities of human skeletal muscle mitochondria : a tool in clinical metabolic research, Nature, 184:1951.
French, J. H., Sheraed, E. S., Lubell, H., Brotz, M. and Moore, C. L., 1972, Trichcopoliodystrophy : report of a case and biochemical studies, Arch. Neurol., 26:229.
Gonatas, N. K., Evangelista, I., and Martin, J., 1967, A generalized disorder of nervous system, skeletal muscle and heart resembling Refsum's disease and Hurler's syndrome, Am. J. Med. 42:169.
Hanzlikova, V., and Schiaffino, S., 1977, Mitochondrial changes in ischemic skeletal muscle, J. Ultrastruct. Res., 60:121.
Haydar, N. A., Hadley, L. C., Afifi, A., Wakid, N., Bolles, S., and Fawaz, K., 1971, Severe hypermetabolism with primary abnormality of skeletal muscle mitochondria : functional and therapeutic effects of chloramphenicol treatment, Ann. Int. Med., 74:548.

Heffner, R. R., and Barron, S. A., 1978, The early effects of ischemia upon skeletal muscle mitochondria, J. Neurol. Sci., 38:295.

Karpati, G., Carpenter, S., Labrisseau, A., and Lafontaine, R., 1973, The Kearns-Shy Syndrome. A multisysem disease with mitochondrial abnormality demonstrated in skeletal muscle and skin, J. Neurol. Sci., 19:133.

Kark, R. A. P., Weinbad, E. C., Blass, D. P., and Engel, W. K., 1972, Oxidative metabolism in small samples of normal and diseased human muscle, Second. Int. Congress on Muscle Diseases Excerpta medica, 99.

Kark, R. A. P., and Buduelli, R. M., 1979, Pyruvate dehydrogenase deficiency in spinocerebellar degenerations Neurology, 29:126.

Kearns, T. P., and Sayre, G. P., 1958, Retinitis pigmentosa, external ophthalmoplegia and complete heart block, Arch. Ophthal. 60:280.

Kuroda, Y., Kline, J. J., Sweetman, L., Nylan, W. L. and Groshong T. D., 1979, Abnormal pyruvate and alpha-ketoglutarate dehydrogenase complexes in a patient with lactic acidemia, Pediatr. Res. 13, 928-931.

Luft, R., Ikkos, D., Palmieri, G., Ernster, L., and Afzelius, B., 1962, A case of severe hypermetabolism of nonthyroid origin and a defect in the maintenance of mitochondrial respiratory control. A correlated clinical, biochemical and morphological study, J. Clin. Inv. 41:1776.

Markesberry, W. R., 1979, Lactic acidemia, mitochondrial myopathy and basal ganglia calcification, Neurology 29:1057.

Morgan-Hughes, A., Derveniza, P., Kahn, S. N., London, D. N., Sherrott, R. M., Land, J. M., and Clark, J. B., 1977, A mitochondrial myopathy characterized by a deficiency in reducible cytochrome b, Brain, 100:617.

Nemni, R., Mitsumoto, A., and Bradley, W. G., 1981, Morphological and biochemical analysis in chronic progessive ophthalmoplegia, J. Neuropathol. Exptl. Neurol., 40:350.

Noll, F., 1974, in "Methoden der Enzymatischen Analyse", Bergmeyer ed. Verlag Chemie, Weinheim p 1521.

Reske-Nielsen, E., Lou, H. C., and Lowes, J., 1976, Progressive external ophthalmoplegia. Evidence for generalized mitochondrial disease with a defect in pyruvate metabolism, Acta Ophtalm., 54:551.

Sahgal, V., Subramani, V., Hughes, R., Shah, A., and Singh, H., 1979, On the pathogenesis of mitochondrial myopathies. An experimental study, Acta Neuropath., 46:177.

Scarlato, G., Pellegrini, G., and Veichsteinas, A., 1978, Morphologic and metabolic studies in a case of oculo-craniosomatic neuromuscular disease, J. Neuropathol., 37:1.

Schapira, Y., Cederbaum, S. D, Cancilla, P. A., Nielsen, D., and Lippe, B. M., 1975, Familial poliodystrophy, mitochondrial myopathy and lactate acidemia, Neurology, 25:614.

Schneck, L., Adachi, M., Briet, M., Walintz, A., and Volk, B. W., 1973, Ophthalmoplegia plus with morphological and chemical studies of cerebellar and muscle tissue, J. Neurol. Sci., 19:37.
Spiro, A. J., Moore, C. L., Prineas, J. M., Strasberg, P. M., and Rapin, I., 1970, A cytochrome-related inherited disorder of the nervous system and muscle, Arch. Neurol., 23:103.
Stumpf, D. A., Mc Cabe, E. R. B., Parks, J. K., and Bullon, W. W., 1979, A mechanism for hyperammonemia in mitochondrial disorders, Neurology, 29:576.
Sulaiman, W. K., Doyle, D., Johnson, R. H., and Jannett, S., 1974, Myopathy with mitochondrial inclusion bodies : histological and metabolic studies, J. Neurol. Neurosurg. Psychiat. 37:1236.
Van Wijngaarden, G. K., Bethlem, J., Meijer, A. E. F. H., Hulsman, W. C., and Feltkamp, C. A., 1967, Skeletal muscle disease with abnormal mitochondria, Brain, 90:577.

INTRACEREBRAL pH REGULATION AND AMMONIA DETOXIFICATION

N.M. Van Gelder

Centre de recherche en sciences neurologiques
Département de physiologie
Case postale 6128, Succursale A
Montréal, Québex H3C 3J7

SUMMARY

   Two regulatory cycles are known which assure the hydrogen and ammonia ion balance in the CNS. The CO_2/bicarbonate system opposes excessive acidoses and may provide an important energy source (H^+) for glia. The glutamate/glutamine system opposes excessive alkalosis and is important in balancing the excitation inhibition modulation of the CNS (glutamate - GABA system). Both systems are critically dependent on two glial enzymes : carbonic anhydrase and glutamine synthetase, respectively. The principal substrates for these enzymes- CO_2 , NH_3 and glutamate- are derived from neurons and become increasingly available with greater neuronal activity.

   Intracerebral pH regulation appears mainly governed by two processes. The first of these entails the participation of the enzyme carbonic anhydrase (EC. 4.2.1.1), now known to be present almost exclusively in the satellite cells of the central gray matter[1,2,3]. The catalytic action of this enzyme highly facilitates the hydration of CO_2 to carbonic acid (H_2CO_3), which in turn dissociates spontaneously and almost entirely into bicarbonate (HCO_3^-) and hydrogen ions. Because of the selective localization of carbonic anhydrase in the satellite cells, bicarbonate ion formation in the brain should be largely confined to these same cells. By analogy with a process occurring in erythrocytes (chloride shift) and at the choroid plexus[4], the bicarbonate ion formed within the glial cell is exchanged with extracellular chloride; this process may be a passive phenomenon or may represent facilitated exchange[4]. The overall action of carbonic anhydrase

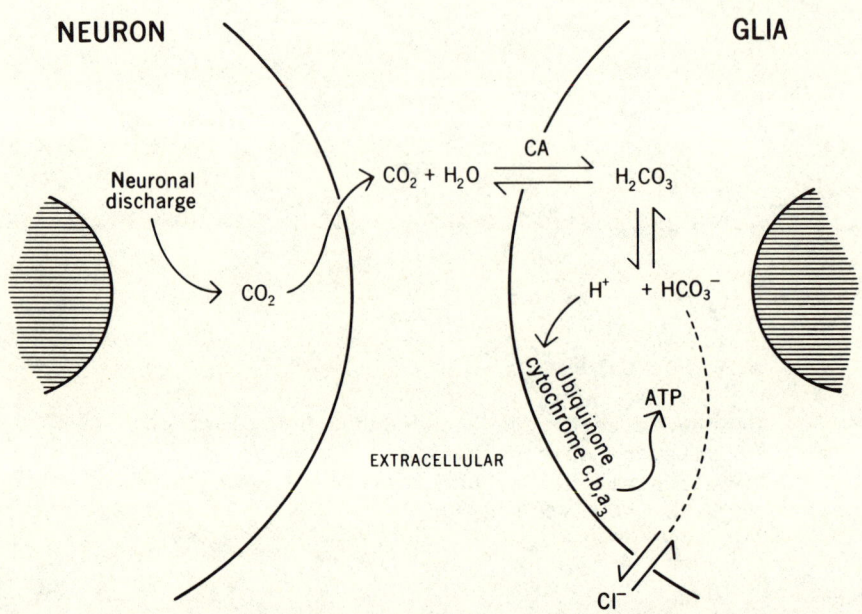

Fig. 1. The action of carbonic anhydrase (CA) in generating hydrogen and bicarbonate ions in glia. Scheme is based on evidence that (i) CA is only found in the satellite cells of the central gray[1,2,3], (ii) intracellular bicarbonate exchanges with extracellular chloride[4], (iii) the existence of a close relationship between glial membrane potential and oxidative metabolism, despite a sparse mitochondrial population[6], and (iv) hydrogen donors elicit the complete cytochrome action spectrum in glia (G. Svaetichin, unpublished).

thus appears to promote interstitial bicarbonate accumulation, coupled to a movement of chloride into non-neuronal cells of the CNS. No energy seems to be required for this phenomenon. Bicarbonate buildup in the extracellular spaces tends to render this environment more alkaline. Conversely, inhibition of cerebral carbonic anhydrase by for example, acetazolamide[5], promotes a decrease of the interstitial pH and increases the partial CO_2 tension. This favours vasodilatation which assures a more rapid elimination of cerebral CO_2 via the blood at the lungs. The effects of carbonic anhydrase inhibition appear therefore largely self-limited.

Inhibition of the enzyme is clinically well documented to

have an anti-seizure action[5]. The effect is usually attributed to
the acidification of the interstitial environment and CO_2 accumulation. Both are reputed to decrease neuronal activity although as
far back as 1965, Svaetichin et al.[6] already suggested that glial
rather than neuronal elements are the most sensitive to changes of
pH, CO_2 tension, or metabolic conditions. It was the contention
of Svaetichin* that the H^+ ions, produced by carbonic anhydrase
from neuronally originating CO_2, are utilized in the retina by glial
cells to drive an energy generating system in the non-neuronal
plasma membrane. The implication would be that many different
adverse metabolic conditions thought to act on neurons, in reality
affect their satellite cells. This extramitochondrial energy system
for which Svaetichin recently also found strong evidence in the
vertebrate brain (unpublished), may represent an ubiquinone-
cytochrome c, b, a_3 complex (Fig. 1). The complex appears directly
driven by hydrogen donors (acetate, succinate, etc.) and is
sensitive to uncouplers of oxidative metabolism e.g. dinitrophenol[6].
The existence of such a system would be compatible with many investigations which indicate that glia efficiently utilize tricarboxylic
acid cycle intermediates (keto-acids) whereas neurons do not[7].
Moreover, the search for the causes of irreversible damage to CNS
functions following short periods of energy deficient states such
as ischemia, hypoxia, etc., so far has been almost entirely focussed
on neuronal elements[8]. Yet the data show that these structures,
with respect to practically all functional and metabolic parameters
studied, rapidly and seemingly entirely recover from such conditions[9].
Damage to the plasma membrane of glia under these adverse conditions
may be more permanent. To quote[6] : "The horizontal cell and other
non neuronal (controller) cells in the retina must strongly depend
on oxidative metabolism, since their membrane potentials and
responses are very sensitive to oxygen lack and to the application
of metabolic agents which interfere with aerobic metabolism. The
carbonic anhydrase, located almost exclusively in glia, is probably
structurally associated with oxidative enzymes". If this observation
can be corroborated further, it will be necessary to assign a greater
importance than is now suspected to the role of glia and carbonic
anhydrase in maintaining brain function (Fig. 1, proposed schema).

The second pH regulating mechanism prevents the intracerebral
pH from shifting too much toward the alkaline side (Fig. 2). Free
ammonia, generated during CNS activation, is the most prevalent
agent tending to alkalinize the cerebral environment[13]. A toxic
effect of excessive ammonia accumulation may be due to a direct
result of alkalosis, or may be due to an action of the ion on cell
membranes and interference with transmembranal monovalent ion
transport. It is nevertheless evident that intracerebral alkalosis
will affect tertiary protein structure and may also affect many

---

*This formidable scientist to my deep regret recently died.

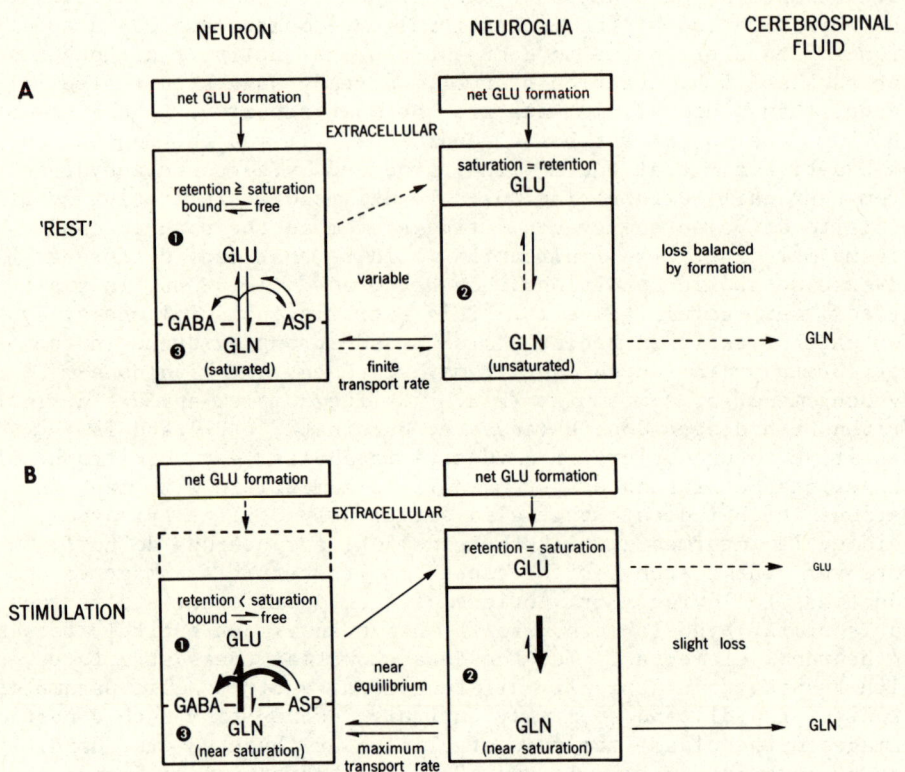

Fig. 2. The compartmentalized metabolism of glutamic acid in the intact CNS[10]. Flow of glutamic acid from neurons (compartment 1) to glia (compartment 2) and back in the form of glutamine into nerve terminals (compartment 3). Compartment 1 may represent the perikarya of neurons where glutamic acid derived from glucose is sequestered in a neutralized form. That portion in exess of this process is released and taken up into adjacent glial elements where it is amidated to glutamine Glial elements, closely interfaced with nerve terminals, release glutamine which on entering the nerve terminals is deamidated to give rise to releasable pools of glutamic acid, aspartic acid and/or GABA (when GAD is present). Stimulation (depolarization) accelerates this process and tends to saturate uptake, retention and enzymic mechanisms so that the amino acids tend to accumulate in the extracellular environment. Taurine influences the compartmentalized metabolism of glutamic acid in some manner, so that less free glutamic acid accumulates extracellularly (from 11, see also 10, 12).

enzyme rates by causing these to operate further away from the normal in vivo pH. This will rapidly disturb delicate equilibrium reactions and enzymic interactions responsible for maintaining a host of complex metabolic cycles within cells as well as between cells.

Two enzymes play a capital role in preventing the accumulation of free ammonia. Glutamic acid dehydrogenase (EC.1.4.1.2), in the presence of α-ketoglutarate, NADH and ammonia, reversibly aminates α-ketoglutarate to glutamic acid (14). Since the bulk of that amino acid seems to be present in neuronal elements rather than in surrounding satellite cells[11], the reaction must predominantly occur in neurons. Moreover, in the presence of ammonia CNS equilibrium conditions strongly favour glutamic acid synthesis rather than its oxidative deamination[14]; the glutamic acid dehydrogenase reaction is considered to be the principal source of glutamic acid in the CNS[13]. The high neuronal content of glutamic acid thus supposes that most of the free ammonia formed in the CNS is also of neuronal origin and, indeed, neuronal excitation does generate free ammonia[15,13]. However, in order for glutamic acid synthesis to serve as an effective detoxification reaction for ammonia, it appears necessary to chelate or sequester the free glutamic acid so formed, since the physico-chemical properties of that amino acid seem to preclude its intracellular accumulation in significant amounts. This is evident from the fact that glutamic acid as such is extremely acidic and, also, very insoluble, so that on its accumulation the cell would be rapidly destroyed. Hence, the high levels of so-called free glutamic acid in the CNS which, incidently, signify an absence or very weak product inhibition of glutamic acid dehydrogenase[14], indicates that in addition to adequate substrate availability, a second regulatory factor must be operating to assure continuous enzyme action. This mechanism must represent a process by which the free glutamic acid formed can be rapidly neutralized or sequestered[11]. Whether the mechanism entails the participation of an intracellular neutralizing ion (Ca^{2+}, K^+) or an intracellular chelating/absorption protein (calmodulin + taurine perhaps), is entirely speculative. It is clear however that the rather steady concentration range of cerebral glutamic acid found throughout most of the species studied[11], implies that the ability to solubilize or neutralize glutamic acid is limited to approximately the 10 mole range (8 - 13 moles). Beyond that maximum capacity, further synthesis of glutamic acid rapidly tends to lower the intracellular pH which in turn would, among other events, cause feedback inhibition to glucose metabolism[16]. But since the cerebral activity accompanying ammonia production may require increased energy, such an indirect feedback control on glutamic acid synthesis would seem counterproductive. The other manner of removing glutamic acid when its ammonia stimulated formation exceeds the intracellular retention capacity (see

17), is by release from the neuron. This in fact is a well documented phenomenon (e.g. 18, 12).

Once glutamic acid is released from neurons, two events assure that, normally, little extracellular glutamic acid accumulates. At the same time, these same mechanisms also aid in the further removal of ammonia. Glia avidly capture glutamic acid where, by the action of glutamine synthetase (EC. 6.3.1.2) in the presence of ammonia and an energy supply (ATP), glutamic acid is amidated to glutamine[19,20]. Glutamine synthetase is only found in glia and glutamine formation therefore also only occurs in these cells[21,22]. Since extracellular glutamic acid is well known to cause strong neuronal excitation, it is evident that the glutamine synthetase reaction by simultaneously "neutralizing" free ammonia and glutamic acid, prevents excessive neuronal hyperexcitability[10,12]. Inhibition of the enzyme by the selective inhibitor methionine sulfoximine, causes strong seizures[23] which are not easily abolished except by direct displacement of the inhibitor from the enzyme (by methionine). Many other conditions, all of which are accompanied by temporary or permanent seizure states[24], may also interfere with the removal of the extracellular glutamic acid. Energy deficiencies will interfere with the active transport of glutamic acid into glia and the ATP requiring glutamine synthetase activity, besides causing a general release of the so-called free amino acids from neurons[18,8]: this retention process is also dependent on energy. Mechanical factors such as a widening of the extracellular spaces (oedema), inadequate or defective maturation of the CNS which normally assures proper neuronal-glial anatomical interrelationships, or scarring, are all situations[24] which may be imagined to lead either to inefficient glial uptake of glutamic acid or, possibly, to a defective or lower glutamine synthetase activity itself. In general, any condition which causes release of glutamic acid to surpass its limited rate of active transport into glia or exceeds maximum glutamine synthetase activity, would lead to extracellular glutamic acid accumulation[12,11]. It should also be noted that GABA in nerve terminals is in part derived from glutamine after that amino acid has been deamidated to glutamic acid by glutaminase (EC.3.5.1.2). According to Kvamme[20], the synaptic enzyme is easily inhibited by ammonia. High ammonia levels (metabolic disease states) may thus not only stimulate glutamic acid synthesis, thereby "flooding" the extracellular environment with strongly excitatory glutamic acid, but may also simultaneously lead to an inefficient GABA synthesis due to lack of substrate and, hence, deficient inhibition in the CNS[20]; inadequate glutamine synthesis which appears rare in nature, would produce a similar effect[15].

In conclusion, it is clear that pH regulation as well as the excitation/inhibition balances within the CNS are importantly dependent on the intimate and precisely arranged contacts between neurons and the surrounding glial network. Furthermore, future investigations into the metabolic causes for cerebral dysfunctions may have to pay closer attention to the fundamental role of glia in the control of neuronal excitability.

ACKNOWLEDGEMENT

Supported by funds from the Université de Montréal (CAFIR), the Savoy Foundation for Epilepsy and the Banting Research Foundation.

REFERENCES

1. Giacobini, E., J. Neurochem. 9 : 169-177 (1962).
2. Mandel, P., Roussel, G., Delaunoy, J.-P. and Nussbaum, J.-L., in : "Dynamic Properties of Glia Cells", edited by E. Schoffeniels, G. Franck, L. Hertz and D.B. Tower. Acad. Press, N.Y., pp. 619-630 (1978).
3. Parthe, V., J. Neuroscie Res. 6 : 119-131 (1981).
4. Woodbury, D.M., Progr. Brain Res. 29 : 297-312 (1968).
5. Woodbury, D.M., "Antiepileptic Drug Mechanisms of Action" edited by G.H. Glaser, J.K. Penry and D.M. Woodbury, Raven Press, N.Y., pp. 617-633 (1980).
6. Svaetichin, G., Negishi, K., Fatehchand, R., Drujan, B.D., and Selvin de Testa, A., Progr. Brain Res. 15 : 243-266 (1965).
7. Tursky, T., Ruscak, M., Lassanova, M. and Ruscakova, D., J. Neurochem. 33 : 1209-1215 (1979).
8. Norberg, K. and Siesjo, B.K., Brain Res. 86 : 45-54 (1975).
9. Patuszko, A., Wilson, D.F., Erecinska, M. and Silver I.A., J. Neurochem. 36 : 116-123 (1981).
10. Berl, S., J. Biol. Chem. 240 : 2047-2054 (1965).
11. Van Gelder, N.M., Can. J. Physiol. Pharmacol. 56 : 362-374 (1978).
12. Van Gelder, N.M., Adv. Biochem. Psychopharmacol. 29 : 115-125 (1981).
13. McIlwain, H., in : "Biochemistry and the Central Nervous System", edited by H. McIlwain, J. & A. Churchill Ltd., London, pp. 114-119 (1959).
14. Chee, P.Y., Dahl, J.L. and Fabien, L.A., J. Neurochem. 33 : 53-60 (1979).
15. Kvamme, E. and Olson, B.E., Brain Res. 181 : 223-233 (1980).
16. Norberg, K. and Siesjo, B.K., Brain Res. 86 : 31-44 (1975).
17. Cummins, J.T., Hamberger, A. and Nystrom, B., J. Neuroscie. Res. 6 : 217-224 (1981).
18. Collins, G.G.S., Anson, J. and Probett, G.A., Brain Res. 204 : 103-120 (1981).
19. Hertz, L., Schousboe, A., Boechler, N., Mukerji, S. and

Federoff, S., Neurochem. Res. 3 : 1-14 (1978).
20. Kvamme, E., In :"GABA - Biochemistry and CNS Function",edited by P. Mandel and F.V. De Feudis, Plenum Publ. Corp., N.Y., pp. 111-137 (1979).
21. Henn, F.A., Goldstein, N.N. and Hamberger, A., Nature 249 : 663-664 (1974).
22. Norenberg, M.D. and Martinez-Hernandez, A., Brain Res. 161 : 303-310 (1979).
23. Meister, A., In : "Enzyme-Activated Irreversible Inhibitors", edited by N. Seiler, M.J. Jung and J. Kock-Weser, Elsevier/ North-Holland Biomedical Press, Amsterdam, pp. 187-209, (1978).
24. Robb, P., Dhew Public. No (NIH) 73-415, Public Health Service Publication # 1357, pp. 4-26 .

CONCLUDING REMARKS

First I would like to point out this was a multidisciplinary meeting, in which both clinicians and basic scientists participated. An improved understanding of the urea cycle was the goal of this symposium. In physiological conditions the urea cycle enables the organism to get rid of ammonia. Malfunction leads to accumulation of ammonia, ammonia containing compounds and other metabolites of the urea cycle. The urea cycle is not an isolated metabolic pathway in the organism; it connects to other pathways, such as those that allow urea metabolism, such as the transformation of ornithine in various metabolites, like proline, and of arginine in monoguanidines. For this last pathway, one remaining question is if guanidinosuccinic acid is a metabolite of urea or of arginine.

These different metabolic cycles and pathways have mostly been studied in man. On the other hand, it is clear that interesting observations can be made in animals, for instance in cats fed an arginine-enriched diet. Experimental models of these metabolic diseases have been presented, and these models can also bring important information.

Three major points have been discussed during the meeting :

1. how can malfunctions of the urea cycle, and its interphasing metabolic pathways, be diagnosed?
2. what are the physiopathological implications of these malfunctions?
3. how can these diseases be treated?

We will now recall the discussion remarks in more details.

1. <u>How can a precise diagnosis be made?</u> There are different steps of diagnosis : screening, precise biochemical analysis, and eventually techniques borrowed from other disciplines. Classical techniques, such as chromatography and biological tests are primarily used for screening. Nevertheless, other methods such

as determination of ammonia, analysis of organic acids in urine, mainly of orotic acid, a.s.o. have been proposed and discussed. One can ask if determination of monoguanidines would provide important information. Some participants in the meeting believe that the so called cystinurias include numerous hyperargininemias that could be diagnosed correctly by analysis of amino acids in peripheral blood, or by determination of guanidines in urine.

Positive screening results require that metabolites and enzymes are determined as fast as possible. The study of metabolites does not appear to present problems. On the other hand, many questions about the enzyme assays remain. Should such assays be made in peripheral blood, erythrocytes or leukocytes? Is liver or musclenerve biopsy indicated? Should assays be made in fibroblasts? Is a simple assay sufficient or should enzyme kinetics be studied? Should only enzymes be identified or should enzyme precursors be studied with the complex but very elegant available methods? What is the value of isoenzyme determination? Can one conclude from enzymatic studies that the heterogeneity of some enzymes implicates that the corresponding disease, as citrullinemia, is also heterogenous? What levels of enzyme activity can be termed pathological : total absence or 50-70 % activity? Are we dealing with absence of enzymes, or rather a decreased enzyme activity due to a gene mutation which can alter the rate of synthesis or/and degradation of the enzyme? The gene mutation probably also influences the regulatory (inhibition or activation) mechanism controlling the active site of the enzyme.

An overview of the communications in this meeting shows that 2 points were virtually not discussed :

a) the morphological changes in diseases of the urea cycle. Morphological anomalies in gyrate atrophy and in Reye's syndrome were shown during the meeting.
b) the incidence, and more general, the epidemiology of urea cycle diseases. It should be mentioned that the first case of argininosuccinic aciduria of Japanese origin was reported during this meeting.

2. The second discussion point is <u>the mechanism by which the symptoms appear in these diseases</u>. Important here is to know which metabolites are toxic and can damage the organism, in particular the nervous system. It should be mentioned that certain metabolic anomalies are tolerated very well, and can be classified as non-disease symptoms. This can also be observed in α-aminoadipic aciduria where, of two siblings presenting the same metabolic anomaly, one is severely oligophrenic and the other is normal. The same is found in histidinemia where the metabolic anomaly often does not give any clinical symptoms.

In diseases of the urea cycle, would increased levels of amino acids, organic acids, orotic acid, monoguanidines or ammonia be looked upon as pathological? Or low levels of urea? Should we think of an autointoxication of the organism, functioning as a vicious circle leading to new intoxications? Ammonia seems to play a central role and could be the toxic product in the various diseases of the urea cycle, but also in many congenital malformations. What could be the link between those two groups? What is the origin of hyperammonemia in diseases where the urea cycle is normal? Can clinical manifestations be the consequence of reactivation of metabolic pathways that are not used in physiological conditions in humans, but are physiological in invertebrates and vertebrates who have no urea cycle? Can we think of inhibition of urea synthesis or of a failure to synthesize enough ornithine to drive the urea cycle? This is seen in cats on a arginine free diet. Another problem is the variation of the symptomatology in some diseases : the clinical condition is sometimes much worse than in others, even though the levels of toxic substances may be the same. How could these cases be identified, and what causes the differences? Why are children with carbamylphosphate deficiency or citrullinemia severely ill, whereas hyperargininemia patients survive? An always returning question is : how can the localisation of toxic symptoms be explained? How can the ophthalmological symptoms in hyperornithinemia be explained? Should assays be made on affected organs to establish these localisations?

3. Lastly therapeutic aspects have been discussed at length during the meeting. We will not recall this problem in detail. Nevertheless, it should be mentioned that the dietary treatments that are proposed are not only restrictive, but sometimes characterized by addition of certain substances to the diet. These supplements can have a direct or indirect effect. Indirect restriction of ornithine by way of an arginine free diet as proposed in gyrate atrophy is an example of a new diet principle.

These are the conclusions we have drawn from these communications. Two essential points that remain to be clarified appeared during the discussion and will be mentioned as conclusion :

a) the methodology for enzymatic assays and the interpretation of the results. It should be remembered that these assays are often made in conditions of pH, temperature and concentrations of metabolites that are not physiological.

b) its seems also to be imperative that toxic substances are identified in order to establish efficient, albeit symptomatic therapies.

A. Lowenthal

PARTICIPANTS

ABE S.
Department of Anesthesia
Okayama University Hospital
Okayama 700
Japan

AKABOSHI I.
Department of Pediatrics
Kumamoto University Medical
School
Kumamoto 860
Japan

AKAMATSU K.
Department of 3rd Internal
Medicine
Ehime University Medical School
Ehime 791-02
Japan

ANDO A.
The 1st Department of Medicine
Osaka University Hospital
Fukushima
Fukushima-Ku, Osaka
Japan

AONO S.
Department of Pediatrics
Osaka City University Medical
School
Osaka 545
Japan

ARASHIMA S.
Department of Pediatrics
Hokkaido University
School of Medicine, North 14,
West 5, Sapporo 060
Japan

BABY T.G.
Department of Zoology
University of Poona
Ganeshkhind
Pune 411007
India

BACHMANN C.
Chem. Zentral labor der Universitätskliniken
Inselspital
CH-3010 Bern
Switzerland

BATSHAW M. L.
The Johns Hopkins University
John F. Kennedy Institute
707 North Broadway
Baltimore, MD 21205
USA

BERGER R.
Department of Pediatrics
University of Groningen
10 Bloemsingel
Groningen
The Netherlands

BLOM W.
Sophia Children's Hospital
Erasmus University
Gordelweg 160, POBox 70029
3000 LL Rotterdam
The Netherlands

BÖHLES H.
Universitätskinderklinik
Loschgestrasse 15
8520 Erlangen
West Germany

BRIAND P.
Laboratoire de Biochimie Génét.
Hôpital Necker-Enfants Malades
149, rue de Sèvres
75730 Paris Cedex 15
France

CATHELINEAU L.
Laboratoire de Biochimie Génét.
Hôpital Necker-Enfants Malades
149, rue de Sèvres
75730 Paris Cedex 15
France

CERRI C.
Department of Neurology
University of Milano
Via F. Sforza 35
20122 Milano
Italy

COHEN B. D.
Bronx Lebanon Hospital
1276 Fulton Avenue
Bronx, N.Y. 10456
USA

DEN TANDT
Department of Pharmacology
Universitaire Instelling Antwerpen
B-2610 Wilrijk
Belgium

FICI
Third Medical Clinic
University of Milan
Via Pace 15
Milano
Italy

FUJIMOTO S.
Department of Public Health and Welfare
Osaka City Office
Osaka
Japan

GRISAR T.
Neurochemical Laboratory
Liège University
17 Place Delcour
B-4020 Liège
Belgium

HAGIWARA H.
Laboratory of Biological Chem.
Tokyo Institute of Technology
Ookayama, Meguroku
Tokyo 152
Japan

HASE Y.
Osaka City Children Health Service Center
Osaka
Japan

HAYASAKA S.
Department of Ophthalmology
Tohuku University
School of Medicine
Sendai 980
Japan

HIRD F. J. R.
Department of Biochemistry
University of Melbourne
Parkville 3078 Victoria
Australia

HOSHINO T.
Pharmaceutical Institute
Keio University Medical School
35-Shinanomachi, Shinjukuku
Tokyo 160
Japan

ICHIBA Y.
Department of Pediatrics
Okayama National Hospital
Okayama 700
Japan

PARTICIPANTS

IKEDA K.
Department of 2nd Internal Med.
Nihon University, Medical School
Itabashiku
Tokyo 173
Japan

ISHIDA M.
The 1st Department of Medicine
St. Marianna University
School of Medicine
Kawasaki
Japan

ISHIKAWA Y.
Department of Health and Human
Services
National Institutes of Health
Bethesda, Maryland 20205
USA

ISODA K.
Sakura National Hospital
Chiba
Japan

ISSHIKI G.
Department of Pediatrics
Osaka City University Medical
School
Osaka 545
Japan

KAMOUN P.
Laboratoire de Biochimie Génét.
Hôpital Necker-Enfants Malades
149, rue de Sèvres
75730 Paris Cedex 15
France

KARCHER D.
Laboratory of Neurochemistry
Born-Bunge Foundation
Universitaire Instelling
Antwerpen
B-2610 Wilrijk, Belgium

KATO T.
Chubu Rosai Hospital
Nagoya
Japan

KATSUNUMA T.
Department of Biochemistry
Tokai University, Medical School
Boseidai
Isehara-city 259-11
Japan

KODAMA H.
Department of Pediatrics
Osaka University Hospital
Fukushima-ku
Osaka
Japan

KUBOTA K.
Department of Pediatrics and
Child Health
Kurume University Medical Center
Kurume 830
Japan

LAMERS W. H.
Anatomical-Embryological Laborat.
University of Amsterdam
Mauritskade 61
Amsterdam 1092 AD
The Netherlands

LEMIEUX B.
Département de Pédiatrie
Centre Hospitalier Universitaire
de Sherbrooke
Sherbrooke, P.Q. J1H 5N4
Canada

LOWENTHAL A.
Laboratory of Neurochemistry
Born-Bunge Foundation
Universitaire Instelling
Antwerpen
B-2610 Wilrijk, Belgium

MARESCAU B.
Laboratory of Neurochemistry
Born-Bunge Foundation
Universitaire Instelling
Antwerpen
B-2610 Wilrijk, Belgium

MATSUDA I.
Department of Pediatrics
Kumamoto University, Med. School
Kumamoto-City
860 Kumamoto
Japan

MATSUDA Y.
Department of Enz. Chemistry
Institute for Enz. Research
Tokushima University, Med. School
Tokushima 770
Japan

MATSUZAWA T.
Department of Biochemistry
Fujita-Gakuen University
School of Medicine
Toyoake, Aichi 470-11
Japan

MIKAMI H.
The 1st Department of Medicine
Osaka University Hospital
Fukushima, Fukushimaku
Osaka
Japan

MOHANA CHARI V.
Research Scholar
Department of Zoology
Sri Venkateswara University Coll.
Tirupati 517502
India

MORI A.
Okayama University
School of Medicine
Okayama 700
Japan

MORI M.
Department of Biochemistry
Chiba University, Medical School
Inohana
Chiba 280
Japan

MURAKAMI T.
Department of Pediatrics & Child Health
Kurume University Medical Center
Kurume 830
Japan

MURAKAMI-MUROFUSHI K.
Department of Biochemistry
Teihyo University
School of Medicine
Tokyo 173
Japan

NAGATA N.
Department of Pediatrics
Kumamoto University
Medical School
Kumamoto 860
Japan

NAKAJIMA S.
Department of Pediatrics
Osaka City University
Medical School
Osaka 545
Japan

NAYLOR E. W.
Department of Pediatrics
State University of New York
352 Acheson Hall-3435 Main St.
Buffalo, New York 14214
USA

NISHIDA N.
Department of Pediatrics
Kansai Medical College
Moriguchi 570
Japan

PARTICIPANTS

NIWA A.
Department of Microbiology
Dokkyo University
School of Medicine
Ibaragi 321-02
Japan

NOPPE M.
Laboratory of Neurochemistry
Born-Bunge Foundation
Universitaire Instelling
Antwerpen
B-2610 Wilrijk, Belgium

NYHAN W. L.
Department of Pediatrics
University of California
San Diego
La Jolla, CA 92093
USA

OCHIAI Y.
Department of Anesthesia
Okayama University Hospital
Okayama 700
Japan

OKADA S.
Department of 3rd Internal Med.
Ehime University Medical School
Ehime 791-02
Japan

OKADA S.
Department of Pediatrics
Osaka University Hospital
Osaka 553
Japan

OKANO Y.
Department of Pediatrics
Osaka City University
Medical School
Osaka 545
Japan

OOTA Y.
Department of 3rd Internal Med.
Okayama University
Medical School
Okayama 700
Japan

OWADA S.
The 1st Department of Medicine
St. Marianna University
School of Medicine
Kawasaki 213
Japan

PLUM
Aldersrogade 43 F IV
DK-2200 Copenhagen M
Denmark

QURESHI I. A.
Centre de Recherche
Hôpital Sainte Justine
Montreal, Que
Canada H3T IC5

ROBIN Y.
Biochimie marine EPHE
Collège de France
11 Place Marcellin-Berthelot
75231 Paris Cedex 05
France

SAHEKI T.
Department of Biochemistry
Kagoshima University
School of Medicine
Kagoshima 890
Japan

SAITO T.
Tohoku University
School of Medicine
Sendai 980
Japan

SAITO Y.
Laboratory of Biolog. Chemistry
Tokyo Institute of Technology
Ookayama, Meguroku
Tokyo 152
Japan

SAKAGAMI K.
Department of Surgery
Okayama University
Medical School
Okayama 700
Japan

SAKAGUCHI Y.
Department of Pediatrics
Kurume University
Medical School
Kurume 830
Japan

SAKIYAMA T.
Department of Pediatrics
Nihon University Surugadai Hosp.
Chiyodaku
Tokyo 101
Japan

SALERNO F.
Third Medical Clinic
University of Milan
Via Pace 15
Milano
Italy

SCARLATO G.
Clinica Neurologica
Via F. Sforza 35
20122 Milano
Italy

SEKI S.
Department of Physiology
Kanagawa Dental College
Yokosuka
Japan

SHIMADA A.
Department of 2nd Internal Med.
Nihon University, Medical School
Itabashiku
Tokyo 173
Japan

SHIMADA T.
1st Department of Medicine
St. Marianna University
School of Medicine
Kawasaki 213
Japan

SHIMOJO S.
Department of 2nd Internal
Medicine
Jikei Medical College
Tokyo 105
Japan

TADA K.
Institute of Neurobiology
Okayama University
School of Medicine
Okayama 700
Japan

TADA K.
Department of Pediatrics
Tohoku University
School of Medicine
Sendai 980
Japan

TANAKA A.
Department of Microbiology
Osaka City University
Medical School
Osaka 545
Japan

TATIBANA M.
Department of Biochemistry
Chiba University, Medical School
Inohara
Chiba 280
Japan

PARTICIPANTS

TERHEGGEN H. G.
Im Jagdfeld 44a
4040 Neuss
West Germany

UCHIYAMA T.
Department of Pediatrics
Chiba University, Medical School
Inohara
Chiba 280
Japan

VAN GELDER N.
Département de Physiologie
Centre de Recherche en Sciences
Neurologiques, Univ. de Montréal
Montréal, Quebec H3C 3T8
Canada

VAN PILSUM J. F.
Department of Biochemistry
University of Minnesota
4-225 Millard Hall
Minneapolis, Minnesota 55455
USA

WALSER M.
Department of Pharmacology
Johns Hopkins School of Medicine
725 N. Wolfe Street
Baltimore, Maryland 21205
USA

WILLMORE J.
4926 N. W. 18 th. Pl.
Gainesville, Fl. 32605
USA

YABUUCHI H.
Department of Pediatrics
Osaka University Hospital
Osaka 553
Japan

YAMAMOTO Y.
Osaka City Children Health
Service Center
Osaka
Japan

YAMASHITA F.
Department of Pediatrics
Kurume University
Medical School
Kurume 830
Japan

YOSHIDA I.
Department of Pediatrics
Kurume University
Medical School
Kurume 830
Japan

YOSHINO M.
Department of Pediatrics
Kurume University
Medical School
Kurume 830
Japan

YUYAMA N.
Department of Physiology
Kanagawa Dental College
Yokosuka 234
Japan

ZOLLNER H.
Karl-Franzens-Universität Graz
Institut für Biochemie
Halbärthgasse 5
A-8010 Graz
AUSTRIA

INDEX

N-acetylglutamate, 144, 153, 155-156, 157, 197-205, 207-216, 217-228, 259, 337
  effect of ammonia, 199
  effect of diet with non-essential amino acids, 259
  effect of glucagon, 217-228
  effect of glutamate, 201
  effect of lactate, 200
  effect of ornithine, 199, 202
  effect of 2-oxobutyrate on synthesis, 157
  effect of propionate on synthesis, 157
  effect of valproate on synthesis, 157
  in methylmalonic acidemia, 337
N-acetylglutamate synthetase, 39-45, 155, 207-216, 224
  activities and kinetics, 42, 155, 208-211, 224
  characterization, 213
  effect of protein on activity, 209
  influence of arginine, 208, 213
  postprandial change in activity 208
  purification, 211-213
N-acetylglutamate synthetase deficiency, 39-45
  arginine, 41, 43-44
  carbamylphosphate synthetase, 42
  citrulline, 41
  glutamate, 41, 43
  glutamine, 41, 43

N-acetylglutamate synthetase deficiency (continued)
  leucine, 44
  lysine, 41
  organic acids, 40, 316
  ornithine, 41
Alanine
  in N-acetylglutamate synthetase deficiency, 41
  in isovaleric acidemia, 142
Ammonia
  in N-acetylglutamate synthetase deficiency, 40, 43
  in argininemia, 113, 122-123
  in argininosuccinic aciduria 96-97, 444
  in blood, 19-27, 89-90, 135-140, 142, 148, 164
  in cirrhosis, 480-481
  in citrullinemia, 65
  detoxification in CNS, 501-508
  influence on N-acetylglutamate content, 198
  in isovaleric acidemia, 136, 142
  in liver, 259
  in methylmalonic acidemia, 138, 334
  in ornithine transcarbamylase deficiency, 346-348
  in plasma, 43, 123, 142, 347, 444, 480
  in propionic acidemia, 136
  in Reye's syndrome, 164
  in serum, 98, 179
  in sparse fur mice, 177
Animal models

Animal models (continued)
  ornithine transcarbamylase deficiency, 173-183, 185-194
  sparse fur mice, 174, 185
Arginase
  activities and kinetics, 65, 79, 88, 114, 143, 168, 300, 304, 336, 342, 357, 364
  in argininemia, 114
  in argininosuccinic aciduria, 88
  in citrullinemia, 65, 79
  developmental profile, 232
  in gyrate atrophy, 357, 364
  in isovaleric acidemia, 143
  in lizard, 304
  in methylmalonic acidemia, 336
  in ornithine transcarbamylase deficiency, 342
  in Reye's syndrome, 168
Arginine
  N-acetylglutamate synthetase, 209, 213
  in N-acetylglutamate synthetase deficiency, 41, 43-44
  in argininemia, 112-114, 122-123, 428
  in argininosuccinic aciduria, 98
  catabolism in invertebrates, 412
  in citrullinemia, 65
  effect of diet with non-essential amino acids, 259
  evolutionary significance, 401
  metabolism
    in invertebrates, 407-417
    in lizard, 307
Argininemia, 111-119, 121-125, 427-434
  arginase, 114
  arginine, 112, 122, 428
  biochemical findings, 111, 122
  clinical findings, 111, 121
  cystinuria, 112
  guanidino compounds, 428-434
  lysine, 112
  ornithine, 112, 122

Argininemia (continued)
  orotic acid, 123, 315
  treatment, 114
Argininosuccinase
  activities and kinetics, 66, 79, 88, 97, 143, 168, 279, 285, 304, 336, 342
  in argininosuccinic aciduria, 88, 97
  bovine brain
    antigenic properties, 286
    cold lability, 282
    molecular weight, 279
    oligomeric structure, 282
    purification, 278
    SH groups, 284
    subunit composition, 280
  in citrullinemia, 66, 79
  gene complementation, 101-110
  heterogeneity, 101-110
  in isovaleric acidemia, 143
  in lizard, 304
  in methylmalonic acidemia, 336
  in ornithine transcarbamylase deficiency, 342
  properties in bovine organs, 279
  in Reye's syndrome, 168
Argininosuccinate synthetase
  activities and kinetics, 64-72, 79-80, 143, 168, 259, 336, 342
  brain, 70
  in citrullinemia, 66, 79
  effect of diet with non-essential amino acids, 259
  fibroblast, 70-72
  gene complementation, 101-110
  heat stability, 81
  heterogeneity, 63-76, 101-110
  immunochemical properties, 68
  in isovaleric acidemia, 143
  kidney, 70
  liver, 64-70, 79-80
  in lizard, 304
  in methylmalonic acidemia, 336
  molecular weight, 80
  in ornithine transcarbamylase deficiency, 342
  in Reye's syndrome, 168

INDEX

Argininosuccinic acid
  in cerebrospinal fluid, 97
  in serum, 89, 97, 444
  in urine, 89, 97, 444
Argininosuccinic aciduria, 83-93,
    95-100, 101-110, 443-449
  arginase, 88
  arginine, 98
  autoradiography, 108
  citrulline, 98
  contingent negative variation,
    85
  electroencephalography, 85, 96
  excretion of guanidinosuccinic
    acid, 443-449
  fibroblast culture, 104
  gene complementation, 101-110
  glutamic oxaloacetic trans-
    aminase, 97
  glutamic pyruvic transaminase,
    97
  lactate, 97
  organic acids, 97
  pyruvate, 97
  signs and symptoms, 83-84, 95-
    96
  therapy, 96
  uric acid, 89-92
Aspartate
  argininosuccinate synthetase,
    66
  diet-induced changes, 250
Autoradiography, 108

Bacterial inhibition assays, 10
Benzoate, 40, 127, 175
Blood urea nitrogen
  in acute renal failure, 461
  in argininemia, 123
  in argininosuccinic aciduria,
    87, 97
  in chronic renal failure, 451

Canaline, 364
Carbamylglutamate, 42
Carbamylphosphate synthetase
  in N-acetylglutamate synthetase
    deficiency, 41
  activities and kinetics, 65, 79,
    143, 155, 168, 224, 336,

Carbamylphosphate synthetase
    (continued)
  activities and kinetics (con-
    tinued), 342
  in citrullinemia, 65, 79
  developmental profile, 230-232
  influence of hormones, 230
  in isovaleric acidemia, 143
  in methylmalonic acidemia, 336
  in ornithine transcarbamylase
    deficiency, 342
  in Reye's syndrome, 168
  synthesis, 269
Cirrhosis
  effect of pyridoxine-2-oxo-
    glutarate, 479-485
  lactate, 480-482
  pyruvate, 480-482
Citrate
  in ornithine transcarbamylase
    deficiency, 345-348
Citrulline
  in N-acetylglutamate synthetase
    deficiency, 41
  in argininemia, 113, 122
  in argininosuccinic aciduria,
    98
  in citrullinemia, 65, 78
  effect of diet with non-essen-
    tial amino acids, 259
  diet-induced changes, 250
  synthesis, in vitro, 31-33
Citrullinemia, 63-76, 77-82, 101-
    110
  arginase, 65, 79
  arginine, 65, 78
  argininosuccinate synthetase,
    65, 79
  autoradiography, 108
  carbamylphosphate synthetase,
    65, 79
  citrulline, 65, 78
  fibroblast cultures, 104
  gene complementation, 101-110
  glutamic oxaloacetic trans-
    aminase, 78
  glutamic pyruvic transaminase,
    78
  heterogeneity, 63-76
  ornithine transcarbamylase,

Citrullinemia (continued)
  ornithine transcarbamylase
    (continued), 65, 79
  orotic acid, 315
Citrullinogenesis, 153-161, 217-228, 248
Contingent negative variations 85
Cystine
  in argininemia, 112

Electroencephalography, 85, 96, 421-422
Enzyme-auxotroph assay, 10

Fibrin polymer preparation, 30
Fluorescent spot test, 11
Fluorography, 268, 272-273

Gas chromatography-mass spectrometry, 431
Gel filtration, 188, 212, 278, 280
Gene complementation
  in argininosuccinic aciduria, 101-110
  in citrullinemia, 101-110
Glucagon
  influence on N-acetylglutamate content, 217-228
Glucose, 480-482
Glutamate
  in N-acetylglutamate synthetase deficiency, 41, 43
  diet-induced changes, 250
  influence on N-acetylglutamate content, 197-205
Glutamic oxaloacetic transaminase
  in argininosuccinic aciduria, 97
  in citrullinemia, 78
  in Reye's syndrome, 165
Glutamic pyruvic transaminase
  in argininosuccinic aciduria, 97
  in citrullinemia, 78
Glutamine
  in N-acetylglutamate synthetase deficiency, 41, 43
  in sparse fur mice, 177

Glycine
  in isovaleric acidemia, 142
Glycocyamine, 413
Gradient centrifugation, 80
Guanidine cycle, 435-441
  influence of hormones, 438
  influence of hydroxyurea, 437
  influence of urea, 437
Guanidino compounds
  N-acetylagmatine, 412
  N-α-acetylarginine, 387, 394, 432, 467
  N-α-acetyl-γ-hydroxyarginine, 467
  in acute renal failure, 459-464
  agmatine, 412, 467
  α-amino-γ-guanidinobutyric acid, 466
  α-amino-β-guanidinopropionic acid, 468
  arcaine, 412
  arginine, 386, 394, 401-406, 412, 423, 427-434, 450, 467
  argininic acid, 423, 432, 467
  audouine, 412
  canavanine, 423, 437, 467
  in chronic renal failure, 449-458
  creatine, 291, 401-406, 412, 423, 450, 467
  creatinine, 386, 394, 423, 449, 461, 467
  guanidine, 385, 394, 423, 468
  guanidinoacetic acid, 385, 391, 412, 423, 432, 449, 462, 468
  guanidinoadipic acid, 467
  γ-guanidinobutyramide, 412, 468
  γ-guanidinobutyric acid, 385, 394, 412, 432, 467
  ϵ-guanidinocaproic acid, 467
  guanidinoethanol, 412
  α-guanidinoglutaric acid, 419-426, 467
  γ-guanidino-β-hydroxybutyric acid, 468
  β-guanidinoisobutyric acid, 412

INDEX

Guanidino compounds (continued)
  guanidinopropionic acid, 385, 394
  β-guanidinopropionic acid, 412, 468
  guanidinosuccinic acid, 385, 391, 432, 437, 443-448, 449, 461-462, 468
  after hemodialysis, 459-464
  hirudonine, 412
  homoarginine, 468
  γ-hydroxy-agmatine, 412
  γ-hydroxyarginine, 412, 467
  β-hydroxy-γ-guanidinobutyric acid, 412
  hypotaurocyamine, 412
  influence on hen egg development, 465-470
  α-keto-δ-guanidinovaleric acid, 412, 430
  lombricine, 412
  methylguanidine, 385, 391, 449, 461, 467
  monoamidinocadaverine, 412
  monoamidinospermidine, 412
  octopine, 412
  opheline, 412
  phascoline, 412
  phascolosomine, 412
  taurocyamine, 385, 394, 412, 467
  thalassemine, 412
  toxicity, 419-426, 449
Guanidino compounds analysis
  analytical procedure, 383
  deproteinization of samples, 381-389, 391-400
  in parallel with amino acids, 429
Gyrate atrophy, 353-358, 359, 361-370, 371-378
  arginase, 357, 364
  ornithine, 356, 359, 371-378
  ornithine keto-acid transaminase, 356-357
  proline oxidase, 357
  1-pyrroline-5-carboxylate dehydrogenase, 357
  1-pyrroline-5-carboxylate reductase, 357

Heterochrony, 229-240
Hippurate, 127, 180
Hydroxyurea, 437
Hyperammonemia
  induced by valproate, 153-161
  in isovaleric acidemia, 141-146
  in neonata with hypoxia, 147-152
  in propionic acidemia, 136, 243
  screening, 19-27
  transient, 331-338
Hyperornithinemia, 353-358

Intracerebral pH regulation, 501-508
Isovaleric acidemia, 138, 141-146
  arginase, 143
  argininosuccinate synthetase, 143
  carbamylphosphate synthetase deficiency, 143
  glycine, 142
  organic acids, 142
  ornithine transcarbamylase, 143

α-Ketoglutarate
  influence on ammonemia, 474-476
  influence on lactacidemia 474-476
  influence on pyruvicacidemia, 474-476
  in ornithine trancarbamylase deficiency, 346-348

Lactate
  in argininosuccinic aciduria, 47
  in cirrhosis, 480-482
  influence on N-acetylglutamate content, 200
  in methylmalonic acidemia, 334
  in ophthalmoplegia plus, 491-497
  in ornithine transcarbamylase deficiency, 347-348
Lactate dehydrogenase, 165
Leucine
  in N-acetylglutamate synthetase deficiency, 44

Liquid anion-exchange chromatography, 344
Liquid cation-exchange chromatography, 382, 387, 424, 429
Loading tests
  ammonia acetate, 90
  ammonium chloride, 259, 343, 347
  arginine, 98, 142, 451-456
  canavanine, 437
  citrate, 345
  creatine, 451-456
  creatinine, 451-456
  glucose, 493-496
  glycine, 142
  guanidinoacetic acid, 451-456
  hydroxyurea, 437
  α-ketoglutarate, 473-478
  monomethylamine, 451-456
  pyridoxine, 482
  pyridoxine-2-oxoglutarate, 479-485
  sodium benzoate, 177
  urea, 437, 451-456
Lysine
  in N-acetylglutamate synthetase deficiency, 41
  in argininemia, 112

Mancini radial immunodiffusion, 58, 188
Methylmalonic acidemia
  N-acetylglutamate concentration, 337
  activities of urea cycle enzymes, 336
  clinical and biochemical findings, 334-335
  lactate, 335
  orotic acid, 316
Microdiffusion of ammonia, 20
Morphology, 166-167, 372-376, 490, 492
Multiple auxotroph assays, 10

Ophthalmoplegia plus
  lactate, 491-497
  pyruvate, 491-497
  treatment of lactic acidemia,

Ophthalmoplegia plus (continued)
  treatment of lactic acidemia (continued), 487-500
  treatment of pyruvic acidemia, 487-500
Organic acids
  in N-acetylglutamate synthetase deficiency, 40, 316
  in argininosuccinic aciduria, 97
  in isovaleric acidemia, 142
  in ornithine transcarbamylase deficiency, 343
Ornithine
  in N-acetylglutamate synthetase deficiency, 41, 43-44
  in argininemia, 112, 122
  diet-induced changes, 248-249
  effect of diet with non-essential amino acids, 259
  effect of intraperitoneal injection of ammonium salt, 260-261
  in gyrate atrophy, 356, 359
  influence on N-acetylglutamate content, 199, 202
  influence on urea synthesis, 259
  serum levels in gyrate atrophy, 356
  toxic effect on retina, 371-378
Ornithine aminotransferase, 306
Ornithine decarboxylase, 306
Ornithine flux, 245
Ornithine ketoacid transaminase
  activities and kinetics, 306, 356-357, 364, 366-367
  in gyrate atrophy, 356-357
  postnatal changes, 366-367
Ornithine-proline metabolism, 247, 355, 361-370
Ornithine transaminase
  effect of diet with non-essential amino acids, 258-259
Ornithine transcarbamylase, 42, 49-50, 53-62, 79, 143, 155, 168-169, 177, 186-193, 224, 232, 304, 336, 342
  activities and kinetics, 49-50, 55-57, 65, 79, 143, 155, 168-169, 177, 304, 336, 342

Ornithine transcarbamylase (continued)
  characterisation and properties, 187-193
  in citrullinemia, 65, 79
  developmental profile, 232
  heterogeneity, 53-62
  immunochemical properties, 54-58
  in isovaleric acidemia, 143
  jejunal mucosa, 49
  leucocytes, 49
  liver, 49, 79
  in lizard, 304
  lymphoid cells, 49
  in methylmalonic aciduria, 336
  in ornithine transcarbamylase deficiency, 342
  precursor, 267-276
  purification, 186
  in Reye's syndrome, 169
  in sparse fur mice, 177-178
  synthesis, 269

Ornithine transcarbamylase deficiency
  animal model, 173-183, 185-194
  arginase, 342
  argininosuccinase, 342
  argininosuccinate synthetase, 342
  carbamylphosphate synthetase, 342
  citrate, 347-348
  α-ketoglutarate, 346-348
  lactate, 347-348
  organic acid metabolism, 341-350
  ornithine transcarbamylase, 177-178, 342
  orotic acid, 178, 315
  pyruvate, 347-348
  succinate, 347

Orotic acid
  in N-acetylglutamate synthetase deficiency, 40, 315
  in argininemia, 123, 315
  in argininosuccinic aciduria, 97, 315
  in biotin dependant multiple carboxylase deficiency, 316
  in citrullinemia, 315
  in isovaleric acidemia, 144

Orotic acid (continued)
  in lysine protein intolerance, 315
  in 3-methyl-3-hydroxyglutaric aciduria, 316
  in methylmalonic acidemia, 316
  in ornithine transcarbamylase deficiency, 178, 315
  in propionic acidemia, 316
  in sparse fur mice, 178
  in transient hyperammonemia, 316
  in urea cycle diseases and hyperammonemia, 313-319
Ouchterlony double immunodiffusion, 55, 68

Paper and thin-layer chromatography, 10, 322, 430
Peptide mapping, 283
Phosphagens
  arginine phosphate, 402, 412
  bonellidine phosphate, 412
  creatine phosphate, 402, 412, 423
  glycocyamine phosphate, 412
  hypotaurocyamine phosphate, 412
  lombricine phosphate, 412
  opheline phosphate, 412
  taurocyamine phosphate, 412
  thalassemine phosphate, 412
Plasma urea nitrogen
  in N-acetylglutamate synthetase deficiency, 40
Polyamines, 377
Precursor carbamylphosphate synthetase
  intracelllular transport, 269
  processing, 274
  synthesis, 268-269
Precursor ornithine transcarbamylase
  inhibition of transport-processing, 272
  intracellular transport, 269
  isoelectric point, 271
  processing, 274
  synthesis, 268-269
Proline oxidase
  activities and kinetics, 306, 357, 364
  in gyrate atrophy, 357

Propionic acidemia, 135-140, 243
  orotic acid, 316
Protein supply,
  in N-acetylglutamate synthetase
    deficiency, 40
  in argininosuccinic aciduria,
    98
1-Pyrroline-5-carboxylate
    dehydrogenase
  activities and kinetics, 306,
    357, 364
  in gyrate atrophy, 357
1-Pyrroline-5-carboxylate reductase
  activities and kinetics, 306,
    357, 364, 366-367
  in gyrate atrophy, 357
  postnatal changes, 366-367
Pyruvate
  in argininosuccinic aciduria,
    97
  in cirrhosis, 480-482
  in ophthalmoplegia plus, 491-
    497
  in ornithine transcarbamylase
    deficiency, 347-348

Renal failure
  acute, 459-464
  chronic, 449-458
Reye's syndrome, 163-170
  clinical findings, 163-164
  glutamic oxaloacetic transaminase, 165
  lactate dehydrogenase, 165
  urea cycle enzymes, 168

Screening
  bacterial inhibition assays, 10
  enzyme-auxotroph assay, 10
  enzyme-multiple auxotroph
    assay, 10
  fluorescent spot test, 11
  hyperammonemia, 19-27
  multiple auxotroph assays, 9

Screening (continued)
  paper- and thin-layer chromatography, 9, 321-329
  SDS polyacrylamide gel electrophoresis, 187, 212, 268,
    272, 280
Sparse fur mice, 173-183, 185-
    194
Succinate
  in ornithine transcarbamylase
    deficiency, 347

Therapy
  immobilization of urea cycle
    enzymes, 29-35
  ketoanalogues of essential
    amino acids, 445
  sodium benzoate, 40, 127, 175
Thin-layer chromatography, 322
Toxicity, 419-426, 449
Transamidinase
  influence of creatine, 291-292,
    294
  influence of hormones, 291-297
  properties, 293-294
  purification, 293

Ultrafiltration, 382, 393
Urea
  diet-induced changes, 250
  effect of diet with non-essential
    amino acids, 258-259
  influence on guanidine cycle,
    435-441
  synthesis, in vitro, 29-35
Urea cycle
  evolution, 407-417
  in myocardium of rats, 241
Uric acid
  in argininosuccinic aciduria,
    89-92

Valproate
  induced hyperammonemia, 153-161
  influence on N-acetylglutamate,
    157